The continent of Europe has a complex geological history of successive tectonic events. Over several thousand million years these have formed the present day configuration of major tectonic provinces. *A Continent Revealed* unravels this history by presenting and interpreting the results of the European Geotraverse (EGT) – a unique study of the continent of Europe and the first comprehensive cross section of continental lithosphere. This illustrated book has been put together by key workers in the EGT project. It uses the wealth of information yielded by the ten years of experiments, study centres and workshops to provide a concise and thought provoking account of the geological processes that created the European continent. It provides a summary of the European Geotraverse, and at the same time a starting point for further work.

This book, along with a comprehensive database from the EGT project – in the form of 25 maps, a descriptive booklet and a CD-ROM – is available in a boxed set (ISBN 0 521 41923 9).

A Continent Revealed

The European Geotraverse

A CONTINENT REVEALED
The European Geotraverse

Edited by

DEREK BLUNDELL,

Royal Holloway and Bedford New College, University of London

ROY FREEMAN AND STEPHAN MUELLER,

ETH Zurich

ILLUSTRATIONS BY SUE BUTTON

Published by the Press Syndicate of the University of Cambridge
The Pitt Building, Trumpington Street, Cambridge CB2 1RP
40 West 20th Street, New York, NY 10011-4211, USA
10 Stamford Road, Oakleigh, Victoria 3166, Australia

First published 1992

A catalogue record for this book is available from the British Library

Library of Congress cataloguing in publication data available

ISBN 0 521 42948 X paperback
ISBN 0 521 41923 9 boxed set

Transferred to digital printing 2003

Contents

Preface   *Page* ix
The authors   xii

1 WHY A TRAVERSE THROUGH EUROPE?   1
D. Blundell and R. Freeman
1.1 Tectonic evolution of a continent   1
1.2 The European Geotraverse   3
1.3 Coordinating the European Geotraverse   4
1.4 Achievements of the European Geotraverse   8

2 MOBILE EUROPE   11
A. Berthelsen
2.1 How far back does plate tectonics go?   11
2.2 A palaeomagnetic kinematic perspective   12
2.3 Decline and fall of an orogen   15
2.4 How Europe's crust evolved   17

3 EUROPE'S LITHOSPHERE – SEISMIC STRUCTURE   33
J. Ansorge, D. Blundell and St. Mueller
3.1 Seismic methods for exploring the crust and upper mantle of Europe   33
3.2 Seismic exploration of the crust along EGT   35
3.3 Seismic exploration of the upper mantle along EGT   60

4 EUROPE'S LITHOSPHERE – PHYSICAL PROPERTIES   71
4.1 Physical properties of the lithosphere   71
E. Banda and S. Cloetingh
4.2 Mechanical structure   80
S. Cloetingh and E. Banda
4.3 Evidence from xenoliths for the composition of the lithosphere   91
K. Mengel
4.4 Integrated lithospheric cross section   102
D. Blundell

5 EUROPE'S LITHOSPHERE – RECENT ACTIVITY   **111**
E. Banda and N. Balling
5.1 Seismicity   111
5.2 State of stress   120
5.3 Recent crustal movements   124
5.4 Recent volcanism   132
5.5 Transient heat flow   135

6 TECTONIC EVOLUTION OF EUROPE   **139**
6.1 Precambrian Europe   139
B. Windley
6.2 From Precambrian to Variscan Europe   153
A. Berthelsen
6.3 Phanerozoic structures and events in central Europe   164
W. Franke
6.4 Alpine orogeny   180
A. Pfiffner
6.5 The fragmented Adriatic microplate: evolution of the Southern Alps, the Po basin and the northern Apennines   190
P. Giese, D. Roeder and P. Scandone
6.6 Sardinia Channel and Atlas in Tunisia: extension and compression   199
D. Roeder
6.7 Recent tectonics of the Mediterranean   202
D. Roeder and P. Scandone

7 GEODYNAMICS OF EUROPE   **215**
D. Blundell, St. Mueller and K. Mengel
7.1 How does geology work?   215
7.2 What drives tectonic processes?   215
7.3 Geodynamic processes in the past?   228
7.4 EGT – the future?   231

References   **233**

Index   **263**

Preface

The European Geotraverse (EGT) project has been a scientific undertaking on an unprecedented scale in the Earth Sciences. Its whole ethos has been founded on the idea that the scale of trying to understand the workings of a continent and its evolution through geological time demanded the combined efforts of a very large number of people with expertise from a wide range of disciplines. Not only would they have to understand each other's point of view and work together in a series of integrated experiments, they would have to produce their findings in such a way that could be understood by all. EGT completed its experimental work in 1990, the culmination of nearly ten years effort. But the work of EGT could not end there. The concluding paragraph of the EGT Final Report to its prime supporter and benefactor, the European Science Foundation (ESF), made clear that 'In many ways, EGT has only just begun'. The wealth of data collected during the EGT experimental programme, together with geological and geophysical data obtained over many years, were reduced to a common reference frame and mapped on to a common scale and projection in order to compare them directly so that the connections between them might become clear. The result of this compilation is the sequence of maps and plates, together with explanatory text and comprehensive reference lists, that form the EGT Atlas which is the complement and companion to this book. To enhance the use of this database, much of the Atlas data is contained on a CD-ROM.

The concept of this book has been to draw all this work together into a coherent account of the tectonic evolution of Europe and the geodynamic processes that have fashioned it. It is, in a sense, the epitome of EGT, having been written in the belief that the whole is greater than the sum of its individual parts. It has also been written with a view to the future, in the knowledge that there is much more still to be gained from further analysis and interpretation of the information gathered during EGT and the hope that the book can serve as a springboard for new research advances. The book is very much a team effort, involving fifteen authors. Whilst each has been identified with particular sections of the book, it might have been added after their names 'with a little help from their friends' because the writing and the ideas have been shared and have benefited from joint efforts. We have attempted to present the book in a unified way. As a vital ingredient, all the diagrams have been computer draughted to a consistently high standard by Sue Button at the University of Leicester, to whom we are immensely grateful.

The EGT project involved too many people for us to mention here and we apologise for the injustice in our not recognising their individual efforts. The length of the list of references at the end of this book is testimony to the great number who have contributed.

EGT was made possible because of the recognition and support that was given so generously by the ESF throughout the eight years formal existence of the project and, indeed, both before and since. There are no words to express adequately the gratitude we owe to ESF. We would like to thank, in particular, Eugen Seibold, the ESF President for much of the period of EGT, and the two officers of ESF who, in turn, gave practical expression to the help that we received, Bernard Munsch and Michele Fratta. Peter Fricker deserves our special thanks. As chairman of the European Science Research Councils Working Group for the European Geotraverse he gave unstinting support and did much to secure the funding for the coordination of EGT from the Research Councils. The management of EGT came through the Scientific Coordinating Committee, chaired by Stephan Mueller, including E. Banda, A. Berthelsen, D. Blundell, P. Giese, A. Hirn, C. Morelli and H. Zwart, whose deliberations were put into practice by the Scientific Secretary, E. Banda, and the Adjunct Secretary, first D. Galson and then R. Freeman. To them fell the major part of the hard work involved in the whole project. For practical purposes the EGT swathe was divided into three segments. Responsibility for the northern segment fell to A. Berthelsen as chairman with first N. Springer and then M. Marker as scientific assistant; for the central segment to P. Giese as chairman with M. Huch as scientific assistant, and for the southern segment to C. Morelli as chairman and D. Polizzi as scientific assistant. Much of the coordination of EGT was devolved to the three segments, particularly the organisation of workshops and study centres. These proved to be especially fruitful in bringing people and ideas together and were greatly enhanced by the willingness of ESF to enable rapid publication of extended abstracts and to disseminate them widely. Financial support and encouragement for this from the Commission of the European Communities was especially valued, through the good offices and continuing interest of K. Louwrier and E. Staroste.

Amongst the 13 experiments making up the Joint Programme of EGT, some involved field campaigns to obtain new observational data whilst others required coordinated efforts to recompile existing data into uniform formats. All required substantial organisational effort. Five major seismic experiments were executed specifically for EGT. EUGENO-S was undertaken by a working group chaired by A. Berthelsen and owed much to the efforts of E. Flüh, S. Gregersen and C.-E. Lund. EUGEMI was coordinated by C. Prodehl and B. Aichroth, EGT-S83 and EGT-S85 were directed by J. Ansorge and C. Morelli and ILIHA was organised through a working group led by A. Udías, A. Lopez-Arroyo and L. Mendes-Victor. Data compilation has been effected primarily by A. Berthelsen, P. Burollet, D. Dal Piaz, W. Franke and R. Trümpy (tectonics), J. Ansorge (seismics), A. Hahn, T. Wonik, A. Galdéano and P. Mouge (magnetics), V. Haak and S.-E. Hjelt (electromagnetics), N. Pavoni, T. Ahjos, S. Gregersen, H. Langer, G. Leydecker, P. Suhadolc and M. Uski (seismicity, focal mechanisms), V. Cermák, N. Balling, R. Schulz and B. Della Vedova (geothermics), E. Klingelé (gravity), D. Lelgeman (geoid), P. Giese (Moho depths), L. Pesonen and M. Westphal (palaeomagnetics), E. Gubler, S. Arca, J. Kakkuri, K. Mälzer and K. Zippelt (recent crustal movements) and D. Gebauer (geochronology).

Primary publication of many of the scientific results from EGT experiments benefited greatly from the agreement by Elsevier Science Publishers BV to publish special issues of *Tectonophysics* devoted to EGT. In all, eight special issues have appeared between 1986 and 1992, identified by the EGT logo. The main editorial effort by D. Galson (Parts 1–2) and R. Freeman (Parts 3–8) was supported by A. Berthelsen, P. Giese, M. Huch, M. von Knorring, H. Korhonen, C.-E. Lund and St. Mueller.

In all such publications, the generosity of time and effort and the judgement of reviewers is critical. This has been particularly so for this book. We are especially grateful to C. Drake, S. Gregersen, R. Hatcher, C. Morelli, G. Panza, R. Trümpy, P. Ziegler and H. Zwart who carefully reviewed an early draft of the complete book and from their detailed suggestions we have been able to make significant improvements. All are very busy and we have greatly appreciated the time and attention that they gave to our work. Individual chapters have also benefited enormously from detailed comments by V. Cermák, C. Doglioni, H. Downes, E. Flüh, A. Hahn, M. Helman, H. Henkel, S.-E. Hjelt, H. Kern, P. Matte, C. Prodehl, S. Schamel, G. Serri and P. Suhadolc. With so much effort by others having gone into this book we hope that it has done them justice.

D. B., R. F. and St. M.

The authors

Dr J. Ansorge, Institut für Geophysik, ETH Hönggerberg, CH-8093 Zürich, Switzerland

Prof. N. Balling, Department of Earth Sciences, Geophysical Laboratory, Aarhus University, Finlandsgade 8, DK-8200 Aarhus N, Denmark

Prof. E. Banda, Institute of Earth Sciences (Institut Jaume Almera), CSIC, Marti i Franquès s/n, 08028 Barcelona, Spain

Prof. A. Berthelsen, Geologisk Institut, Afdeling for almen Geologi, Øster Voldgade 10, DK-1350 Copenhagen K, Denmark

Prof. D. Blundell, Department of Geology, Royal Holloway and Bedford New College, University of London, Egham, Surrey TW20 0EX, England

Prof. S. Cloetingh, Institute of Earth Sciences, Vrije Universiteit, De Boelelaan 1085, 1081 HV Amsterdam, The Netherlands

Prof. W. Franke, Geologisch-Paläontologisches Institut, Senckenbergstrasse 3, D-6300 Giessen, F R Germany

Dr R. Freeman, Institut für Geophysik, ETH Hönggerberg, CH-8093 Zürich, Switzerland

Prof. P. Giese, Institut für Geophysik, Freie Universität Berlin, Rheinbabenallee 49, D-1000 Berlin 33, F R Germany

Prof. K. Mengel, Geochemisches Institut, Universität Göttingen, D-3400 Göttingen, F R Germany

Prof. St. Mueller, Institut für Geophysik, ETH Hönggerberg, CH-8093 Zürich, Switzerland

Prof. A. Pfiffner, Geologisches Institut, Universität Bern, Balzerstrasse 1, CH-3012 Bern, Switzerland

Dr D. Roeder, Anschutz Overseas Corporation, 2400 Anaconda Tower, 555 Seventeenth Street, Denver, Colorado 80202, USA

Prof. P. Scandone, Dipartimento di Scienze della Terra, Università degli Studi, Via S Maria 53, I-56100 Pisa, Italy

Prof. B. Windley, Department of Geology, University of Leicester, Leicester LE1 7RH, England

1 Why a traverse through Europe?

D. BLUNDELL AND R. FREEMAN

1.1 TECTONIC EVOLUTION OF A CONTINENT

Piecing together the geological evolution of a continent is rather like a detective investigation. Various pieces of evidence provide clues as to what might have happened, but these can be assembled in a variety of ways. Various theories, based on certain geological mechanisms, are put forward to test the evidence. Europe has a history of geological activity and continental evolution spanning over 3500 million years (Ma) to the present day and is one of the best places in the world to discover how a continent evolves. The geology of Europe has been studied intensively for well over a century by examining outcrops of rocks at the surface, so that the surface geology is probably better known than anywhere else in the world. In contrast, knowledge of what the geology is like beneath the surface is limited to information from boreholes and indirect evidence, principally from geophysical measurements.

Since the theory of plate tectonics came to prominence in the mid 1960s, a mechanism has become understood which explains how global tectonic processes take place at the present day. It is known that, on a global scale, the outer region of the Earth called the lithosphere, which includes both crust and upper mantle, acts as a more rigid layer above a more plastic layer of the upper mantle, called the asthenosphere. The lithosphere is divided into a dozen or so major plates which move relative to each other, interacting and deforming, mostly around their perimeters. Direct evidence of plate movements has been recorded in oceanic crust for the past 200 Ma but beyond that time no oceanic crust exists intact to tell the tale. As a consequence of the success of plate tectonics, the structure of the present oceanic regions appears to have become better understood and attention can now move from these relatively youthful features to focus on the continental regions, which contain what remains of the main time span of the Earth's history. Continental evolution within the last 200 Ma is firmly established within the framework of lithospheric plate interactions. The geodynamics of continental evolution therefore take place on a lithospheric scale, extending several hundred km below the surface, so that to understand the processes of continental evolution requires a knowledge of the geology of the whole lithosphere and, indeed, the underlying asthenosphere. But the evolutionary history of the continental lithosphere has been long and complex, with successive thermal and deformational episodes superimposed, and our knowledge of the processes involved has remained obscure. Furthermore, it is fundamental to discover whether plate tectonics have operated not just during the past 200 Ma but over the full 3500 Ma (3.5 Ga) timespan of continental evolution, so that the geology can be interpreted within the context of plate tectonics theory.

Rigidity depends on a number of factors, the most important of which include the composition of the rock and its temperature. New lithosphere, created through upwelling of magma from the mantle to the surface at the mid-ocean ridges, cools and thickens with time as heat escapes, forming an outer boundary layer to the Earth. A temperature gradient exists through the lithosphere to its base, where the temperature equals that of the asthenosphere. Because of its reduced rigidity, the asthenosphere mixes by convection currents to keep it uniform in temperature at around 1330°C. The thickness of the lithosphere can be estimated through knowledge of the surface heat flow and the thermal conductivity of the rocks to find the temperature gradient, from which the depth at which a temperature of 1330°C is reached can be calculated. Certain other factors, such as the amount of heat produced within the lithosphere from the decay of naturally radioactive elements within the rocks, have to be taken into account in making the calculations. Lithosphere determined in this way is sometimes called the 'thermal lithosphere'.

Rigidity is one of the factors that controls the speed of propagation of seismic waves through the Earth. A reduction in rigidity, other things being equal, results in a reduction in seismic wave velocities. This is particularly noticeable for S-waves and surface waves. The base of the lithosphere can therefore be determined by locating the top of a low velocity zone within the upper mantle which can be identified as the asthenosphere. Lithosphere determined in this way is sometimes called the 'seismic lithosphere'. In Chapter 4 we shall also introduce the term 'mechanically strong lithosphere' (MSL) as a measure of the thickness of that part of the lithosphere which is mechanically strong.

Although these measurements may seem simple enough, and are reasonably easy to determine in the relatively uniform composition and structure of the oceanic regime, they are far more difficult and complex for the continental lithosphere, and uncertainties in estimating lithosphere thickness are in the order of tens of kilometres. The highly variable composition, structure and thickness of continental crust ensures that the behaviour of continental lithosphere is far from uniform, and makes it difficult to determine the properties of the underlying mantle. To comprehend these complexities there is a fundamental need to find a better understanding of the tectonic evolution of the continental regions of the Earth.

Having worked on the geology and deep structure of the Alps and central Europe for a number of years, Stephan Mueller at Zürich recognised that the best place to study the tectonic evolution of a continent is Europe, because it is made up of a number of tectonic provinces ranging in geographical succession from the oldest Precambrian areas of Scandinavia to the currently active area of the Mediterranean. In 1979, he and his colleague Giuliano Panza published a contour map of the thickness of the lithosphere across Europe, based on an analysis of seismic surface waves. It showed considerable variability, but could it be related to surface geology? Mueller set about appraising the state of knowledge about the lithosphere of Europe. It was apparent that whilst surface geology might be relatively well known, evidence at depth was, to say the least, fragmentary. A number of geophysical experiments had been carried out, but data quality was variable and survey data from neighbouring countries were incompatible, having been reduced to different datums. Information was patchy and parochial, and a tradition had evolved in which there was virtually no communication between geologists and geophysicists. There was no way that the clues could be fitted together. But what an extraordinary opportunity there would be if a concerted effort could only be made across Europe to bring the right people together and create a coherent knowledge of the lithosphere across a whole continent and a 3.5 Ga timespan. The scale of the effort needed to bring this to fruition, however, was truly daunting.

1.2 THE EUROPEAN GEOTRAVERSE

Mueller turned for help to one of the foremost Alpine geologists, Rudolf Trümpy, who was at the time President of the International Union of Geological Sciences (IUGS). During the 26th International Geological Congress at Paris in 1980, Trümpy and Mueller discussed the problem informally with Eugen Seibold, then President of the German Research Association (Deutsche Forschungsgemeinschaft), and Peter Fricker, Secretary General of the Swiss National Science Foundation. They needed to find a mechanism that could bring together a large group of Earth scientists from a wide variety of disciplines from every country in Europe and persuade them of the value of working together to a common purpose. More than that, this international group would need to be sufficiently motivated and credible to convince their various national scientific funding agencies, the Research Councils, to provide the money to support the range of experiments that would be needed.

These wise men recognised that the European Science Foundation (ESF) could provide the ideal way forward. Centred in Strasbourg, the ESF is an internationally recognised organisation supported by most of the national research councils of European countries. They suggested that Mueller should initiate a proposal for a coordinated research programme and obtain the willingness of the ESF to provide initial help to launch the idea. Under the aegis of the ESF, Mueller called together a Working Group of national representatives of European Science Research Councils (ESRC) which met in Zürich in February 1981 to discuss how best to set up and manage an interdisciplinary scientific programme to investigate the lithosphere of Europe. The goal of the project was to develop a three-dimensional picture of the structure, properties, and composition of the continental lithosphere of Europe as a basis for understanding its nature, evolution and dynamics.

From this meeting emerged the concept of a continental geotraverse of lithosphere proportions along which coordinated experiments could provide consistent information across each of the tectonic provinces to link them together. Recognising the need to examine geological structures in three dimensions, the geotraverse was conceived as a swathe, rather than a line, 4600 km in length, 200–300 km in width, and 450 km in depth, from northern Scandinavia to central Tunisia. It was designed to encompass the succession of tectonic provinces from the oldest known Archaean (3.5 Ga) to those active today. The merit of the EGT swathe is that it includes one of the widest possible ranges, along a single continuous path, of processes in which continental crust is built up, maintained, and destroyed. Importantly, because the provinces occur in succession geographically as well as in time, there is the opportunity to follow the progression of tectonic activity through time.

Extending from the northern tip of Scandinavia southwards to North Africa (Figure 1-1), the European Geotraverse is located to encompass the Archaean nucleus in the northernmost part of the Baltic shield, the Proterozoic, Paleozoic and Cenozoic provinces of northern and central Europe that have been added on to this nucleus, and the active transition zone between the Eurasian and African plates in the Mediterranean region. The broad aim of the Geotraverse is thus to secure an understanding of how the continental lithosphere of Europe formed and reacted to changing physical and geometric conditions through successive Precambrian, Caledonian, Variscan and Alpine tectonic episodes. It was clear that an understanding of these processes would require detailed knowledge of the structure and dynamics of the whole lithosphere, including not only the crust, but also the underlying mantle which is intimately bound up with it. The large scale of the geotraverse was required both to provide lateral continuity of information across the major structural elements of Europe and to achieve a deeper view of variations within the lithosphere. It was intended that the results of the investigations should be integrated to produce a north–south section through the crust and upper mantle of Europe, providing the basis for a reconstruction of the evolutionary development of the various tectonic provinces and their mutual interaction, and leading on to a better understanding of the dynamics of the lithosphere–asthenosphere system.

The guiding principle behind the project was that progress in understanding the continental lithosphere evolution and dynamics could only be achieved through a well defined programme of linked experiments involving international collaborative effort and drawing on a large number of Earth scientists with the widest possible extent of experience and knowledge. To be effective it was essential to integrate a broad range of techniques in a number of disciplines, in which geophysics would feature on the experimental side to gain information on the deep structures and on dynamic and kinematic problems, but geological and geochemical expertise would be very important at the stage of interpretation. Although certain techniques were proposed for the full length of the Geotraverse to provide continuity and depth of information, most of them were to be applied selectively so that the combination of methods utilised would be that most appropriate to the particular problem under investigation. Vital to the concept of the Geotraverse was that it should be carefully directed and managed.

1.3   COORDINATING THE EUROPEAN GEOTRAVERSE

At the ESRC Working Group meeting in February 1981, it was agreed to draw up a project proposal to the ESF. Later that year, the Working Group invited ideas from all of the countries represented for specific projects which could make up the programme of experiments for the EGT Project.

The Working Group set up a Scientific Coordinating Committee (SCC), chaired by Mueller, to be responsible for the direction, organisation and coordination of EGT. The SCC provided the scientific management to EGT and became the driving force for the Project. The SCC received various ideas generated by the Working Group and from them formulated an integrated programme, requiring international collaboration. This 'Joint Programme' of 13 experiments formed the backbone of the EGT Project. These are set out in Figure 1-1 and Table 1-1. They include field experiments involving the collection of new data, laboratory experiments and analysis, and compilation of data from existing surveys in compatible form.

For practical purposes the Geotraverse was divided into three segments each managed by a Segment Chairman, who was a member of the SCC. These were chosen to cover the following regions:

(a) The Precambrian Baltic Shield of Fennoscandia (age *ca.* 3100–600 Ma) with its border regions, including the Caledonides (age *ca.* 600–400 Ma). Segment Chairman: Asger Berthelsen.

(b) The Variscan realm of Central and Western Europe (age *ca.* 400-230 Ma). Segment Chairman: Peter Giese.

(c) The Alpine–Mediterranean region (age 230 Ma to present). Segment Chairman: Carlo Morelli.

In designing the Joint Programme, particular attention was given to the border zones between the segments.

With the agreement of the Working Group this programme was put forward as a formal proposal to ESF for approval. It was adopted as an 'Additional Activity' by the ESF Assembly in November 1982 to begin on 1 January 1983 and to last up to seven years. This support proved vital. ESF provided the necessary standing and environment to facilitate international collaboration, and in giving support and encouragement in the capacity that its name implies. It provided the key funding to coordinate the entire project, through subscriptions from ESRC members, which enabled the SCC to appoint a Scientific Secretary and an Adjunct Scientific Secretary committed to the project and three part-time Scientific Assistants, one for each segment, as well as the means to hold regular meetings. It also provided the secretariat support for the ESRC Working Group and the SCC, and produced the Workshop Proceedings. Throughout the Project the encouragement given by ESF was of enormous benefit.

The SCC devised a strategy for carrying out the EGT experiments, concentrating on each of the segments in turn according to the experimental time schedule. Each major experiment was preceded by a workshop and followed by a 'Study Centre' from which stemmed publication of the work. This proved remarkably effective.

EGT workshops

Workshops were instigated to assemble, categorise, present and review all the available data relevant to the particular problem prior to each of the major international experiments. Publication and distribution of Workshop Proceedings within six months, under the auspices of ESF and financial support from the Commission of the European Communities, has been a particularly valuable feature.

Experimental programme

Large-scale seismic refraction experiments involved multinational teams working together in the field, coordinating their efforts to predetermined time schedules, recording uniformly to give their data to agreed centres for processing and analysis. Other experiments and the compilations usually involved dedicated specialist groups coordinating their efforts especially for EGT.

The network of Earth Science Study Centres

To assemble scientists and data ready to undertake the interpretation, use was made of the ESF Scientific Network Scheme set up in 1985 to hold a series of Earth Science Study Centres. Each was located where 60 or so scientists could live and work together without distraction for a 2-week period of intensive analysis, interpretation, synthesis, debate and preparation of initial drafts for publication. Their great strength has been the international, multidisciplinary mix of younger and more experienced scientists. To those fortunate enough to take part in one it was an enthralling and exhilarating experience which cemented many a lasting friendship.

To get the best out of the Study Centres, the most effective scheme entailed:

    (a) Preparatory meetings to get data assembled in standard form;

    (b) The main Study Centre;

    (c) Follow-up meetings of small groups to finalise geological models a write up the results.

Table 1-1 The EGT 'Joint Programme' and Data Compilations (nos. refer to Figure 1-1)

No.	Programme	Field Experiment	Study Centres, Workshops
1	Multidisciplinary studies of the evolution of the Baltic Shield	FENNOLORA: Fennoscandian Long-Range Project [1979], POLAR Project [1987]	First Workshop [1983], Second [1986] and Fifth [1990] Study Centres: Integrated interpretation of geophysical and geological data
2	Multidisciplinary studies along a south Scandinavian east-west traverse	Heat-flow, magnetotelluric, seismic reflection surveys across the Protogine and Mylonite Zones [1985-9]	Data and interpretations presented and discussed at the First Workshop and Fifth Study Centre
3	Multidisciplinary study of the contact zone between Precambrian and Hercynian Europe	EUGENO-S Network (EGT Northern Segment - Southern Part) [1984]	First Study Centre[1985]: Synoptic interpretation of the EUGENO-S network of seismic lines
4	Deep seismic sounding of the lithosphere, Central Segment of the EGT	EUGEMI: EGT Central Segment Profile [1986] (Baltic Sea to the Alps)	Interpretation combined with results from DEKORP 2-S and 2-N at the Fourth [1989] and Fifth Study Centres
5	Synoptic geological and geophysical studies of border regions between different tectonic units of Hercynian age, Central Segment	EUGEMI Profile DEKORP Profiles Various magnetotelluric studies	Third Workshop [1986] and Fourth Study Centre: crustal-scale balanced cross sections applied to the DEKORP 2-N profile
6	Multidisciplinary studies as well as synoptic geophysical surveys in the Southern Alps, Po Plain, and Northern Apennines	EUGEMI Profile, EGT-S83 and EGT-S86: Central Alps-Po Basin-Northern Apennines Profile [1983, 1986]	Results combined with the Swiss NFP20, French ECORS, and Italian CROP seismic reflection traverses at the Second Study Centre [1985]
7	Deep seismic sounding of the lithosphere in the Southern Segment of the EGT	EGT-S83: Western Alps, Northern Apennines, Ligurian Sea to Sardinia Channel Profile, EGT-S85: Sardinia -Tunisia Profile	Gross crustal structure defined at the Second Workshop [1985] and Third Study Centre [1988], further developed at the Fifth Study Centre
8	NARS: Network of Autonomously Recording Seismographs	Broadband seismology experiment along a transect from southern Sweden to the Alboran Sea [1983-4]	Results presented at the Third, Fourth [1988] and Fifth [1988] Workshops
9	ILIHA: Iberian Lithosphere Heterogeneity and Anisotropy project (an EC 'Stimulation Action')	Broadband Seismology Subproject Deep Seismic Sounding Subproject [1988-91]	ILIHA Working Group meetings, Fifth Workshop

EGT-Wide Programmes:

	Mapping of the lithosphere-asthenosphere system along the EGT by seismological techniques	E.g. dispersion of surface waves and tomography (P-wave delay) studies [1983-90]	Results first presented at the Third Workshop, major theme at the Fourth Workshop; Atlas Map 2: Moho depths
	Mapping of the resistosphere and conductosphere along the EGT	Magnetotelluric studies in Fenno-scandia, Germany, Switzerland and Sardinia	Results presented and discussed at the Sixth Workshop [1989] and the Fourth and Fifth Study Centres; Atlas Maps 11 & 12: Magneto-telluric and magnetovariational data
	Geomagnetic observations along the EGT	Compilation of national surveys	Atlas Map 10: Magnetic anomalies
	Integrated geothermal studies along the EGT	Compilation of national surveys	Atlas Map 13: Heat-flow density

Data Compilations

	Tectonics	Compilation of geological fieldwork	Atlas Map 1: Tectonics
	Seismicity	Compilation of national catalogues	Atlas Map 3 & 4: Historical and instrumental seismicity
	Focal Mechanisms	Compilation of national catalogues	Atlas Map 5: Focal mechanisms
	Recent vertical crustal movment	Compilation of national surveys	Atlas Map 6: Recent vertical crustal movement
	Geoid undulations	Compilation of international data sets	Atlas Maps 7 & 8: Geoid undula-tions and gravity disturbance vector
	Bouguer gravity	Compilation of national surveys	Atlas Map 9: Bouguer gravity anomalies
	Palaeomagnetic results	Compilation of catalogues	Atlas Map 14: Drift of Fennoscandia

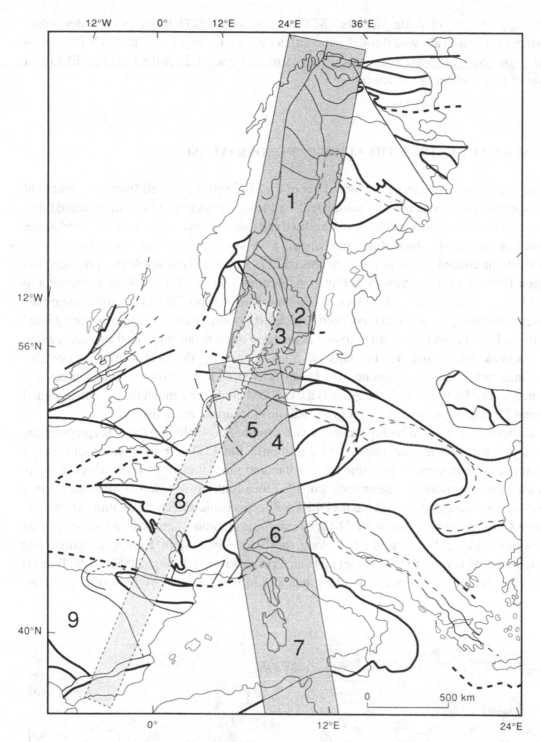

Figure 1-1. Location map of 'Joint Programme' EGT Experiments. Numbers refer to experiments listed in Table 1.

Symposia and presentation of results and ideas

The SCC has arranged for results to be presented in symposia dedicated to EGT at most of the relevant international meetings in recent years. Through the good offices of the publishers, Elsevier (Amsterdam), it arranged for original publications, internationally refereed to ensure scientific quality, to be gathered together in eight Special Issues of *Tectonophysics* and has encouraged publication of approved work under the EGT logo in other appropriate scientific journals. SCC has also made EGT known to a wider public whenever occasion allowed through short articles in magazines, newspapers, TV interviews and so on. *New Scientist* even published a cartoon (Figure 1-2) at the time that EUGEMI (Figure 1-1) was being carried out.

1.4 ACHIEVEMENTS OF THE EUROPEAN GEOTRAVERSE

During the seven years of the formal life of the EGT Project the full Joint Programme of 13 experiments was successfully completed. To do this, a workforce of several hundred Earth scientists from over 14 countries was mobilised and their efforts coordinated. Six workshops were held and their proceedings published. Five study centres were held from which publications ensued, many in the eight special EGT issues of *Tectonophysics* produced between 1986 and 1992. Primary publications stemming from EGT work are continuing to appear in a variety of journals. Over 20 Diploma, Masters and PhD theses containing EGT material as their primary data have been completed. A large part of the information gained from the EGT Project has been prepared in the form of an Atlas, published as a companion to this book, which includes 14 maps and 5 plates covering the EGT swathe. It contains explanatory text and a comprehensive list of references to the primary data sources and related work. Included with the Atlas is a CD-ROM containing much of the data in digital format for reading into a computer (a PC will do) for further analysis.

The total cost of coordinating the seven-year programme, including the Atlas preparation, came to about FF7 million (around $1.2 million). But EGT attracted Research Council funding from many countries some ten to a hundred times greater in value which actually enabled the experiments to be undertaken. EGT also contributed towards stimulating other allied research programmes, such as the DEKORP deep seismic reflection profiling programme across Germany and the Swiss NFP20 deep seismic programme across the Alps, one profile of which was deliberately aligned along the EGT. In eastern Europe a comparable geotraverse, known as EU-3, was set up from Czechoslovakia through USSR to the Barents Sea. The earlier experience of working together in EGT assisted in the formation of the

Figure 1-2. Cartoon by Bill Tidy published in New Scientist (No 1539, 18 December 1986), reproduced with permission. The 'genius behind the European Geotraverse' was, in this case, the EUGEMI Working Group led by C. Prodehl and B. Aichroth who organised the experiment.

BABEL Working Group, composed of British, Danish, Finnish, German and Swedish scientists who jointly succeeded in getting funded and carrying out in 1989 a combined normal incidence and wide-angle deep seismic reflection survey across the Baltic Shield. A similar experiment in the Mediterranean region called STREAMERS has similar origins in EGT. Perhaps the most lasting value of the EGT Project lies in the inheritance it leaves for others, resulting from the interdisciplinary collaboration of so many scientists from many countries. A way of working together has been established which has set the pattern for the future.

But for all the activity and goodwill that has been generated, all the experiments and the publications,what of the scientific achievement of the EGT? Has it achieved its goals? Has it provided the key evidence in the detective investigation of the evolution of a continent? Do the clues make much better sense now that they can be brought together? Have genuine new discoveries been made? It is the main purpose of this book to answer these questions and to persuade you, the reader, that useful scientific advances have been made. The authors have worked together as a team to explain the various facets of the science, reviewing the work of their colleagues, as far as they can, to create a coherent story of the EGT so far. We are well aware that there are many deficiencies in this story but we hope that these will simply serve to stimulate you to put them right. The Atlas and data on CD-ROM provide you with the means to do so.

Chapter 2 sets the scene with a brief review of the major provinces of Europe and their tectonic evolution, broadly based on palaeomagnetic evidence. The key element is their mobility. Chapter 3 discusses the structural framework of the lithosphere of Europe determined largely from various lines of seismic evidence. It begins to show something of the physical properties of the lithosphere. This aspect is taken further in Chapter 4 which reviews geophysical evidence of various kinds about the physical characteristics of the rocks and the physical conditions pertaining within various regions of the lithosphere. Complementary to the geophysical information is the direct evidence obtainable from samples from deep in the crust and upper mantle brought up to the surface by volcanic activity as xenoliths in igneous rocks. Combining the geophysical and geological information leads to a synthesis of the European Geotraverse in the form of a cross section of the geology of the whole lithosphere, across the whole continent, as it is at present. Chapter 5 reminds us that the geology of Europe is very active at present, including earthquakes and volcanoes, uplift and subsidence. The analysis of present-day processes gives valuable insight into how they may have acted in the past. Chapter 6 makes use of the information assembled in the previous chapters to interpret the evolution of Europe through successive orogenic periods. This provides the story of an evolving continent, while Chapter 7 attempts to reveal the underlying geodynamic mechanisms and the forces that control the way that continental geology works.

It is now for you to judge for yourself how far we have succeeded in our aims and to continue the story for yourself.

2 Mobile Europe

A. BERTHELSEN

Present day Europe forms part of the large Eurasian plate which is surrounded by 12 large and at least as many small plates. This plate configuration, where curving Alpine fold belts and island arcs wind along the convergent borders, is relatively young, of Late Mesozoic–Cenozoic origin. In earlier geological times quite different plate configurations existed. Some of the former plate borders can still be recognised as deep scars, called sutures, in the continental crust, other plate borders have been obliterated. Because all the pre-Mesozoic oceans that once surrounded Europe have also been lost (consumed by subduction), the answers as to how Europe was formed and assembled must be sought in the continental lithosphere. To look for these answers was a prime aim of the European Geotraverse Project.

In this introductory review of Europe's tectonic evolution, we focus on *when, where, and how* the crystalline basement and folded cover sequences of Europe's Precambrian and Phanerozoic fold belt were formed and assembled. Europe's growth started about 3.5 Ga ago, in Archaean time, in the northeastern part of the Baltic Shield, and since then the growth continued episodically. Along the Geotraverse, the European crust becomes younger and younger, roughly speaking, in a southwards direction up to the present plate border at the Sardinian channel in the Mediterranean.

2.1 HOW FAR BACK DOES PLATE TECTONICS GO?

During the 1980s there was a growing recognition among geoscientists of the intimate relationship between plate tectonics and continental growth since early geological time. Seafloor spreading, subduction of oceanic lithosphere, formation of accretionary wedges at leading edges, ascent of calc-alkaline melts in magmatic arcs, development of fore-arc and back-arc basins, telescoping and docking of arc terranes to form continental nuclei, continental collision, escape tectonics and wrenching, formation of successor basins, rifting and break-up of continents, renewed sea floor spreading and so on, are processes that appear to have been functioning since the formation of the first Archaean continents (Friend *et al.* 1988, Hoffman 1989, Kröner 1991, Park 1991, Windley Chapter 6.1). We have been further encouraged to take this view by recent evidence (BABEL Working Group 1990) of deep seismic reflection images of an Early Proterozoic (1.9 Ga) collision zone which looks so similar to modern ones that it appears to have resulted from the same processes that occur in plate tectonics at the present.

Notwithstanding this uniformitarian view of plate tectonics, we must admit that the Earth's physical conditions, not least its thermal regime, have changed markedly since the Early Archaean. Even though the mechanisms and kinematics of the earliest plate tectonics might have been similar to modern ones, the compositions of their rock products must have changed with time. When the first Archaean lithosphere formed, the Earth's heat production was 2–3 times greater than that at present (McKenzie and Weiss 1975). This allowed ultramafic komatiitic (Mg-rich) melts to ascend at the mid-oceanic ridges instead of modern mid-ocean ridge basalts (Arndt 1983). Therefore the bulk composition of the subduction-derived calk-alkaline rocks of primitive arc terranes was also slightly more basic than that of younger igneous rocks. This, naturally, does not preclude the likelihood that, with regard to their chemical composition, trace element content and their calc-alkaline associations, supposedly mantle-derived granitoid members of Archaean age are very similar to younger granitoids. One important difference, however, is the lower potassium content of Archaean crustal rocks and this probably explains why early formed continental crust became less heat-productive than subsequently accreted crust. The high Archaean geothermal gradient did not preclude the development of a thick continental lithosphere (Haggerty 1986, Groves *et al.* 1987). Surplus heat possibly escaped by increased transient heat flow towards neighbouring komatiitic ocean areas. The early plate tectonics that governed Archaean crustal evolution were responsible for the formation of extensive greenstone and granite-gneiss terranes. By 2.0 Ga, the thermal regime of the Earth had cooled so much that more modern-looking plate tectonics took over. But only truly modern plate tectonics are blessed with a complete oceanic record.

Understanding petrological and geochemical processes, and how rock products have changed with time, is essential for an initial understanding of Archaean and Early Proterozoic plate tectonics, but this only allows us to propose idealised or generalised plate tectonic models. It does not provide a picture of the actual kinematics of the ancient plates involved. With no knowledge about the kinematics, plate tectonic interpretations of Archaean and Proterozoic crustal domains are bound to be as speculative as less mobile alternative models, such as the ensialic orogeny model (Kröner 1981, Martin and Elder 1983). But recent advances in palaeomagnetism have changed all this (Kröner 1991).

2.2 A PALAEOMAGNETIC KINEMATIC PERSPECTIVE

Thanks to modern advances in palaeomagnetism, a breakthrough has occurred in plate tectonics during the 1980s. By now, a much clearer picture of the last 600 Ma's plate kinematics is emerging, casting new light on the successive assemblage of Europe's Caledonian, Variscan and Alpine crustal domains. Throughout this book we use 'Variscan' to describe the Late Palaeozoic orogeny in Europe and for the resultant structures, irrespective of their trend. We reserve the term 'Hercynian' to describe the coeval worldwide orogenic events.

The pre-600 Ma assemblage history is still difficult to resolve with the same amount of detail as for the last 600 Ma, but important progress has been achieved all the same. In the crystalline basement most palaeo-poles of extra-European origin have been obliterated by metamorphism accompanying docking or plate collision. Palaeo-poles formed during and after the arrival and amalgamation of foreign terranes to Precambrian Europe are, however, well

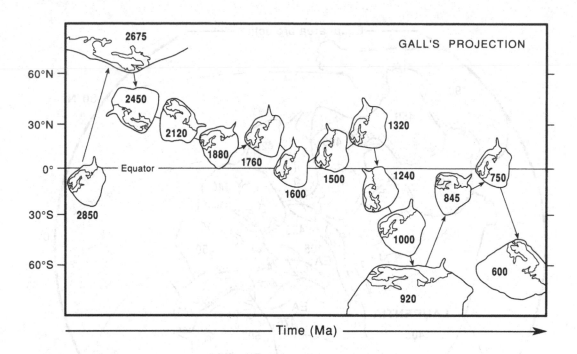

Figure 2-1. Precambrian drift history for the Baltic Shield portrayed in a Gall's projection. The time scale is arbitrary as E–W drift components cannot be shown. Ages are in Ma. This figure was kindly prepared by Trond Torsvik and is based on the work of Elming et al. (1992) and Torsvik et al. (1992).

preserved in the Baltic Shield and the shields of other continents. They show that both Archaean and Proterozoic Europe continental blocks were drifting around the globe, following a drift path different from those established for other continents. Admittedly the story is not complete, but all the same it provides substantial support for Archaean and Proterozoic plate tectonics. This progress would not have been achieved without the assistance of modern radiometric dating methods which make it possible to determine the age of a given palaeo-pole with sufficient precision.

Figure 2-1 shows a palaeomagnetic reconstruction of the Archaean and Proterozoic Europe's plate movements. In this type of diagram which, to non-palaeomagnetists, is more digestible than apparent polar wander paths, the palaeomagnetic latitude is plotted against an arbitrary time scale, and rotation (angle between palaeo-North and present North) is indicated. Not shown are the palaeomagnetically unresolvable longitudinal components in the drift movement.

Where unmetamorphosed cover rocks of pre-docking or collision age have their palaeo-poles preserved, the resolution becomes greatly increased. Paying due respect to constraints supplied by palaeoclimatic relations and distribution of fossil fauna and flora provinces, it is sometimes possible to trace the origin of a terrane back to its 'birth place' and to follow its later track and drift experiences, including its docking and post-docking history. Figure 2-2 shows an attempted reconstruction of the Late Cambrian to End Silurian drift of the terranes and continents which were assembled to form Caledonian Europe. Note how Baltica rotated anticlockwise through 180° during its northward wandering. The longitudinal positions shown are hypothetical.

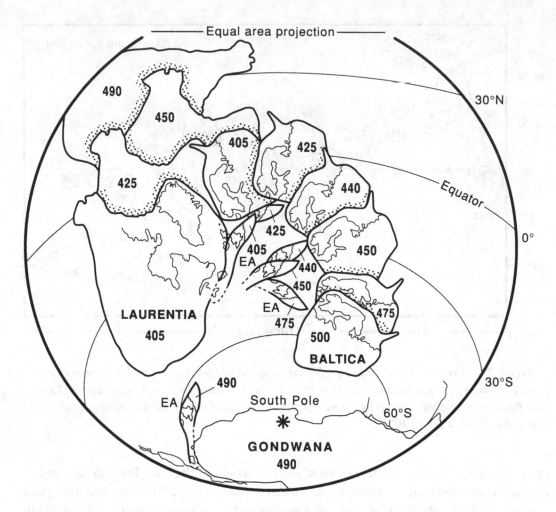

Figure 2-2. Drift of Baltica, Laurentia, and Eastern Avalonia (EA) during the Early Palaeozoic. Ages are in Ma. Note Baltica's anticlockwise rotation after 520 Ma. Gondwana-derived terranes arriving during the succeeding Variscan evolution are not shown. This figure was kindly prepared by Trond Torsvik and is based on the work of Torsvik and Trench (1991), Trench and Torsvik (1991) Torsvik et al. (1990a,b, 1991 and 1992).

2.2.1 MODERN ART – A EUROPEAN COLLAGE

So what is the present state of the art in terrane tracking? Admittedly, we are unable to apply strictly the principles laid out by Coney (1980) to distinguish and delineate all suspect terranes. For this, our data are not sufficiently detailed. In some cases they only relate to basement rocks, in others only to cover rocks. But we know that the crystalline crust of some terranes has undergone a prolonged and complex evolution comprising a number of amalgamation and separation stages before eventually they were permanently welded to Europe. Considering the very long time span encompassed by the EGT, errors in tracking may eventually cancel out overall, so that the collage patterns established for the major orogenic divisions can be regarded as reasonably representative. Figure 2-3 shows our present conception of Europe's division into terranes and crustal domains. The boundaries between the individual terranes and domains are shown where they outcrop at surface or subcrop under younger cover.

Because of the marked contrast in surface and subsurface geology between the geologically old and young regions in Europe (as will become apparent in Chapter 3), a division into terranes and domains of comparable origin and crustal significance must be interpretative. It depends largely on the availability of geophysical data, or rather a range of geophysical information, that has now been produced along the EGT. In the high-grade metamorphosed Precambrian crust of the Baltic Shield, a terrane boundary can be located by means of a prominent geoelectric anomaly, whilst in the southern segment of EGT, deep seismic profiling and tomography serve to outline the young to recent plate boundaries at comparable crustal and lithospheric levels in the Mediteranean region.

2.2.2 HOMEMADE AND SUSPECT TERRANES

Only limited parts of Europe's continental crust bear the imprint 'Made in Europe'. Much was formed elsewhere and was imported from 'abroad' as suspect (or displaced) terranes. On their way to Europe, these terranes travelled long distances as single or composite terranes, micro-continents or as part of a large continent, changing from high to low, or from northern to southern latitudes and back again, before eventually joining up with Europe. Not all crustal terranes that came to harbour in Europe remained there. Major break-up events at the Archaean–Proterozoic transition, during the Late Proterozoic to Early Palaeozoic, and the Mesozoic caused crustal losses, but in the long run Europe gained in size with time.

A striking feature in Figure 2-3 is the similarity of the overall terrane structure between the so-called old and stable Precambrian crust of NE-Europe, and the Phanerozoic crust of western and southern Europe. This, we think, is a consequence of the similar mechanics and kinematics of the ancient, early and modern plate tectonics. Differences between these two parts of Europe we relate to a change with time in the petrological products, as well as to different degrees of consolidation and levels of erosion exposed at surface.

It should be recalled that no simple relation between crustal age and depth of erosion exists. Large areas of the Baltic Shield have only been denuded to 5–10 km depth since the Early Proterozoic whereas other areas formed at the same time, but in different tectonic environments, suffered much more advanced exhumation, to depths of about 25 km or more. In the Variscan domain, deeply eroded terranes are also found side by side with flysch basins which almost escaped erosion.

2.3 DECLINE AND FALL OF AN OROGEN

From their surface and subsurface geological expression, the Caledonian, Variscan and Alpine domains crossed by the EGT form an almost ideal evolutionary series to illustrate the decline and fall in time of an orogen.

The North German–Polish Caledonides, the oldest Phanerozoic orogen crossed by the EGT, have long since collapsed and are now buried under a thick cover of Late Palaeozoic and younger sediments. The conclusive evidence for the former existence of a Caledonian fold belt from England across the North Sea to northern Germany and Poland (Figure 2-3) was not produced until a number of deep boreholes drilled in search for oil and gas were found

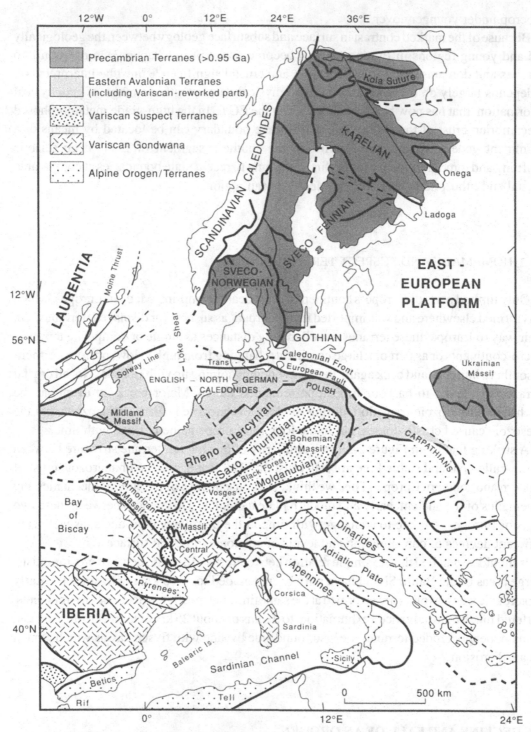

Figure 2-3. The 'terrane collage' of Precambrian and Phanerozoic Europe, a simplified sketch. Sutures and orogenic fronts are shown as bold lines, internal borders as thin or thin broken lines. Note that the size and shape of the terranes do not change significantly with time (approximate direction of younging is from north to south).

to terminate in Caledonian basement (Frost *et al.* 1981, Ziegler 1982).

In the Variscan domain, surficial to mid-crustal orogenic structures and a number of deep-reaching sutures can still be traced from one Variscan massif to another, and have been drilled to moderate depths beneath intervening younger basins. Based on surface geological studies,

Kossmat (1927) provided the first valid tectonic zonation of the Variscan crustal domains. It emphasised first of all the post-collisional setting. Recent reviews of Variscan geology are presented by Franke (1989a, Chapter 6.3), Matte (1991) and Ziegler (1988, 1990). In the Alpine domain, seismicity demonstrates that plate convergence is still going on, young flysch and molasse basins are widespread, the nappe roots and the crystalline rocks are hardly exposed as yet, and the crustal structure has become further complicated by the development of local oceanic basins of Tertiary to Recent age in the Mediterranean which are probably about to close as a consequence of further plate covergence between Europe and Africa. The pioneering studies and syntheses of the structure of the Alps date back to the last half of the 19th century and the dawn of the 20th (Escher, von der Linth, Heim, Schardt, Lugeon, Argand, Termier, etc). Recent reviews are provided by Trümpy (1990), Laubscher (1989), Ziegler (1990) and Pfiffner (see Chapter 6.4).

In this comparison between the states of degradation of orogens of different ages, the Precambrian has so far not been mentioned. This is because the post-orogenic evolution of the Baltic Shield followed a different line, or rather several different lines. There are various reasons for this which will be discussed in the following chapters. One important circumstance has been the almost constant freeboard of the shield since the Mid Proterozoic, a feature that is probably related to the particular three-layered crustal structure in most of the shield's Archaean and Early to Mid Proterozoic orogens, with a lowermost, high velocity layer (V_p >7.0 kms^{-1}, see Figure 3-4) of dense mafic material which has functioned as a ballast load, stabilising the shield's thick crust.

2.4 HOW EUROPE'S CRUST EVOLVED

We will now briefly review how Europe's continental crust was formed and assembled, beginning with the Precambrian. For ease of reading, we have hyphenated composite terms such as Sveco-Fennian and Rheno-Hercynian, even though this is not the common practice.

2.4.1 A DIFFICULT START: ARCHAEAN EUROPE

Europe's crustal evolution started *ca.* 3.5 Ga ago in present Russian Karelia, north of Lake Onega. Throughout the middle and most of Late Archaean time, accretion of greenstone and granite gneiss terranes was maintained (Windley, Chapter 6.1). Towards the close of the Archaean, a continental nucleus of Archaean Europe had evolved, probably of considerable size. However, this nucleus no longer exists as such. It became rifted and broke apart at the Archaean–Proterozoic transition (2.5 Ga) and the fragmented terranes separated.

2.4.2 REASSEMBLAGE AND GROWTH OF PROTEROZOIC EUROPE

During the Early Proterozoic (2.0–1.9 Ga) some or all of the dispersed Archaean terranes were reassembled, and a collisional fold and thrust belt was formed, trending between NW–SE and E–W, through the Kola Peninsula in the far north. Thus the Kola–Karelian orogenic domain originated, comprising a number of reworked Archaean crustal terranes stacked together with belts of newly formed Early Proterozoic sedimentary and magmatic material. The tectonics of this *ca.* 1.9 Ga old thrust belt are very similar to modern collision belts. Between 2.0 and 1.5 Ga, juvenile Early to Mid Proterozoic crustal terranes consisting of 2.2–1.5 Ga old arc-type crust were telescoped on to the formerly passive margin southwest of the reassembled Archaean; first the Sveco-Fennian orogenic domain, then the Old and Young Gothian domains. Proterozoic Europe was growing. This lateral crustal growth off the SW margin of the Karelian was characterised by accretion of juvenile terranes.

Interpreting the FENNOLORA refraction and wide-angle seismic profile, Guggisberg and Berthelsen (1987) argued that the deep crustal and sub-crustal seismic structure of the Baltic Shield probably still images these Proterozoic events, and recently the BABEL Working Group (1990) published a deep reflection seismic section which convincingly depicted a 1.9–1.86 Ga collisional suture (see Figure 3-6). The seismic memory of the lithosphere of the Baltic Shield appears to be like that of old people: it recalls best what happened during childhood and youth. It is interesting to note that the present heat flow in the Baltic Shield is 40–50 mWm^{-2} in its northeastern part where Archaean crust predominates, but it increases southwestwards to 60–70 mWm^{-2} in the Proterozoic crustal domains (Balling 1989, Chapter 4.1). Despite the influence of the thicker lithosphere in the north, it is tempting to explain part of this change in heat flow as being due to a lower potassium and rare-earth element (REE) content in the Archaean than in the Proterozoic crust, the memory of which lingers on.

During the Early Proterozoic, the North America–Greenland continent was Europe's close neighbour. Side by side, the two continents shared a common active margin, developing similar age zonations (Hoffman 1989). Around 1.5 Ga ago, Proterozoic Europe had probably reached its maximum size. Intrusions of local dyke swarms and rapakivi massifs then testified that it had cooled and stiffened, and had been converted into a craton. When the Grenville orogeny was about to start in adjacent North America, *ca.* 1.35 Ga ago, Proterozoic Europe left its neighbour and drifted away for a while (cf. the change in Baltica's drift in Figure 2-1). Clockwise rotation during the drift gave rise to rifting and downfaulting of Jotnian redbeds, and the Central Scandinavian Dolerite Group was intruded (Gorbatschev *et al.* 1987).

2.4.3 THE SVECO-NORWEGIAN DETOUR

When Europe again approached North America, this time from a different angle and in southern latitudes, the Sveco-Norwegian orogeny (1.1–0.95 Ga) began in present southwestern Scandinavia. It culminated when the two continents collided, and continued plate convergence caused peeling and thrust stacking of pre-existing (1.8–1.5 Ga old) crust in adjacent parts of the craton. With the close of the Sveco-Norwegian orogeny, Mid Proterozoic Europe had probabaly decreased in size, but it was fringed to the west by a lofty Sveco-Norwegian mountain belt which formed a worthy counterpart to the Grenville thrust belt of North America (Berthelsen 1987, 1990, Gower 1990).

Before long, the drift path took a sharp turn towards equatorial latitudes and the thickened crust of the Sveco-Norwegian orogen suffered gravitational collapse. In this way, 0.92 Ga old granulite facies rocks were probably brought close to the surface and dyke swarms were emplaced along the eastern margin of the orogen. Erosional leveling progressed rapidly and a post-Sveco-Norwegian peneplane was formed as an extension of the Mid Proterozoic erosional surfaces that had been preserved in other parts of the craton. No Sveco-Norwegian molasse has been preserved but, apart from this, the Sveco-Norwegian tectonic evolution bears great ressemblance to that of the younger Phanerozoic orogens.

Interestingly enough, the crustal structure of the Sveco-Norwegian orogen resembles Phanerozoic Europe's rather than those of the Early to Mid Proterozoic laterally accreted crustal domains. No high-velocity lower crustal layer is found, except under the orogen's easternmost part, where 3-layered Sveco-Fennian Gothian crust appears to underlie a west-dipping shear zone (Green *et al.* 1988).

With the decline of the Sveco-Norwegian orogen, the European Proterozoic craton drifted towards equatorial latitudes. Associated rifting caused the Vättern graben system to form in southern Sweden where up to 1 km of 0.85–0.7 Ga old fluviatile to marine clastic sediments are downfaulted close to the eastern border of the Sveco-Norwegian orogen and unconformably overlying the marginal shear zone.

2.4.4 BREAK-UP OF PROTEROZOIC EUROPE AND FORMATION OF BALTICA

Around 0.75 Ga ago, Proterozoic Europe began to drift again towards high southern latitudes where the Gondwana continent was being assembled during the Pan-African and Cadomian orogenies (0.65–0.55 Ga). Africa, South America, Australia and India all belonged then to Gondwana. Proterozoic Europe remained at high southern latitudes during Cambrian time but at the turn of the Early Ordovician those parts which now constitute the Baltic Shield and the basement of the East European Platform broke off and drifted away as an independent plate, 'Baltica', predestined to become the backbone of future Caledonian Europe. What we here call Baltica was named Fennosarmatia by Stille (1929), and could also be called Ancient Europe or the Russio-Baltic Platform. Baltica is preferred for the sake of brevity, and for consistency with the palaeomagnetic reconstructions.

Other parts, maybe as much as half of the original Proterozoic Europe, had become welded on to Gondwana, strongly influenced by the Cadomian orogeny. They were left behind, at least for the time being. Originally they may have comprised crustal domains formed between 1.5–1.35 and 1.1–0.9 Ga when Proterozoic Europe was situated next to North America, and when the two continents shared a continuous active margin.

After the Cadomian orogeny, deep rifted sedimentary basins developed across those parts of Proterozoic Europe which by then had become a part of Gondwana. During the Late Ordovician, they experienced, like neighbouring regions of Gondwana, a major glaciation, the 'Saharan' glaciation. Meanwhile, a wide ocean, the so-called 'Tornquist Sea', was opening between the ice-covered relics and the northward drifting and anticlockwise rotating Baltica, Figure 2-2.

2.4.5 BALTICA'S DIVERSIFICATION

Most of Baltica was covered by shallow epicontinental seas during the Cambrian and Ordovician and today, shelf-type Cambrian and Ordovician sediments still fringe the Baltic Shield. In those parts of Baltica that are now known as the East European Platform, sedimentation continued with little interruption throughout the Palaeozoic, persisting in places even into Mesozoic times. Beneath the East European Platform, Baltica's Precambrian basement now generally lies at depths of 1–3 km except in the Ukranian massif where it is exposed at surface and in the Voronech uplift where the cover is so shallow that large-scale open-pit mining of Archean and Proterozoic banded iron formations can be performed.

The eastern border of the East European Platform is defined by the Urals which formed in Carboniferous–Permian time when the eastern platform margin collided with newly formed island arcs and the Kazakstan plate. Only the western part of the ensuing collisional belt, the Uralides, is now exposed in the Urals; the eastern portions are hidden under the (hydrocarbon-producing) sedimentary basins of western Siberia.

2.4.6 BALTICA COLLIDES WITH LAURENTIA

In contrast with the East European Platform, the northwestern part of Baltica underwent a different geological evolution. During the Early Palaeozoic whilst Baltica's eastern margin (prior to the developement of the Uralides) was situated on the tectonic trailing edge of Baltica, the northwestern margin was acting as the leading edge.

The margin was originally passive when Late Proterozoic Europe separated from North America, Greenland, and NW-Scotland, which were at that time united as a single continent, 'Laurentia'. It remained a passive margin during the latest Precambrian and Cambrian when Baltica (as part of Proterozoic Europe) paid a visit to Gondwana. Upon its departure, however, its anticlockwise rotation 520–500 Ma ago made it face and approach Laurentia (formed from the union of Scotland, Greenland and North America). The ocean in between, the so-called 'Iapetus Ocean', was closing whilst Baltica drifted towards milder climates. This resulted in the formation of island arc terranes sheltering backarc basins along the mutually approaching continental marigins.

After initial obduction of ophiolitic sequences and the collision of arc terranes, Baltica collided with Laurentia and the two continents soon united into one super-continent, 'Laurussia'. The timing of the collision is still a matter of debate. Some researchers emphasise the occurrence of marine Silurian sediments in the Scandinavian Caledonides and claim that collision did not take place until during the Late Silurain 'Scandian orogeny' (see also Figure 2-2). Others explain this orogeny by continued post-collisional plate convergence and consider the collision (and ensuing peak of eclogite facies metamorphism) to be of Late Ordovician age, allowing marine conditions to survive in places. Because palaeomagnetic data cannot constrain the E–W component of movements, it is possible to put forward kinematic models, with quite different drift paths, for both an early and a late collision between Laurentia and Baltica. According to the early collision model, which we prefer, a much thickened crust was being formed around 430 Ma ago in the Scandinavian collision belt (Bucher-Nurminen 1991). This promoted uplift with gravitational nappes spreading on to the adjacent foreland of Baltica. In the Early Devonian, strike-slip escape and extensional collapse took over. Coeval with the Devonian collapse in the Scandinavian Caledonides, the crust of the Baltic Shield was forced into a gentle dome-like peripheral bulge that became the core of the so-called 'Old Red Continent' of Devonian Europe.

The present-day relief of the Scandinavian and Scottish Caledonides stems from a Late Tertiary uplift in response to coeval rifting in the North Sea and the Atlantic (Ziegler 1988).

2.4.7 ENTER AVALONIA: THE CALEDONIAN TRIPLE JUNCTION

During the Early Silurian, a third partner joined the Caledonian scene in Northwest Europe. A Gondwana-derived terrane, Eastern Avalonia (from here on just called Avalonia) was approaching the united continents of Laurentia and Baltica which had already collided, docking sidewise along the south-facing Laurussian margin. Significant strike-slip and transpressional movement along the Avalonia–Laurussia suture persisted to the end of the Silurian and, together with post-docking plate convergence absorbed within Avalonia, it caused the rise of yet another Caledonian mountain belt, here called the English–North German–Polish Caledonides to distinguish it from the term used previously, 'Mid-European Caledonides', which also included the Caledonian-deformed parts of Variscan Europe (cf. Ziegler 1988, 1990).

The northwest European Caledonian triple junction had been established with its centre lying in the northern part of the present North Sea (Figure 2-4). Its western leg, the Avalonia–Laurentia suture, corresponds to the Solway line (marked 1 in Figure 2-4) across the British Isles (Matthews and the BIRPS Group 1990, Klemperer and Hurich 1990); whilst the eastern leg, the Avalonia–Baltica suture, is presumably indicated by the Trans-European fault, shown as 2 in Figure 2-4 (EUGENO-S Working Group 1988). These two legs, which together form the Avalonia–Laurussia suture, are still detectable as deep seismic reflection images in the lower crust. The northern and oldest leg of the triple junction, the Laurentia–Baltica suture, is more difficult to trace because not only has there been large-scale post-collisional shortening, but also later strike-slip movements, and finally crustal extension and rifting, have been superimposed.

2.4.8 THE ENGLISH–NORTH GERMAN–POLISH CALEDONIDES

The English–North German–Polish Caledonides stretched from southern Ireland, Wales and England across the central and southern part of the North Sea to northern France, Belgium, Holland, Northern Germany and Poland (Figure 2-4). Their overall structure is outlined by deformed Caledonian, low to medium-grade metamorphosed accretionary belts that surround at least two Cadomian (and older) basement massifs with a barely disturbed Early Palaeozoic cover. One is the Midlands massif of southern England (Pharaoh *et al.* 1987) and the other is the Lüneburg massif (Figure 2-4) which is situated to the south of the Elbe Lineament in northern Germany and southern Poland.

The former existence of a Caledonian mountain belt in North Germany and Poland during the Late Silurian is also documented by the large quantities of detritus that were shed northwards over adjacent parts of Baltica where a deep foreland basin developed during the Late Silurian (EUGENO-S Working Group 1988). The southernmost part of this foreland basin was probably cannibalised when the north-verging Caledonian thrust front encroached on Baltica. The deep crustal Avalonia–Baltica suture, we believe, was not outlined by the Caledonian front (Figure 2-3). Presumably it is to be found more to the south at the Trans-European fault close to the Baltic coast of Germany (EUGENO-S Working Group 1988), as indicated in Figure 2-4. Together with the Solway line across the British Isles, the Trans-European fault appears to cut the overall structure of the English–North German–Polish Caledonides discordantly, which would suggest that important strike-slip movement along the suture has occurred since Avalonia's first contact with Laurussia.

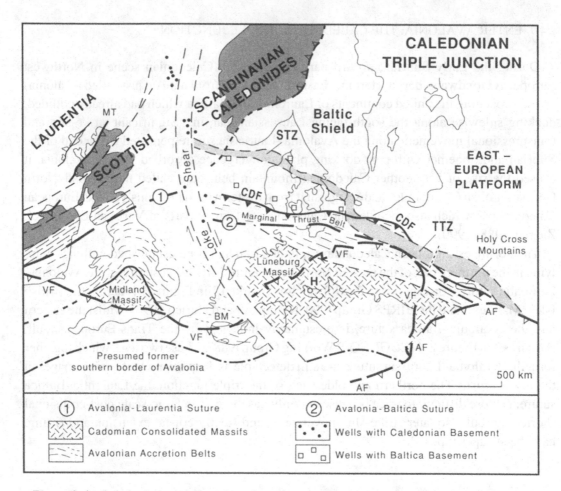

Figure 2-4. Caledonian triple junction of northwestern Europe. The Laurentia–Baltica Suture is hidden within the Scottish–Scandinavian Caledonide Belt. BM: Brabant massif. VF: Younger overlapping Variscan front. AF: Alpine front. H: Locality in Harz where late-Variscan gabbro intrusion has brought Cadomian gneiss to surface.

In this connection it is interesting to find that whilst most metamorphic ages determined from drill core samples fall into the range 450–440 Ma (Late Ordovician), others are as old as 530–490 Ma (Late Cambrian–Early Ordovician), and some are as young as 420–400 Ma (Late Silurian – corresponding to the final emplacement and rise of the North German–Polish mountain belt and the development of the northern deformation front). When leaving Gondwana and drifting towards the colliding Laurussia, an active margin was apparently developed in front of the Avalonian terranes, while an extensional passive margin was developed at Avalonia's rear. In other words, the English–North German–Polish Caledonides were brought to their present position in Europe as a 'ready-made' terrane collage, whose overall structure became but little modified by later Caledonian overprints. On faunal evidence, it has been argued that Baltica and Avalonia could not have been widely separated and then again brought into juxtaposition by long-range drift or strike-slip displacements because they share a common Early Cambrian 'Baltic' trilobite fauna (Bergström 1984). However, in the light of the recently established drift paths for Baltica, these relations become more of a support to 'strike-slip' speculators than an obstruction.

The existence in the central North Sea of a curved, N–S to NW–SE trending shear zone called the Loke shear has recently been postulated by Berthelsen This shear appears to cause a dextral offset of the suture of, perhaps, 150–200 km, and of the Caledonides to the south.

The Loke shear is believed to have been initiated in Early Devonian time. To the west of it, Caledonian deformation was active throughout the Early Devonian (Soper and Woodcock 1990), whilst to the east it ceased at the very end of the Silurian. The Loke shear apparently influenced the Mid-Devonian palaeogeography of the North Sea region (cf. Ziegler 1982, pl. 7) and it probably predestined the later development of the Central-Viking Graben rift system. The crustal extension accompanying this later development may account for a considerable part of the otherwise exceptional width of the shear.

2.4.9 WHOLESALE LITHOSPHERIC TERRANES OR AN OROGENIC FLOAT?

The greater part of the English–North German–Polish Caledonides is now covered by the thick sedimentary sequences of the Variscan foreland basin, the Southern Permian basin, the Mesozoic North German–Polish basins and the Tertiary North Sea basin. Outcrops are only found in the British Isles, in the Brabant massif (BM in Figure 2-4), in the Ardennes of France and Belgium and in the Holy Cross Mountains of Poland. In between, geophysical and scattered borehole information is all that is available. Caledonian granites are known to occur locally in the basement (Lee *et al.* 1990) but there is no evidence of the presence of any major subduction-related calc-alkaline batholiths – neither due south, nor due north of the presumed Avalonia–Laurussia suture as might be expected if large quantities of oceanic lithosphere had been subducted prior to the suturing. Considering how far Avalonia had travelled before docking, this obviously calls for an explanation.

One way to explain it would be to assume that the convergent boundary between the Avalonian plate and the Laurentia–Baltica plate system was located not too far off Avalonia, adjacent to the wide Tornquist Sea. Subduction of oceanic lithosphere under the leading edge of the Avalonian plate would then have caused the construction of accretionary wedges and primitive arc terranes. With time, as Avalonia drifted, and the site of subduction shifted to new positions, the arc terranes would have been telescoped on to Avalonia. Such a model would explain the ready-made structure of Avalonia prior to its docking, and because the wide Tornquist Sea would have been consumed in time shifting, short-lived subduction zones, no major magmatic arcs or batholiths would have formed. An alternative explanation could be that the Avalonian massifs and separating accretionary slate-schist belts actually formed an 'orogenic float,' made up of subcrustally detached terranes. In this case, the subduction zone would have dipped towards the Laurussian continent, but as additional terranes were added to the 'float' the subduction zone would have been situated successively further and further away from the continent, as in the model suggested by Oldow *et al.* (1989) for parts of the North American Cordillera.

Whilst the first of these models results in the production of a wholesale lithospheric collage, where upper lithospheric structures concur with those of the lower lithosphere, the 'orogenic float' model produces disharmonic upper and lower lithospheric structures. In Chapter 6, we will return to this issue, which may also be relevant to Variscan crustal evolution.

2.4.10 VARISCAN SEQUEL: THE PRE-COLLISIONAL SETTING

The docking of Avalonia at the southern border of Laurussia both concluded the Caledonian and heralded the Variscan development, which came to an end with the assemblage of a Late Palaeozoic supercontinent, which the inventor of the continental drift theory, Alfred Wegener, named 'Pangaea'.

During the Early and Mid Devonian, when the North German–Polish Caledonides had collapsed, thick clastic sequences derived from the Old Red Continent in the north spread southwards over the rifted and attenuated crust of Avalonia's Gondwana-facing margin. Today the clastics and overlying carbonates, together with younger flysch, form part of Kossmat's Rheno-Hercynian zone. In this zone, weakly metamorphosed Rheno-Hercynian sequences are detached from their original basement and have been thrust over the more or less imbricated units of the post-Cadomian cover of the southern part of the Lüneburg massif. Let us turn back to the pre-collisional set-up, although, in doing so, we move into the field of speculation. Strong post-collisional tectonic events have overprinted and obliterated most of the early features or buried them deep down in the crust. Sparce palaeomagnetic data available from the vicinity of the EGT, a tenuous biostratigraphy, DECORP deep seismic profiles, and geochemical hints as to the geotectonic origins of the more or less metamorphosed magmatic rocks of either known or unknown age can all be used in support of quite different models, from the more fanciful (e.g. Frank *et al.* 1977) to the more prosaic. A mobile model such as the one presented in Figure 2-5 presumes that the proto-Rheno-Hercynian southern margin of Avalonia was originally separated by a fairly wide ocean from the proto-Saxo-Thuringian terrane which remained adjacent to Gondwana up to the close of the Ordovician. Most of this ocean is assumed to have been consumed during the Silurian at an intra-oceanic, north dipping subduction zone, at the leading edge of which a volcanic arc was being formed in pre-Devonian time. At the beginning of the Devonian (Figure 2-5a–b), back-arc spreading was initiated north of this subduction zone, behind the pre-Devonian arc. Towards the close of the Early Devonian, the proto-Saxo-Thuringian terrane, which travelled northwards along with the subducting plate, finally collided with the arc and overrode it (Figure 2-5b). Later upthrusting of parts of this arc may account for the presence of Silurian high-grade metamorphic rocks within the Saxo-Thuringian. The cessation of northward subduction enabled olistostromes with Ordovian and Silurian sediments from the front of the overriding terrane to move into adjacent parts of the young back-arc basin. However, subduction was soon resumed, but with an opposite polarity (Figure 2-5c). Back-arc generated oceanic crust was subducted under the proto-Saxo-Thuringian terrane, converting its northern part into an Andean-type magmatic arc, the so-called mid-German Crystalline high (Figure 2-5d). Meanwhile Middle Devonian pelagic shales and radiolarian cherts were deposited in the remainder of the back-arc ocean north of the subduction zone. In turn, they were overlain by Upper Devonian to Lower Carboniferous greywacke turbidites, supplied into the closing back arc basin by the rising magmatic arc in the south. With the closure of the back-arc basin and the Saxo-Thuringian terrane's collision with Avalonia (Figure 2-5e), slices of Devonian oceanic crust and sediments scraped off it were telescoped on to Avalonia's margin as the Lizard–Giessen–Harz nappes (*ca.* 330 Ma ago). Following this, upthrust and retrogressed rocks from the accretionary wedge north of the 'mid-German Crystalline high' formed the 'Northern Phyllite zone' in front of the then deeply eroded volcanic arc with abundant Devonian–Early Carboniferous calc-alkaline plutonics. The present southern part of the Saxo-Thuringian domain corresponds to the rifted passive margin of this originally Gondwana-derived terrane. On this margin at the rear of the terrrane,

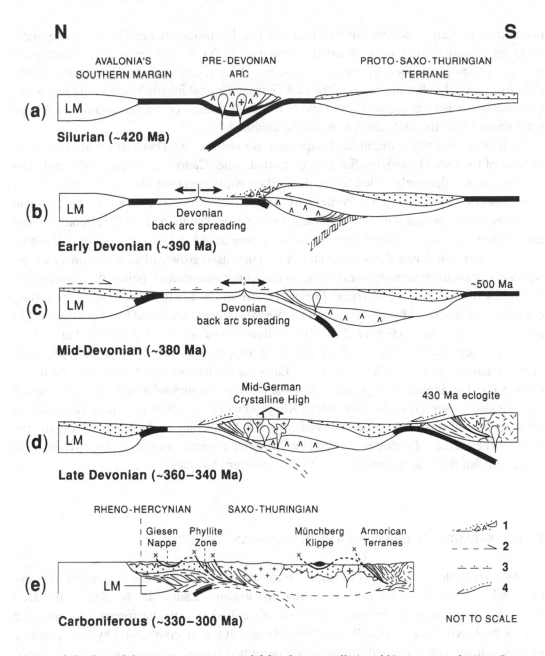

Figure 2-5. Fanciful tectonic cartoon model for the pre-collisional Variscan evolution. Pre-Devonian oceanic crust in black, Devonian in white. LM: Lüneburg massif. 1: Late Early Devonian olistostromes. 2: Clastic influx from north. 3: Pelagic shales and chert. 4: Advancing flysch turbidites. See also an alternative model in Figure 6-14.

sedimentation continued through Devonian to Early Carboniferous time, until other so-called Armorican terranes arrived from the south (see below). Admittedly, evidence for such a mobilistic model is meagre, much of it stored away at deep crustal levels. The occurrence in the southern Saxo-Thuringian zone of glacio-marine deposits related to the Saharan glaciation of Gondwana is dubious evidence which could also be accounted for by letting icebergs, and not micro-continental terranes, drift northwards. Therefore, several research-ers (see also Chapter 6.3) prefer a model where the basement of the Saxo-Thuringian zone is considered to be a part of Avalonia that was split off when Ordovician–Silurian rifting gave way to seafloor speading and a narrow Devonian ocean started to form between the split off part and the remainder of Avalonia (cf. Figure 6-11). In this model, pre-Devonian magmatic rocks are considered to have originated in continental rift environments, and pre-Devonian metamorphism is ascribed to crustal extension. The formation of the Devonian to Early Caboniferous calc-alkalinc rocks of the mid-German Crystalline high is not only explained by southwards B-type subduction of the (narrow) Devonian ocean but additional A-type subduction of continental lithosphere is also assumed.

Whichever way that it might have happened, we venture to go back again in time. With the end of the Late Ordovician Saharan glaciation, other Cadomian-influenced Gondwana terranes, here collectively called 'Armorica', rifted away from Gondwana. The bulk of their crust was of Cadomian origin, carrying remnants of 2.45–2.1 Ga old rocks. Post-Cadomian cover sequences included typical Saharan-type tillites and contained clastic zircons derived from Cadomian and older crystalline rocks, some even as old as 3.8 Ga (Gebauer *et al.* 1989).

Northward drift during the Silurian and Devonian caused growth of an accretionary wedge with high-pressure metamorphism in front of the Armorican terranes before these eventually collided with the growing Variscan Europe. Close to the EGT, in the Bohemian massif, collision was preceded by subduction of *ca.* 500 Ma oceanic crust and by obduction of 430 Ma old eclogitic rock (Figure 2-5c–d). Collision was followed 330–320 Ma ago by northwestward tectonic transport of a pile of nappes with inverted stratigraphic and metamorphic sequences. They now rest on Early Carboniferous flysch and older sediments of the Saxo-Thuringian zone. Remnants of such 'exotic' nappes with eclogite in their upper part are also preserved in the Münchberg Klippe situated *ca.* 100 km north of the collision suture (Figure 2-5e). Armorican derived crust also forms part of the Black Forest and the Vosges, the northern Central massif and the Armorican massif but syn- and post-collisional strike-slip faults render correlations with these western terranes less clear.

2.4.11 ASSEMBLAGE OF PANGEA: THE VARISCAN OROGENY

Having despatched the suspect Armorican arc terranes as forerunners, the remaining entity of Gondwana itself started to drift northwards towards Laurussia, heralding the Late Paleozoic assemblage of Pangaea. Gondwana's collision with the forerunners occurred during the Devonian between 380 and 360 Ma ago. It was accompanied by high pressure metamorphism, and it welded further newly deformed crustal terranes to Europe; those of the Iberian Indentor, southernmost France and some of the Variscan massifs that were later on caught up in the Alpine orogeny. South of these terranes, an African foreland bordered the evolving Varsican belt. Plate convergence continued for almost another 100 Ma, turning the Variscan belt into a very wide Alpine-type orogen. The post-collisional convergence produced large-scale crustal stacking down to present lower crustal levels, overprinted by 'Hercynian-type' low pressure metamorphism, syntectonic flysch deposition, and prolonged emplacement of S-type granites, whereby the tectonic zonation became accentuated. Indentation of 'African' promontaries in Iberia and the Bohemian massif caused the formation of two major syntaxial bends as well as large-scale lateral strike-slip escape, parallel or oblique to the tectonic zonation. This post-collisional orogenic activity migrated with time from the centre towards the marginal parts of the belt, the final thrusting and folding affecting the Late Carboniferous coal-bearing molasse basins which extend from southern Wales (Britain) to Silesia (east of Bohemian massif). These foreland basins were partly overridden by allochthonous Rheno-Hercynian units. Through this prolonged evolution, the 700–800 km wide Variscan belt finally acquired its present characteristic bilateral symmetry in cross section, with thrusts and fold structures verging towards the two forelands, as already visualised by Kossmat (1927).

This extremely mobile evolutionary picture, with a large amount of crustal shortening, forces us to presume that a deep crustal root, if not several roots, must have been developed. How can this be reconciled with the fact that the Moho has been mapped under the European Variscan crust along a number of deep reflection and refraction seismic profiles at a constant depth of 30 km (see Figure 3-9). This apparent paradox will be discussed by Mengel in Chapter 4.3 and again in Chapter 7.3. Degradation of the Variscan fold belt, including extensional collapse along low-angle detachments, faulting and fragmentation into resistant massifs and other parts that became buried under younger sedimentary basins have been continuing since Permo-Carboniferous time, when subsequent bimodal volcanicity dominated in the orogen. But this degradation could hardly, on its own, account for the removal of a deep crustal root.

2.4.12 EAST–WEST CONNECTIONS: ORIGIN OF THE TORNQUIST ZONE

When post-collisional N–S convergence in the Variscan belt of central, western and southern Europe had ceased in the Carboniferous, Gondwana no longer acted as a firm vice in the south. A system of large dextral strike-slip faults with connecting pull-apart structures developed to accommodate the continuing convergence and collision of the Uralides in the east and the Appalachians in the west. The distribution of subsequent widespread Permo-Carboniferous magmatism was largely controlled by this faulting (Ziegler 1982).

Deep seated dextral strike-slip faulting giving rise to magmatism also influenced the Caledonian-assembled foreland north of the Variscan orogen, and it even affected the adjacent southwestern part of Baltica. Here, in the Danish–Scanian area and around the southern Baltic Sea, Upper Silurian clastics derived mainly from the North German–Polish Caledonides had been deposited on top of the older, Early Palaeozoic sediments of platform facies that covered the Precambrian basement of Baltica. This Late Silurian foreland basin and its substrate were cut up in fault-limited segments when a WNW-widening splay of deep faults branched off the Avalonia–Laurussia suture at the Trans-European fault in northwestern Poland. Out of this splay, and the suture southeast of it, grew the Tornquist fault zone.

After this, the more westerly part of the Caledonian suture at the Trans-European fault in northernmost Germany and the eastern North Sea partly lost its tectonic integrity. Permo-Carboniferous and later tectonic activity now became diverted into the southwestern part of Baltica. The Permo-Carboniferous activity followed a branch across Scania in southern Sweden that continued due west of the Swedish west coast, joining up to the north with the Skagerrak–Oslo rift system. In Scania, a dense swarm of dolerite dykes was intruded more or less parallel with the fault zone. Mesozoic rifting, however, changed the trend and made it cross northern Jutland and enter the northern North Sea. Finally, Cretaceous–Tertiary inversion reshaped the Sorgenfrei Tornquist zone (STZ in Figure 2-4) into what we know today (EUGENO-S Working Group 1988). The Sorgenfrei Tornquist part of the Tornquist zone can therefore rightly be considered a tectonic 'parvenu' that does not outline Baltica's former boundary. Its development, where trespassing the already weakened southwestern part of Baltica, was caused by intraplate tectonics emanating from the squeezed and much tightened Variscan belt, the distal effect of the Mesozoic breakup of Pangaea, and the subsequent Alpine collision.

On the other hand, the Teisseyre Tornquist part of the Tornquist zone between the Baltic coast of Poland and the Carpathians is coincident with the Trans-European fault and Avalonia–Laurussia suture. This part, therefore, is of more noble and ancient origin. None the less, the crustal structure of the Polish Teisseyre Tornquist zone (Guterch *et al.* 1986) and that of the Sorgenfrei Tornquist zone due south of Scania (BABEL Working Group 1991) appear to show a great many similarities.

2.4.13 FINAL BREAKUP OF PANGEA: THE TETHYAN EVOLUTION

With the entire assemblage of Gondwana and Laurussia locked together into the single continent of Pangaea, its massive continental entity extended, in Early Jurassic times, virtually from one pole to the other, giving the world a very different configuration of continent and ocean from the present. It could not last. During the Jurassic the southern half of Pangea began to shear eastwards relative to the north and a huge sinistral strike-slip system developed as the two halves wrenched and rifted apart. A new ocean, the Tethys, opened in the east and grew in size. In the west, early Jurassic transtension produced rifting between Africa and America and, as the continents split apart, led to the opening of the central Atlantic by the Late Jurassic. Ziegler's (1990) reconstruction of the continents for the Late Jurassic, Figure 2-6a, clearly shows these events. Within the strike-slip zone between the major continents new rifted margins developed and extensional and pull-apart basins abounded. In this dominantly transtensional regime small continental blocks fragmented and broke away from the two major ones, forming a collage of microplates, and new seas opened between them which connected through to the Tethys.

In Mid to Late Jurassic times, western Europe had been feeling the influence of stresses from the opening of the central Atlantic, which resulted in an updoming of the central North Sea and then rifting. These stresses predominated over the transtensional stresses to the south where the region between Europe and Africa was fringed with passive margins and carbonate dominated shelf seas. The Piemont ocean representing the main gap between the two continents, had grown to somewhere between 100 and 500 km by the Late Jurassic, more of which in Chapter 6.4.

2.4.14 ARRIVAL OF THE ADRIATIC PLATE: EO-ALPINE COLLISION

Inevitably, within this sinistral strike-slip system, microplates rotated anticlockwise and collided with each other as well as growing and moving apart. At the beginning of the Cretaceous, the Adriatic microplate translating eastwards with Gondwana, broke away and, rotating anticlockwise, collided with Variscan Europe and the southwestern margin of the East European Platform, thus initiating the Alpine orogeny.

As this part of the Piemont ocean was consumed, a south-dipping subduction zone was active during the Early Cretaceous along the northern margin of the Adriatic microplate. With collision, during the Mid Cretaceous (Figure 2-6b), an eo-Alpine chain developed along the Eastern and Western Alps and the Carpathians, marked by a change in sedimentation from

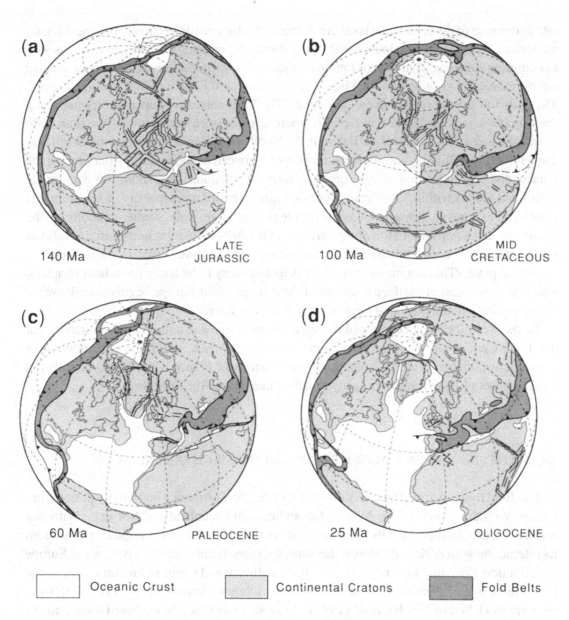

Figure 2-6. Plate tectonic reconstructions by Ziegler (1990), reproduced with permission, for
(a) the Late Jurassic, (b) the Mid Cretaceous, (c) the Palaeocene and (d) the Oligocene.

carbonates to the accumulation of clastic wedges. The mountain chain was characterised by
west to northwest-verging thrusting, and concomitant E–W strike-slip movements which
allowed eastward tectonic escape. High-pressure metamorphic overprints accompanied the
eo-Alpine events.

Meanwhile, to the west of Europe during the Early Cretaceous, Atlantic rifting had
migrated northwards as far as Labrador and Greenland, generating E–W tensional stresses
across western and central Europe. Within this north Atlantic regime the Iberian continental
block had separated from North America and, perhaps influenced by the sinistral strike-slip
regime to the south, rotated anticlockwise and translated southeast.

However, the eo-Alpine continental collision created a fundamental change in stresses
affecting central and western Europe which from then on were largely compressional, first
N–S and later NNW–SSE. These were strengthened towards the end of the Cretaceous by
the onset of convergence between Africa and Europe, probably brought on by the increasing

rate of opening of the central and southern Atlantic. By the end of the Cretaceous the Adriatic microplate had rotated anticlockwise through about 35°, creating dextral strike-slip as well as compression along its northern front in collision with Variscan Europe. Collision created a progressive imbrication and stacking of crystalline-cored Penninic-Austroalpine nappes. The collision front spread both east and west. The Carpathians produced a response along the Tornquist zone and its various splays, where stresses were transmitted more than 1000 km northwards to invert Mesozoic Basins in the Norwegian–Danish Basin and the North Sea. The Helvetic Shelf along the southern margin of Variscan Europe was uplifted during the Palaeocene and converted into a foreland basin. Collision, translation and rotation of individual blocks within this overall orogenic regime created a complexity of structure that is difficult to unravel. It was responsible for the present large-scale 'wedge-structure' in the deep crust of the Alps where forceful protrusion of the Adriatic plate at mid-crustal levels has peeled the upper crust with its sedimentary cover off the lower crust in the descending European plate. This is a major part of the detective story to be taken up in later chapters, when the revelation of the deep structure of the lithosphere of Europe begins to make sense of the geological activities that have been observed at surface.

To the west, Iberia had continued its eastward migration and anticlockwise rotation, but the Late Cretaceous onset of convergence between Africa and Europe resulted in a convergence of Iberia with the southern passive margin of France. The timing of collision and the onset of the Pyrenean orogeny matched that of the Alps (Figure 2-6c).

2.4.15 THE FINAL ROUND: ALPINE BELTS AND THE MEDITERRANEAN

The final round began effectively 40 Ma ago after North America had finally parted from Europe with the growth of the Atlantic Ocean between Greenland and Norway. With that influence gone, Europe from then on has been dominated by the stress regime created from its interaction with Africa. At around this time, the convergence between Africa and Europe set off a new Eocene–Miocene phase of Alpine collision and major mountain uplift, nappe emplacement and development of the Molasse foreland basin through crustal flexure, recently modelled by Sinclair *et al*. (1991). At around this time, the eastward translation of Africa relative to Europe ceased and began to reverse so that dextral strike-slip began to develop, resulting in a concentration of compression and crustal shortening in the Western Alps and the onset of transtension and subsidence of the Pannonian Basin to the east (Figure 2-6d). Dextral transpressive movement within the Pyrenees slid Spain west relative to France and rotated it slightly clockwise in the final stage of the orogeny. Transtensional stresses in western Europe permitted the development of the Rhone–Rhine–Eger rift system from Late Eocene to Oligocene time.

Within the Mediterranean region, the collage of microplates that had evolved through the Jurassic and Cretaceous had created a complexly looped chain of arcuate structures. Where the western arc of the Alps meets the Mediterranean, it switches polarity and continues, facing eastward, in the Apennine along the Italian peninsula, and, in an east-facing loop, across the straits of Messina and Tunis. In Maghrebide North Africa, the south-facing orogen continues with the Tell and the Sahara Atlas. This chain contains no more ophiolites, and the red desert sandstones in its African foothills show that it was formed in the shoaling west end of the Tethyan embayment. Finally, in the High Atlas, it reaches stable Africa where, for lack of sediment, its sole thrusts are within, not above the basement. From the Rif of Morocco

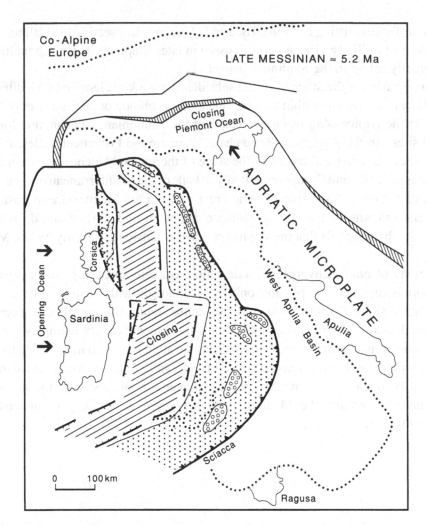

Figure 2-7. Sketch map reconstructing the position of the Adriatic microplate in the Late Messinian (5.2 Ma), the anticlockwise rotation of Corsica and Sardinia and the opening and closing of oceanic basins, after Patacca and Scandone (1990).

across the Straits of Gibraltar and into the Betic Cordillera of Andalusia, the orogenic chain loops once more around the Alboran sea, again with ophiolites and Alpine-looking units of sedimentary carbonate. On the fringe of this loop, the Mediterranean orogen has dumped the chaotic terrane of the pre-Rifian and pre-Betic thrust sheets out into the Atlantic. Along the Iberian south coast, the north-facing Betic Cordillera can be followed into the Balearic islands, and may possibly loop through Corsica back into the Alps.

It is not surprising that such a complex of tectonic units should act as a system which is internally more or less independent of its surroundings. Caught between Africa and Europe, Eocene–Oligocene N–S convergence resulted in the closure of the Ligurian-Alboran Sea and movement between the SW margin of Iberia and the Corsica–Sardinia block. When in the Miocene this convergence became directed NW–SE and dextral movement increased, crustal shortening became concentrated on the Appenines and the western and Southern Alps. The Corsica–Sardinia block rotated anticlockwise, crustal shortening occurred in the Betic Cordillera of SE Spain and a rapid opening of small oceanic basins ensued, including the West Mediterranean and Tyrrhennian basins. This is seen in Figure 2-7 in relation to the situation for the Adriatic microplate in the Late Messinian. Movement of Africa and Arabia relative to Europe continues to the present day, exemplified by dextral movement along the Anatolian fault system in Turkey, rifting of the Red Sea and of grabens transecting the Calabrian arc to the south and east of Sicily. Indeed, as discussed in later chapters, strike-slip faulting may well be currently active in the Sardinia Channel.

Extensional basins in the stage of initial subsidence are a key element of Mediterranean tectonics. They occur nested within the orogenic loops as oblong or circular sites of backarc spreading. In the Alpine–Zagros chain, we count the Pannonian, Aegean, and Menderes extensional sites. In the Apennine–Betic chain, there are the Provencal–Balearic and the Tyrrhenian sites. Extensional basins in the stage of thermal subsidence occur on the pre-Alpine cratons of Africa and Europe, essentially outside of the Mediterranean province. They also surround the early Mesozoic oceanic crust of the East-Mediterranean basin. The Provencal–Balearic site of backarc spreading contains an undisturbed and deep layer of Messinian salt which signals that the site had reached extensional stability by late Miocene time.

The interplay of plate convergence, indentation, backarc spreading, and tightening of orogenic loops is kinematically possible only by involving the asthenosphere. It mushrooms above subducted slabs of lithosphere, and, at shallow depth, the forces of its contrasting rheology and density are capable of powering the movements of mountains. We examine these implications further in Chapter 7. This tectonic style may also have been alive when some of the older provinces of Europe consolidated, but here in the Mediterranean, as we show in Chapters 4 and 5, we can measure these movements and estimate the stress fields. With this insight, perhaps the Mediterranean may give us the key to understanding lithosphere dynamics.

3 Europe's lithosphere – seismic structure

J. ANSORGE, D. BLUNDELL AND ST. MUELLER

3.1 SEISMIC METHODS FOR EXPLORING THE CRUST AND UPPER MANTLE OF EUROPE

Methods for exploring the Earth's interior which follow the passage of seismic waves through the ground are adapted according to the scale on which the Earth is viewed. The resolution of the seismic method utilised governs the clarity of the image of the Earth's structure and the extent of information on physical properties. In general, the deeper the investigation, the lower the resolution and the more blurred is the image. Normal incidence seismic reflection techniques provide the best resolution, particularly in a vertical sense, and deep seismic reflection profiles have yielded spectacular views of the detailed structure of the crust and upper mantle to depths of 60 km in recent years. EGT has been able to take advantage of the work of a number of deep seismic reflection profiling programmes, particularly BIRPS (UK), CROP (Italy), DEKORP (Germany), ECORS (France) and NFP 20 (Switzerland). However, this method gives poor information on seismic velocities, for which wide-angle reflection and refraction experiments are much better suited (Giese *et al.* 1976). These too, provide good resolution and strong control on the properties of the crust and upper mantle but to reach depths of 200 km needed to explore the full thickness of the lithosphere, quite elaborate and large-scale experiments have to be conducted along profiles at least 1500 km in length, firing several tens of explosions with dynamite charges measured in tons into large arrays of seismometers spaced 2–3 km apart along the entire length of the profile. The cost and organisational effort involved limits the number of such experiments but along EGT six major experiments of this kind were carried out: POLAR Profile (P), FENNOLORA (F), EUGENO-S (E), EUGEMI (C), EGT-S86 (A) and EGT-South (S) (Figure 3-1). Of these, only FENNOLORA was sufficiently long and comprehensive to allow the lower part of the lithosphere to be mapped out in any detail. In fact, FENNOLORA was able to map the upper mantle to a depth of 450 km.

To explore the upper mantle on a broader scale, other seismic techniques are needed which sacrifice something in resolution but, by making use of earthquakes as natural sources of seismic energy, can gain in coverage by recording large numbers of seismic wave paths. The development of digital recording and data processing has allowed detailed analysis to be undertaken in recent years of P-waves, S-waves and surface waves (usually the Rayleigh waves) which can also yield S-wave velocities. The large number of seismic observatories spread across Europe has meant that seismic waves from earthquakes have been recorded

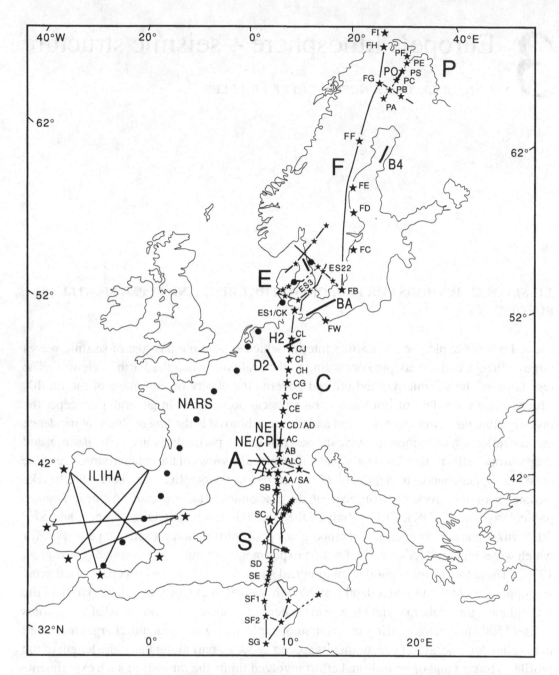

Figure 3-1. Location map of EGT seismic experiments POLAR Profile (P), FENNOLORA (F), EUGENO-S (E), EUGEMI (C), EGT-S86 (A) and EGT-South (S). Stars: shotpoints. Thin lines: refraction profiles. Thick lines: deep reflection profiles; BABEL line 4 (B4), BABEL line A (BA), DEKORP-2N (D2), North German basin (H2), NFP-20/CROP eastern and southern traverses (NE–NE/CP). Lithosphere Heterogeneity and Anisotropy Project (ILIHA). NARS array stations shown as filled circles. References are in the text.

along a huge number of intersecting pathways through the crust and upper mantle beneath Europe. With such a high density of sampling, it is possible to use the same technique as applied in medicine to scan the human body by ultrasound or X-rays, and invert the travel time information mathematically to create a three-dimensional image of the interior. The technique is known as seismic tomography and has been advanced to a highly sophisticated level for P-waves and applied to Europe and the EGT by Spakman (1988).

Surface waves travel with different velocities for different frequencies dependent on the distribution of elastic properties and density within the Earth and therefore arrive at a recording seismometer station as an extended wavetrain in time. The purpose of the analysis is to make use of the complete signature of that wavetrain, both in amplitude and phase, to determine the Earth structure that had modified the initial impulse at the source (the earthquake) into the wavetrain observed and recorded at the seismometer. Waveform inversion, as the technique is known, requires the analysis of many wavetrains recorded from a dense network of seismic wave pathways through the same volume of the Earth and, in consequence, computer analysis of a large dataset. In practice, limitations are imposed on the analysis either to make it more tractable or to cope with limitations in the dataset that result from earthquakes which may not have been located optimally for the purpose of the analysis. Making use of the large number of long period seismometers recording in observatories all over Europe, Panza *et al.* (1980b) created by triangulation a network of Rayleigh wave pathways which were inverted to map the lithosphere–asthenosphere system to depths of about 240 km. Over the past decade, Panza, Mueller, Calcagnile and Suhadolc have elaborated on this work to great effect (Suhadolc *et al.* 1990). The EGT project was also fortunate to be able to take advantage of the development by Nolet and his colleagues at Utrecht of the NARS (Network of Autonomously Recording Stations) portable, 3-component seismic stations capable of recording digitally across a wide range of frequencies, so-called broadband instruments (Nolet *et al.* 1986), that record P-waves, S-waves and surface waves equally well. These were initially deployed in an array of 18 stations set out along a great circle across Europe from Göteborg to Malaga (Figure 3-1), approximately in line with, and at suitable ranges from, earthquakes in the island arcs of the northwestern Pacific. In this way, travel paths for most of the distance through the Earth's interior were the same for all the stations but the differences between them were just those beneath the stations across the array. Thus the analysis yielded information about the nature of the crust and upper mantle beneath the array, across Europe just to the west of the main swathe of EGT. Later, the NARS stations were moved to record for a year across Spain and Portugal as part of the ILIHA project, the full results from which have yet to emerge.

3.2 SEISMIC EXPLORATION OF THE CRUST ALONG EGT

Seismic normal incidence, wide-angle reflection and refraction surveys are the techniques best suited to probe the detailed structure of the Earth's crust and lower lithosphere. Energy sources for deep penetration are explosives on land or offshore, vibrators on land or offshore, and large airgun systems at sea. All the data are processed digitally and displayed originally in the form of record sections as shown in the example of Figure 3-2 by Behrens *et al.* (1986) for wide-angle reflection and refraction data and on Figure 3-8 after Dohr *et al.* (1983) from the northwest German Basin for normal incidence reflections. Normally the time scale for wide-angle reflection and refraction record sections is reduced with a velocity which is appropriate for the depth range investigated, e.g. 6.0 kms^{-1} for the crust. Figure 3-2 shows as example an airgun survey in the Kattegat Sea close to the Swedish coast recorded at offsets giving a range of reflections from near-vertical to wide-angle. A single instrument at a location with a rather low noise level recorded a sequence of airgun shots while the ship was travelling southwards. A number of wave trains can be seen on the record section. P$_g$ denotes P-waves refracted at the top of the crystalline basement with a velocity of 6.0–6.1 kms^{-1} A second phase, P$_i$, is observed from 65 km distance onwards with a velocity between 6.6 and

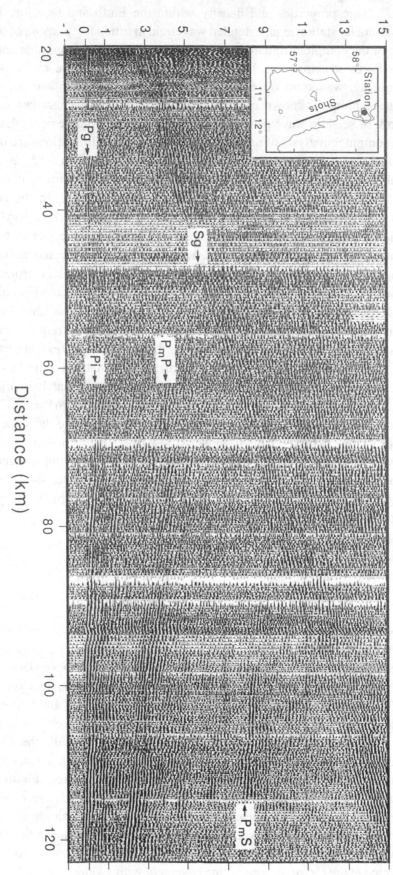

Figure 3-2. Reduced seismogram section from a profile shot offshore and recorded on land in southern Sweden to illustrate different seismic phases propagating through the crust and the data quality obtainable from an airgun source.

6.8 kms^{-1} which has propagated through the middle or lower crust. This, and others like it, helps to identify boundaries within the crust. Phase P$_M$P is reflected from the Moho, taken to be the crust–mantle boundary, and is identified by its curved shape on the record section. The S$_g$ label indicates shear waves through the crystalline basement and P$_M$S is most likely a P-to-S conversion at the Moho. These phases provide the basic information for the interpretation and modelling which leads eventually to the velocity–depth distributions shown in the following paragraphs. Their travel times and amplitudes are interpreted by creating model cross sections of the crust and upper mantle divided into regions with characteristic seismic velocities. Seismic wave paths are calculated, for example by ray tracing, to match the observed travel times for sets of record sections from shots where the waves traverse the same regions in opposite directions. Synthetic seismograms are calculated to match the amplitudes as well as the travel times of the observed signals.

Although the seismic experiments along the EGT were not carried out sequentially from north to south but were completed according to financial and organizational conditions, the description that follows of the seismic exploration of the crust proceeds from north to south, so that it covers progressively younger tectonic regions. Only representative crustal sections are described, but full details can be found in the references here and in the EGT Atlas.

3.2.1 NORTHERN SEGMENT

POLAR Profile

In 1985 a 440 km long NE–SW oriented deep seismic refraction survey was organised as part of a multidisciplinary research project, the so-called POLAR profile (P in Figure 3-1), with the aim of determining the Archean to Early Proterozoic crustal structure in northern Norway and Finland (Windley Chapter 6.1, Freeman *et al.* 1989; for experimental details and results see Luosto *et al.* 1989). Densely spaced recordings of explosions at 9 shotpoints provided ample data to derive a detailed cross section of the velocity–depth distribution under this profile. The seismic refraction and wide-angle reflection data were compiled by Luosto and Lindblom (1990).

Figure 3-3a shows the crustal cross section derived from these data. The velocity distribution becomes increasingly complex along the profile from northeast to southwest. Over a distance range of about 80 km around shotpoints PF and PE the crust is composed of four layers with average P-wave velocities of 6.1 kms^{-1} for the upper crust, 6.4 kms^{-1} for the middle crust, 6.6 kms^{-1} and 7.1 kms^{-1} for the mid–lower and lower crust, respectively, and with thicknesses of about 10 km for each of the upper three layers and 15–20 km for the lowermost layer. The Moho depth decreases significantly southwestwards from 47 to 40 km. Further south the upper crust in particular becomes internally more structured. A high-velocity (6.4 kms^{-1}) layer dips NE from the surface as part of the Lapland Granulite belt to a depth of 10 km at about 50 km north of shotpoint PD in the Inari Terrane. A similar high-velocity body with 6.5–6.6 kms^{-1} is found further southwest in the Archean basement of the Karelian Province between shotpoints PB and PA surrounded by material with normal near-surface velocities of 6.0 kms^{-1} found elsewhere on the profile. This southwestern section of the profile across the Lapland Granulite belt and the Karelian Province is underlain by a low-velocity zone with 6.15 kms^{-1} velocity between 8 and 14 km depth.

The two layers in the middle crust are separated by only a small velocity increase. They have an almost constant thickness along the entire profile except for the last 70 km in the

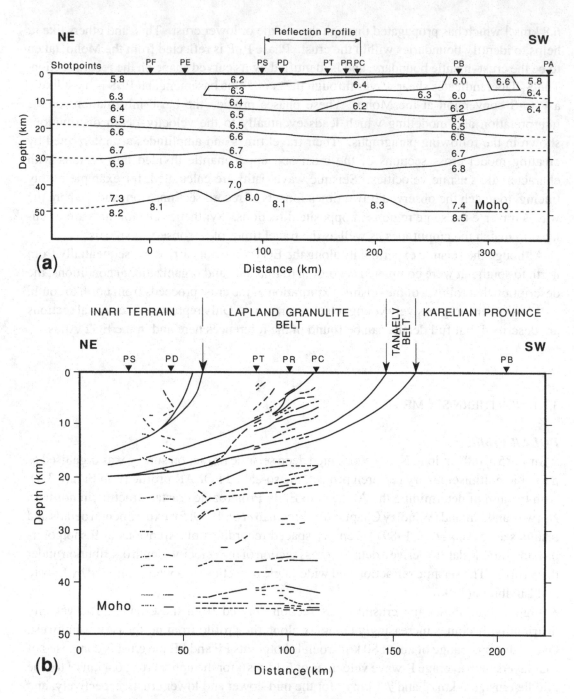

Figure 3-3. Seismic structure cross sections of the northern Baltic Shield derived from the POLAR Profile data (for location see Figure 3-1)

(a) Cross section of main profile

(b) Cross section based on a line drawing of a deep reflection profile of part of POLAR Profile (Figure 3-3a) and integrated gravity and magnetic modelling.

extreme southwest, where both layers come closer to the surface with the lower one decreasing in thickness to only 5 km. Strong lateral variations occur in the lower crust beneath the 180 km long central part of the profile. The sections further northeast and southwest are more representative of a normal shield structure with a crustal thickness of 47 km and P-wave velocities of 6.9–7.3 kms[-1] between the top and bottom of the lower crust. In the centre of the profile, velocities in the lower crust are reduced to only 6.8–7.0 kms[-1], accompanied by a clear reduction to 10–14 km thickness, i.e. an updoming of the Moho to 40 km. Strong undulations of the crust-mantle boundary of the order of 4 km over a distance range of only 30 km are observed near the northeastern end of this anomalous lower crust, which are unusual in areas of Archean age. Underneath this same area of the lower crust the P-wave velocity in the upper mantle immediately below the Moho has a significantly lower value of 8.0 kms^{-1} than the ones of 8.2–8.5 kms^{-1} on either side.

According to Marker *et al.* (1990) the Tanaelv belt and Lapland Granulite belt form the most prominent Early Proterozoic thrust belt in the northern Baltic Shield. Therefore this part of the POLAR profile was the target of a deep reflection survey which was carried out simultaneously with the main crustal seismic experiment in 1985 (Behrens *et al.* 1989). The area surveyed lies between shotpoint PC and 20 km NE of shotpoint PD, as shown on Figure 3-3a, and has a total length of 82 km. Figure 3-3b shows a line drawing of the significant reflectors over a depth range of 43 km, that covers the entire crust. The reflectors from the upper crust are dipping NE to a depth of more than 15 km. This made it possible to trace the southwestern base of the Lapland Granulite belt deeper into the crust towards the northeast. Initial problems to determine the detailed shallow transition to the Inari Terrane because of the scarce reflection data were later solved by Marker *et al.* (1990) using a combined gravity and magnetic model that is consistent with an electromagnetic model produced by Korja *et al.* (1989). The deeper reflectors in Figure 3-3b at 34 km depth can be related to the top of the lower crust and a band of reflections between 40 and 43 km to the crust–mantle transition. This result represents an excellent example of solving a tectonic problem by integrated modelling of several independent sets of data.

FENNOLORA

Although carried out in 1979 as a collaborative European research project within the International Geodynamics Project, the experience of the Fennoscandian Long-Range seismic experiment (FENNOLORA) was one of the main cornerstones on which the EGT was based when it began in 1981. It was the aim of FENNOLORA (Guggisberg and Berthelsen 1987) to obtain much fuller information about the velocity–depth distribution in the lithosphere and asthenosphere down to a depth of 400 km than was available hitherto from other methods. This required multiple reversed coverage of the profile with closely spaced observations (3 km for the crustal survey) of a large number of dynamite explosions at appropriate intervals over a distance of 2000 km. These conditions could only be met in Scandinavia where most of the explosions could be fired offshore. In addition, the traverse (Figure 3-1) was located almost entirely on tectonic units of Precambrian age (Windley Chapter 6.1) with the assumption that the velocity structure of the lithosphere would not change too much along the profile. A spacing of about 300 km for the explosions was chosen so that the structure of the crust could be determined sufficiently well in order to ascertain its influence on those seismic phases which penetrated the upper mantle. All seismic wide-angle reflection and refraction data were compiled by Hauser *et al.* (1990).

The main profile (F in Figure 3-1) lies between the North Cape in Norway, with the most northerly shotpoints FI and FH offshore in the Barents Sea, and southern Sweden, with the southernmost shotpoints FB off the coast and FW onshore in eastern Germany to bridge the Baltic Sea. The experiment and data processing have been described in detail by Lund (1983) and Guggisberg (1986). The northernmost part (Figure 3-1) of the profile is located on the Caledonian nappes which are thrust over the Archean terranes. About 200 km south of shotpoint FG the profile crosses from the Archean to the Proterozoic Sveco-Fennian units and

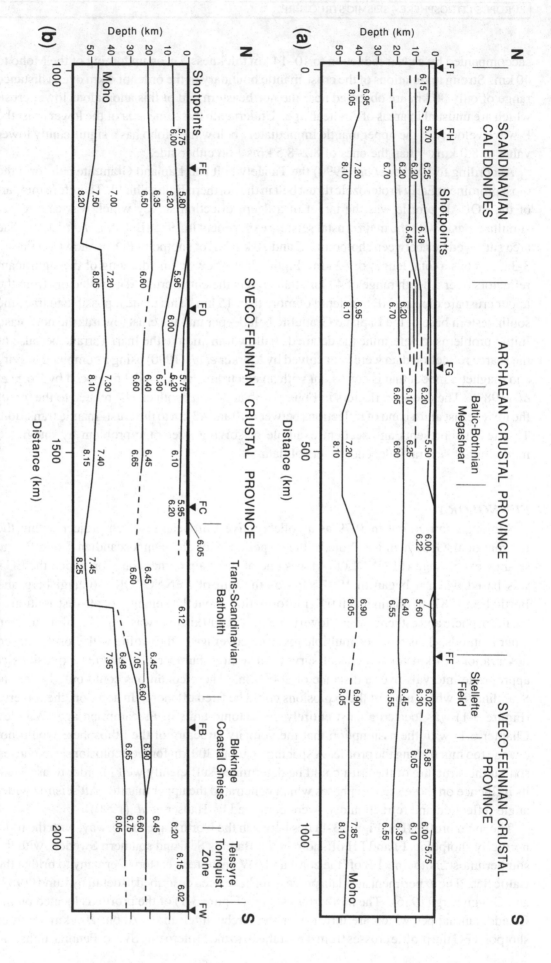

Figure 3-4. Seismic structure cross sections of the Baltic Shield derived from interpretation of FENNOLORA (for location see Figure 3-1).

about half way between shotpoints FC and FB it traverses the Gothian granites. Still further south the Teisseyre Tornquist zone is crossed north of shotpoint FW which is located on the Caledonian orogen of eastern Germany. Detailed two dimensional interpretations of the crustal P-wave data giving similar results were presented by Guggisberg (1986), Lund (1990), Stangl (1990) and Guggisberg *et al.* (1991) for the complete profile, while Clowes *et al.* (1987), Hauser (1989) and Hossain (1989) have worked on the southern and central part. Kullinger and Lund (1986), Hauser (1989), Hossain (1989), and Stangl (1990) have also presented interpretations of S-wave data. The structure discussed here is based on the interpretation by Guggisberg (1986) and Guggisberg *et al.* (1991). The cross section is displayed in Figure 3-4.

Overall the crust is divided into an upper and a lower crust with only one major continuous interface along the entire profile. Superimposed on this basically two-layered crust are substantial lateral variations of the velocity distribution and layer thickness (see Figure 3-4). Because of the large distance between the shotpoints the observations are only reversed for depths greater than about 15 km. Therefore the apparent uniformity of the uppermost crust is rather questionable and information from surface geology was used to extrapolate between the distance ranges covered by seismic observations. The upper crust has an average thickness of 20 km in which the P-wave velocity rises from about 6.0 kms^{-1} to 6.4 kms^{-1}. Close to the surface in the upper 5 to 10 km, sedimentary basin areas have a lower velocity of only 5.75 kms^{-1}. This holds especially south of shotpoint FH, where the profile crosses the Caledonides. This area has an overall low-velocity upper crust down to 20 km depth with only thin intercalated high-velocity layers of 6.25 and 6.4 kms^{-1}. The boundary between the exposed Archean province and the Caledonian nappes lies about halfway between shotpoints FH and FG. It seems as if the Caledonian orogeny has influenced the Archean upper crust. In the area around shotpoint FG the profile traverses the Baltic–Bothnian megashear zone which is expressed by a highly inhomogeneous upper crust that incorporates interleaved high- and low-velocity layers. Further south towards the transition to the Proterozoic regime near shotpoint FF the upper crust becomes quite homogeneous with a smooth increase of the velocity from 6.0 to 6.4 kms^{-1}. Further south between shotpoints FF and FE the upper crust consists of two layers with lower average velocities of 5.8 kms^{-1} and 6.1–6.35 kms^{-1}. Along the section from FE to FC two zones of lower velocity are dipping southwards from the surface into the upper crust with surface velocities of 5.75 kms^{-1} and 5.9–6.0 kms^{-1} at their deepest location with thicknesses between 5 and 8 km. The velocity at the base of the upper crust remains 6.45 kms^{-1}. South of FC towards FB the upper crust shows no internal structure. The velocity increases smoothly from 6.05 to 6.45 kms^{-1}. South of FB nothing can be said about the structure of the upper crust because no data are available from this experiment.

The lower crust is quite uniform along the profile. After a sharp velocity increase from 6.3–6.4 kms^{-1} to 6.55–6.7 kms^{-1} at its upper boundary it increases gradually to 6.9–7.3 kms^{-1} near the base of the crust. Beneath the Caledonian orogen the velocity jumps from 6.2 to 6.7 kms^{-1} at the transition to the lower crust. Only under the Gothian orogen in the south and beneath the adjacent Baltic Sea does the lower crust develop an internal layering with a high-velocity layer of 6.9 kms^{-1} on top of a zone with only 6.48–6.65 kms^{-1} just above the Moho. The average thickness of the lower crust decreases from 20 km north of the Gothian orogen to 16 km beneath the Gothian part of the shield and to only 11 km further south beneath the Baltic Sea.

Changes in the velocity contrast at the Moho are more pronounced. In areas with a flat lying Moho north of FE the P-wave velocity rises from about 6.9 kms^{-1} in the lower crust to 8.1 kms^{-1} in the uppermost mantle. South of FE the difference is slightly smaller. There are

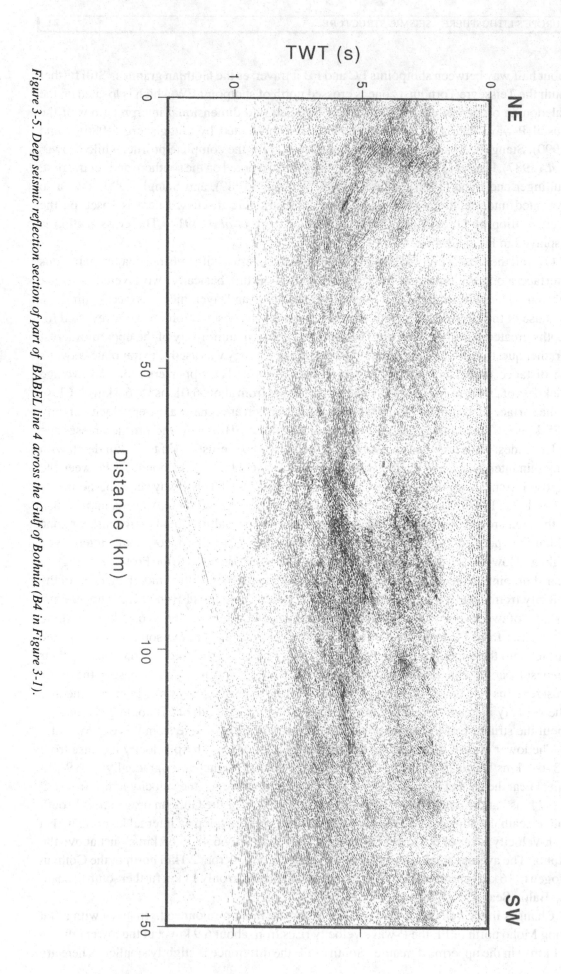

Figure 3-5. Deep seismic reflection section of part of BABEL line 4 across the Gulf of Bothnia (B4 in Figure 3-1).

two areas where the crustal thickness appears to increase sharply by 10–12 km accompanied by a more gentle rise of the Moho towards the north. In the areas between shotpoints FE and FD and beneath the transition from the Gothian to the Sveco Fennian orogen from FC to FB, the velocity in the lowermost crust gradually increases downwards to 7.4–7.5 kms^{-1} followed by a jump to 8.25 kms^{-1} below the Moho The maximum Moho depth of 55 km is reached 50 km south of shotpoint FC. Outside these anomalous areas the average crustal thickness is 38 km south of the sharp increase between FC and FB and 45–47 km north of shotpoint FE.

These pronounced sharp changes in the crustal thickness have raised considerable controversy because they cannot be seen in the Bouguer gravity anomaly (see Chapters 4.1.2 and 4.4) and they are in sharp contrast with the other crustal interfaces. Guggisberg (1986) and Clowes *et al.* (1987) have found independently the same sharp lateral transition using a ray-tracing interpretation. It was also found in an earlier preliminary one-dimensional interpretation by Prodehl and Kaminski (1984). Interpreting the gravity data to be consistent with the seismic model, Henkel *et al.* (1990) have concluded that this lowermost crust would have to have such a high density that there is effectively no contrast at the Moho, in which case it is probably of eclogitic composition. The cause of the sharp changes proposed in the Moho topography still remains to be resolved.

The structure within the crust under the FENNOLORA profile discussed here has no pronounced sharp discontinuities. The transition to lower crustal units occurs less sharply than in Central Europe. Low-velocity zones were only found in the uppermost crust. In the Archean section north of FF, which is undisturbed by the Caledonian overthrust, the average velocity in the upper crust is higher than in Proterozoic crust further south. The depth and transition from the upper to the middle or lower crust is similar to that found under the POLAR profile. But Luosto *et al.* (1989) found a further interface in the lower crust which led to a three-layer crust. Stangl (1990) also proposes in his interpretation one more interface in the lower crust under FENNOLORA on the basis of amplitude calculations for P- and S-waves although the velocity contrast in both models is not very big (0. 2 and 0. 1 kms^{-1} in the POLAR and FENNOLORA profiles, respectively).

The FENNOLORA project was aimed more at the principal features of the entire lithosphere and not so much at the fine details of the crust. Hence questions regarding the mechanisms of convergent plates of Archean to Proterozoic age cannot be answered from these data. However, normal incidence reflection profiling can help to solve this problem. The BABEL Working Group (1990) has run a detailed marine seismic reflection survey in the Baltic Sea and the Gulf of Bothnia parallel to the FENNOLORA line using BIRPS' techniques and data processing for deep crustal reflection profiles (Warner 1986). Figures 3-5 and 3-7 show two sections from this survey located as indicated in Figure 3-1.

Figure 3-5 shows a migrated seismic section across a presumed collison zone in the Skellefte Field of the Sveco-Fennian orogen over a distance of 150 km, about 100 km south of FENNOLORA shotpoint FF and 170 km to the southeast of it. Main features to be noted are the bivergent crustal reflectors dominantly dipping southwest in the NE half and dipping northeast in the SW half of the section with the deepest high-amplitude reflections identified as the Moho. This can be interpreted as the still very clearly recognizable image of a collision zone in which a Proterozoic plate has been subducted under the crust and Moho to the northeast, of which the latter can be resolved on the FENNOLORA data (Figure 3-4, south of FF). The pattern of reflectors resembles very closely that of the Phanerozoic and recent active continental subduction zones, such as the Alps, that is described in Section 3.2. Figure 3-5 should be compared with the mirror image of Figure 3-11, allowing that the subduction is in the opposite sense. There is undoubtedly much more to come from the interpretation of the BABEL data, particularly from the observations of wide-angle reflections recorded on fixed land stations, which is still at an early stage.

EUGENO-S

The important stretch of the FENNOLORA profile (Figure 3-1) from southern Sweden across the southern Baltic Sea and the Caledonian front to eastern Germany crosses the highly complicated area of the Sorgenfrei Tornquist zone. The deep structure and lateral variations of this area were impossible to reveal from these few data. So, an EGT seismic research project was conceived and carried out in 1984 to determine the structure of the transition zone by means of a grid of seismic lines from which a more complete picture and better understanding of the deep seismic structure in three dimensions could be constructed, the EUropean GEotraverse NOrthern Segment, Southern part, i.e. the EUGENO-S Project (E in Figure 3-1). A complete description of the experiment and basic results were published by the EUGENO-S Working Group (1988). This first publication was followed by a series of individual papers which are dedicated to specific aspects of the area. Only those can be cited here which bear directly on the north–south profile extending the EGT southwards to the EUGEMI project on the central segment (EUGEMI Working Group 1990). Most of the seismic wide-angle reflection and refraction data were compiled by Gregersen *et al.* (1987).

Profile 1 of the EUGENO-S project (Figure 3-6) provides the shortest and most direct connection between FENNOLORA and EUGEMI along the EGT central segment. It starts on the Precambrian Sveco-Norwegian basement in southern Sweden (shotpoint ES22), crosses the Kattegat platform, the Sorgenfrei Tornquist zone (STZ), the Danish basin, the Ringkøbing-Fyn high (RFH), the Caledonian front, and the Trans-European fault (TEF) near shotpoint ES1, which coincides with shotpoint CK of EUGEMI (see Figure 6-7). The section then runs due south as part of EUGEMI to the southern end of the North German basin. Thybo *et al.* (1990) have presented the latest interpretation of the available seismic data on that profile, incorporating as much information as possible from the other profiles as well, and Thybo (1990) has subsequently extended the cross section to the south by adding all the information derived from the EUGEMI project on the central segment (Aichroth *et al.* 1992, Prodehl and Aichroth 1992) to the southern end of the North German basin. The summary given in Figure 3-6 follows this work.

There are two major sedimentary basins along the profile. From north to south, the Danish basin and the North German basin, separated by the Ringkøbing-Fyn basement high, have sediment thicknesses of at least 10 km. Details of the sediments in the North German basin are presented by Pratsch (1979). Brink *et al.* (1990, 1992) have summarised the most important features of the late and post-Variscan development of this basin. Average seismic velocities for the basin structure shown in Figure 3-6 were derived from the EUGENO-S and EUGEMI data.

Very roughly the crustal structure along the profile can be divided into three main blocks. In the north we see the relatively uniform homogeneous crustal structure already known from FENNOLORA with a thin low-velocity layer in the upper crust and a continuous velocity increase in the lower crust from 6.5 to at least 6.9 kms^{-1} to reach the Moho at 45 km depth. Next to the south, in the Sorgenfrei Tornquist zone, the low-velocity layer disappears and the vertical velocity gradient in the upper crust increases towards the Danish basin, and the crust is two-layered (excluding the sediments with 4.4–5.7 kms^{-1} in the basin), characterised by velocities of 6.2–6.7 kms^{-1} in the upper crust, and 6.9–7.1 kms^{-1} in the lower crust. The Moho rises southwards in steps to 32 km under the Sorgenfrei Tornquist zone and to only 29 km

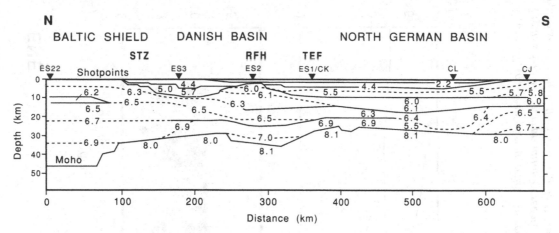

Figure 3-6. Seismic structure cross section from the southern part of the Baltic Shield to the North German basin derived from refraction and wide-angle reflection recordings along EUGENO-S line 1 (for location see Figures 3-1 and 6-7).

depth under the basin. Beneath the third block to the south, the Ringkøbing-Fyn high has only a very thin sedimentary cover above a crystalline crust consisting of three layers with velocities ranging from 5.9–6.2 kms[-1] in the upper crust, 6.4–6.6 kms[-1] in the middle to 6.9–7.1 kms[-1] in the lower crust. A sharp drop of the Moho down to 34 km accompanies the basement high, whereas the subsequent rise towards shotpoint ES1/CK is less well constrained. This transitional zone is the suggested location of the Caledonian front (see Chapter 2.4.8).

Brink *et al.* (1992) have discovered a further deepening of the basement to 13 km about 40 km south of shotpoint ES1/CK in the northern part of the North German basin, coinciding with another regional rise of the Moho with a block type structure to 26 km. The crust changes here to two layers beneath the basin and the velocities are decreasing towards the south with values of 2.2–5.9 kms[-1] in the basin, 6.0–6.1 kms[-1] in the upper crust and laterally decreasing from 6.9 to 6.3–6.4 kms[-1] in the lower crust. At the southern end the section ties well into the crustal model presented in the following paragraph (Prodehl and Aichroth 1992).

Figure 3-7 shows the results of a deep seismic reflection profile from the BABEL survey across the complicated fracture zone of the Tornquist Teisseyre zone southeast of the EUGENO-S research area (BABEL Working Group 1991). The traverse lies between southernmost Sweden and the Island of Bornholm (BA in Figure 3-1), about 50 km to the northwest of the trace of the FENNOLORA profile across the southern Baltic Sea between shotpoints FB and FW (Figure 3-1). Figure 3-7a displays the seismic section migrated so that the locations of the reflectors better reflect the actual geometry. The structure of sedimentary basins within the Tornquist Teisseyre zone is highly fractured and broken up into a series of horsts and grabens with the basement apparently coming close to the surface. The upper crust contains a series of distinct northeast dipping reflectors. These are followed by a band of subhorizontal reflections which represent the lower crust and are comparable to observations in other areas of central Europe, such as those seen in Figure 3-10 from the Rhenish massif. This band is clearly uplifted in the centre under the Tornquist Teisseyre zone with a slight shift to the northeast from where it deepens to the Baltic Shield in accordance with the general crustal structure found there by Clowes *et al.* (1987) and with the very detailed simultaneous recordings of airgun shots by onshore fixed stations over long offsets as shown in Figure 3-2. Figure 3-7 also shows reflections from below the Moho which confirm the fine-scale inhomogeneity of the upper mantle as derived from the FENNOLORA data and described in Section 3.3. Geometrical relationships between the sediments and faulting within and

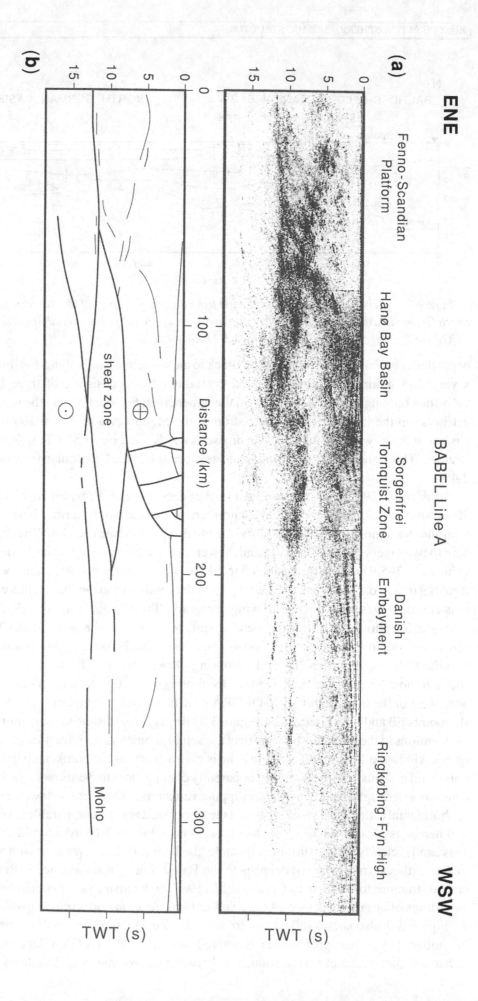

Figure 3-7. (a) Deep seismic reflection section of part of BABEL line A across the Sorgenfrei Tornquist zone near Bornholm, southern Baltic sea (for location see Figures 3-1). (b) Interpretation of record section showing transfer of faults at mid-crust to a region of distributed shear in the lower crust, and in turn to a dipping shear zone in the upper mantle (Blundell and BABEL Working Group 1992).

adjacent to the Tornquist Teisseyre zone indicate strike-slip movements and inversion structures. These have been used by Blundell and BABEL Working Group (1992) to interpret the fault patterns at depth as detaching at mid-crust and Moho levels, where strength is least (see Chapter 4.2.1), and transferring down into the mantle as shown in Figure 3-7b. Preliminary models of the wide-angle reflection data indicate a lowermost crust with high seismic velocity (>7.0 kms^{-1}) continuing across the Tornquist Teisseyre zone to the south as far as the Ringkøbing-Fyn high, consistent with the view expressed in Chapter 2.4.12 that the Sorgenfrei Tornquist is not the edge of the Baltic Shield and is a feature that developed only in post-Caledonian times.

To supplement the structural picture, Figure 3-8 presents an example of a crustal reflection survey by Dohr *et al.* (1983) from the North German basin at the southern end of the section shown in Figure 3-6 just north of shotpoint CJ and subparallel to the EGT. The main reflecting features shown in Figure 3-8 are:

(i) the base of the Zechstein between 0.5 and 1.0 s TWT,

(ii) the almost completely transparent upper crust,

(iii) the sharp boundary to the lower crust at 6 s TWT,

(iv) the missing band of reflections, which is often observed in Central Europe and is replaced here by several strongly layered reflectors which vary laterally in depth and extent,

(v) and the clear band of reflections at the Moho.

Note how similar the strong subhorizontal reflection banding in the lower crust is to that near the southern end of the BABEL section of Figure 3-7. This reflectivity pattern is also typical of the offshore areas around Britain (e.g. BIRPS and ECORS 1986). In contrast, note how strongly this character of the reflectivity north of the Variscan deformation front differs from that shown in Figure 3-10 which traverses the Variscan Rhenish massif. The reflection patterns can also change over much shorter distances within the same area as shown by Wever *et al.* (1990).

3.2.2 CENTRAL SEGMENT

EUGEMI

The central segment of the EGT the EUropean GEotraverse MIddle segment (EUGEMI) extends over a distance of 825 km from the Baltic Sea near Kiel to the northern margin of the Alps south of Lake Constance (shotpoints ES1/CK to CD/AD in Figure 3-1). It traverses the Caledonian and Variscan tectonic units of western Germany which are covered in the north by the sediments of the North German basin and in the south by those of the Tertiary Molasse basin. The seismic refraction and wide-angle reflection survey of 1986 was described in detail by the EUGEMI Working Group (1990) and all the data were compiled by Aichroth *et al.* (1990). A detailed description of the modelling is given by Aichroth *et al.* (1992) and by Prodehl and Aichroth (1992). The northern part of this section across the North German basin has been described in the previous section.

The crustal section presented here covers the area between the southern end of the North German basin and the northern front of the Alps (Figure 3-9). Only a very schematic picture of the upper crust with the sedimentary cover is given in this short presentation. The base of the sediments appears clearly in the velocity–depth distribution as the level at which a P-wave velocity of 6.0 kms^{-1} is reached. Beneath the North German basin this is at a depth of 10 km.

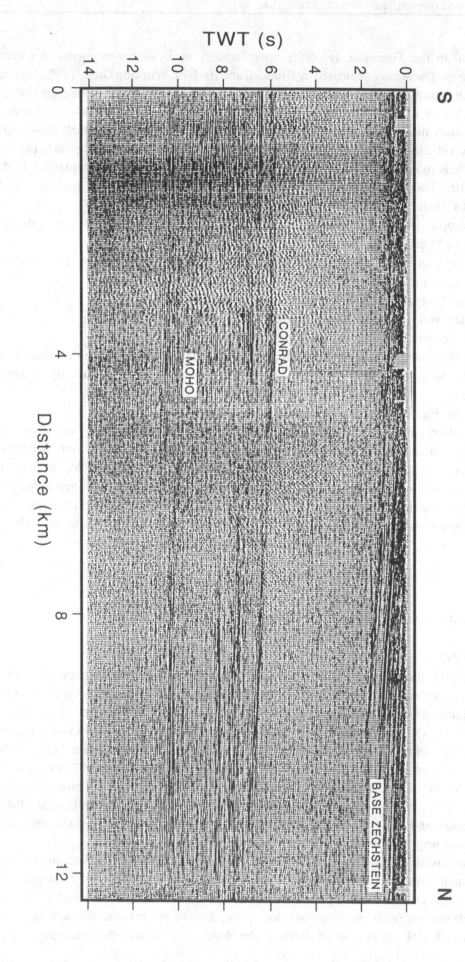

Figure 3-8. Seismic reflection section in the north German basin showing a region of strong, sub-horizontal reflectors in the lower crust marked by the Conrad as its upper limit and the Moho as its base (for location see H2 in Figure 3-1), after Dohr et al.(1983), reprinted by permission.

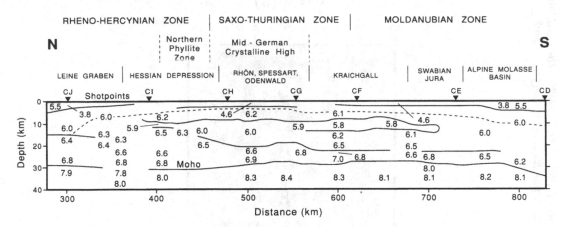

Figure 3-9. Seismic structure cross section of the Variscan units of Germany derived from interpretation of EUGEMI (for location see Figure 3-1).

Under the mid-German crystalline rise it lies at 3 km depth, from where it deepens to 12 km under the northern margin of the Alps.

The middle crust is characterised by a zone which varies both in thickness and velocity along the profile, with a significant reduction of the velocity in the centre from 6.2 kms^{-1} to 5.8 kms^{-1}. No such reduction could be identified under the northern and southern ends of the profile. Beneath the North German basin this may be due to the poorer quality of the data. The average thickness of the low-velocity zone of 4 to 5 km in the south increases significantly to about 10 km in the central part of the profile. This area coincides with the NNE–SSW trending aborted branch of the central European rift system which extends from the Odenwald across the volcanics of the Rhön mountains, and dies out further north under the Hessian depression (Figure 3-9).

The internal structure of the lower crust also varies considerably. In the northern third of the section the velocity increases continuously from 6.3 kms^{-1} to 6.9 kms^{-1} at the Moho over a depth range of 12 to 14 km. The thickness of this layer decreases continuously towards the south to reach a value of 5 km at the northern margin of the Alps. Concurrent with that thinning, the velocity increases southwards to reach an average value of 6.8 kms^{-1} under the northern end of the Molasse basin. The transition to the thin high-velocity layer above the Moho occurs at about 23 km depth. Southward from there the crust does not exhibit any significant internal differentiation and the average velocity becomes as low as 6.2 kms^{-1} in the lower crust. Thus it is impossible to trace a clear continuous mid-crustal Conrad discontinuity. There is little or no crust with P-wave velocity above 7.0 kms^{-1}.

The Moho lies at a depth between 29 and 30 km under the northern two thirds of the profile. Its depth increases to 34 km at the northern margin of the Alps. The velocities in the uppermost mantle change significantly from a minimum value of 7.8 kms^{-1} in the north to as much as 8.4 kms^{-1} under the southern end of the mid-German crystalline rise, from where it decreases again to a more normal value of 8.1 kms^{-1}.

This two-dimensional crustal cross section is documented in three dimensions by a wealth of both older and more recent high-resolution wide-angle reflection and refraction data (for a complete reference list see Prodehl and Aichroth 1992) as well as a number of densely observed normal-incidence deep reflection surveys of the German DEKORP Project. A petrological interpretation of this section was described by Franke *et al.* (1990b): see also Chapter 6.3, Figures 6-18 and 6-20.

As is clear from Figure 2-3, the central segment of the EGT traverses obliquely across the

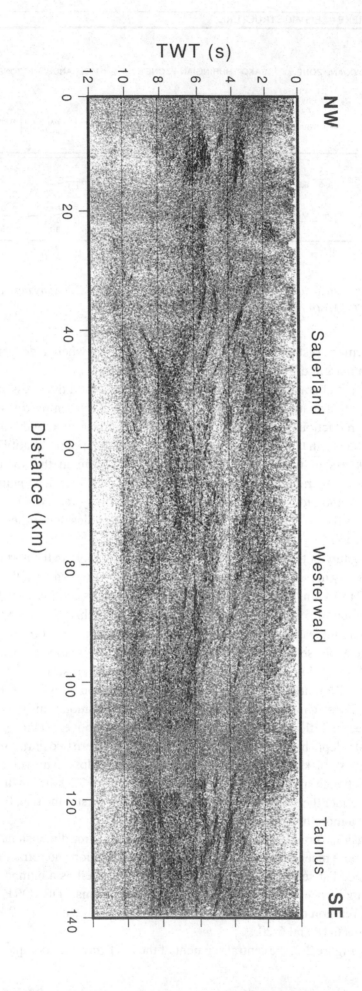

Figure 3-10. Deep seismic reflection section of part of DEKORP line 2N across the northwestern Variscan units of Germany (for location see D2 in figure 3-1); from DEKORP-Atlas (Meissner and Bortfeld 1990) Section 19, reprinted with permission from Springer Verlag, Heidelberg, Germany.

strike of the main tectonic units. The NNW–SSE oriented deep seismic reflection profiles of DEKORP-2S and DEKORP-2N (DEKORP Research Group 1985, Franke *et al.* 1990a), however, cross the tectonic units at right angles and thus reveal a more representative picture of the main characteristics of the crustal structure. Figure 3-1 (D2) shows the location of the main portion of the DEKORP-2N reflection line relative to the EGT and Figure 3-10 reproduces the reflection profile DEKORP-2N across the Rhenish massif (Meissner and Bortfeld 1990). The upper crust to a TWT of 3 s is characterised by reflections which can be related to tectonic and stratigraphic features visible at the surface (Franke *et al.*, 1990a). It is accompanied by a general increase of the velocity with depth as derived by Giese *et al.* (1990) from wide-angle observations along the same profile. The TWT interval between 3 s and 5 s is relatively clear of reflections. It coincides with a zone of reduced velocity shown by the same authors and the mid-crustal electrical conductor compiled by Volbers *et al.* (1990) from magnetotelluric studies along the same line (see Figure 4-6). Reflectivity in the lower crust increases to a broader band between 5 s and 7.5 s followed by another less reflective, low-velocity zone. Immediately above the Moho the reflectivity becomes more pronounced again and is represented in the velocity structure schematically by a 3 km thick zone with thin high-velocity lamellae. The Moho itself as determined by seismic refraction measurements coincides very well with the base of the reflecting crust.

3.2.3 SOUTHERN SEGMENT

The Alps

The whole of the southern segment of EGT from the northern margin of the Alps to Tunisia, referred to as EGT-South, was covered in three major seismic experiments, EGT-S83 (S), EGT-S85 (S) and EGT-S86 (A), located in Figure 3-1. We continue to describe the results from north to south, and begin with the Alps. This survey of the crustal structure covers geographically the north–south section between the Molasse basin and the Po Plain across the Central Alps of Switzerland (A in Figure 3-1). Two major long-term research programmes have added a wealth of new data from which a fully revised structural picture of the Earth's crust under the Alps could be developed. One data set was acquired with the EGT-S86 survey together with EUGEMI along the EGT central segment between Kiel in northern Germany and Genova at the Ligurian coast. Aichroth *et al.* (1990) and Maistrello *et al.* (1991) have compiled all the data and technical parameters. The second set of data was provided by the eastern and southern normal-incidence deep reflection profiles across the Central Alps of Switzerland as part of the Swiss special research programme NFP-20 (Frei *et al.* 1989, Pfiffner *et al.* 1990, Bernoulli *et al.* 1990, Valasek *et al.* 1991)

All the EGT data were obtained along profiles oriented more or less perpendicular to the tectonic strike of the Alps, which in such a structurally complex area may lead to severe spatial aliasing problems in the case of the seismic wide-angle data. Therefore, the reliability of any interpretation of these data depends heavily on the velocity and structural information obtained from seismic refraction profiles along the tectonic strike. This vital information was provided by a third set of data, collected over many years and interpreted by several authors prior to the EGT (see ETH Working Group on Deep Seismic Profiling 1991, and Holliger and Kissling 1991 for a more complete list of references).

The system of reversed and overlapping profiles from shotpoints CE (northern boundary of the Molasse basin), CD/AD (northern boundary of the Helvetic zone), AC (Insubric line)

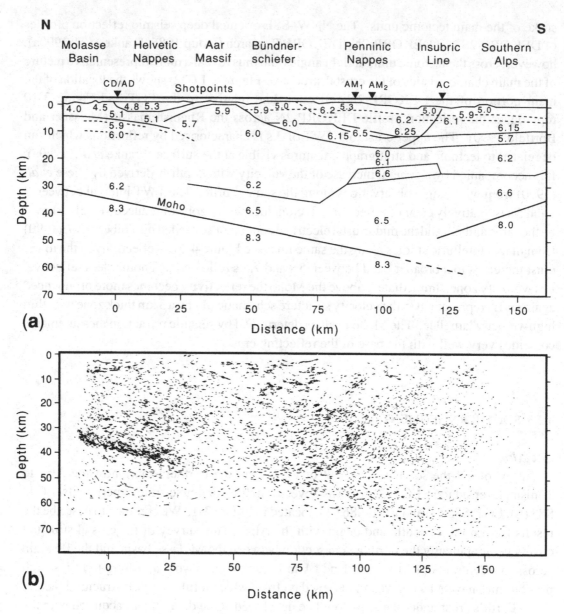

Figure 3-11. Seismic structure cross sections of the Alps derived from
(a) interpretation of EGT-S86, after Valasek (1992). For location see A in Figure 3-1
(b) depth migrated deep seismic reflection section constructed from the eastern and southern traverses of NFP-20, extended to the south by CROP lines. For location see NE–NE/CP in Figure 3-1.

and AB (southern boundary of Po basin), together with the structural constraints provided by the older profiles, has led to the P-wave velocity–depth cross section shown in Figure 3-11a between the Molasse basin and the Po basin (Ye 1992). In the north the cross section joins with the crustal structure derived by Aichroth *et al.* (1992), described in the previous section for the EGT central segment, and in the south it links with a rather similar extension of the crustal structure to the Ligurian Sea proposed by Buness and Giese (1990) shown in Figure 3-12. Figure 3-11b shows the migrated line drawing of the composite reflection sections of the eastern and southern NFP-20 reflection traverses. A smoothed crustal velocity structure from Figure 3-11a was used for the migration (Valasek 1992). The most uniform crustal structure is found beneath the southern Molasse Basin and the northern part of the Helvetic

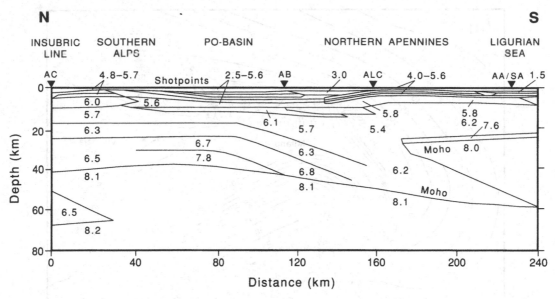

Figure 3-12. Seismic structure cross section from the Southern Alps to the northern Appenines derived from interpretation of EGT-S86 and EGT-South data (for location see Figure 3-1).

zone. The section between shotpoint AD at the northern margin of the Alps and shotpoint AC at the Insubric line shows pronounced lateral structural variation. Major elements in the upper crust are the relatively low velocity in the Aar massif compared with the well-documented higher velocities in the metamorphic crystalline nappes of the Penninic zone derived from the observations of shots at AM_1, AM_2 and AC. The upper and middle crust beneath the Aar massif seems to be devoid of significant reflections. It separates the highly differentiated uppermost crust consisting of the Helvetic nappes with well constrained structural geometry in the north (Pfiffner *et al.* 1990) from the Penninic units in the south. The horizon dipping southward from the Aar massif derived from wide-angle reflection data (Figure 3-11a) was not observed by Stäuble and Pfiffner (1991) in the normal-incidence reflection data and is therefore questionable. A high velocity layer (6.5 kms^{-1}) under shotpoints AM_1 and AM_2 can be correlated with an interface within the pile of Penninic nappes found on the migrated reflection traverse (Pfiffner 1990). Another high-velocity layer with a velocity of 6.6 kms^{-1} observed at about 20 km depth under the Penninic zone coincides with the north-dipping broad band of high reflectivity in the same depth range and position seen on Figure 3-11b. This may be explained as the upper boundary of the Adriatic lower crust which has been driven northward as a wedge between the European upper and lower crust. From about 25 km south of shotpoint AD the lower crust is distinguished by a transitional layer of about 10 km thickness with an average velocity of 6.5 kms^{-1} on top of the south-dipping Moho. This layer is characterised by an increasing reflectivity with depth which is most intense in the region between 3 and 6 km above the Moho. To the south the reflectivity of this lower crustal transition zone ceases abruptly even though from refraction observations the Moho is continuous. Holliger and Kissling (1991) suggest that this feature is caused by scattering in the complex upper crustal geology of the Penninic domain. The north-Alpine reflection character appears again in the deepest part of the crustal root zone north of the Insubric line. This implies that the European lower crust is a pre-Alpine, possibly Variscan, feature and extends relatively undisturbed from the northern Molasse basin to about 120 km south of shotpoint AD where it is subducted beneath the Adriatic microplate.

Shotpoint AC is located almost exactly on the Insubric line. Therefore, nothing can be said

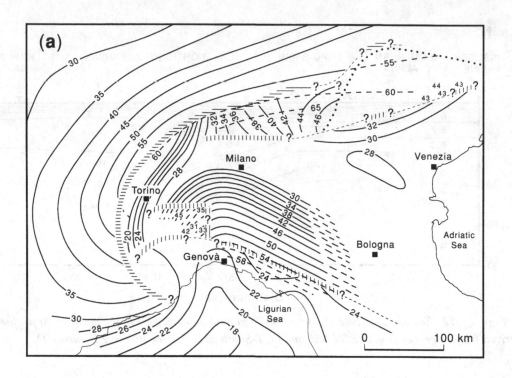

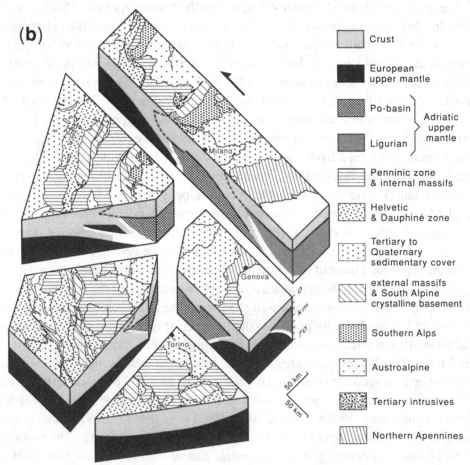

Figure 3-13. Moho surfaces in the region of the Alps
(a) shown as a contour map of Moho depth
(b) shown in a 3-dimensional view, looking northeast.

about the seismic nature of this fault from the refraction data. However, this geological feature at the surface is perfectly imaged on the normal-incidence seismic section by the strong north-dipping reflections observed after migration and can be traced clearly to a depth of 17 km (Figure 3-11b). Further north its counterpart, the Penninic front, is equally well displayed where it is outcropping at the southern margin of the Aar massif.

The next portion of the profile ranges from the Insubric line well into the Po Plain. The low velocity layer in the upper crust under the Southern Alps (5.8 kms^{-1}) had already been discovered by Deichmann *et al.* (1986) from an earlier survey. This low velocity layer is coincident with a band of high reflectivity on the normal incidence profile (Figure 3-11b). Perhaps it represents the stack of a southward oriented thrust complex as suggested by Roeder (1989a). The transition to the lower crust occurs at about 19 km under the centre of the Po Plain and dips northward to reach 24 km depth under the Southern Alps. The velocity increases at this boundary to 6.6 kms^{-1}.

The Moho dips gently to the south from 29 km beneath the northern end of the Molasse basin (Aichroth *et al.* 1992) to 35 km under shotpoint AD at the northern end of the Helvetic overthrust. The greatest clearly determined depth of 56 km is reached some 105 km south of AD. A very steep offset in the Moho has to be assumed below shotpoint AC with a vertical throw of up to 11 km between the lower, S-dipping Moho from the north and the upper, with a much steeper dip towards the north (Figure 3-13b). The Moho lies at 33 km depth under the southern margin of the Southern Alps. The northern European Moho is clearly defined by the sudden termination of the high reflectivity, refracted and wide-angle reflected observations. In contrast, the base of the Adriatic crust is much less well defined on the reflection data, whereas the refraction and wide-angle reflection data are just as good as further north. There is now general agreement that the European plate is subducted under the Adriatic lithosphere where the Alpine crust is thickest (e.g. Valasek *et al.* 1991). However, it is unclear how deep the European Moho can be followed as an interface into the lower lithosphere and how well the corresponding velocity–depth distribution is controlled by the presently available data. Buness and Giese (1990) have derived, from later P-wave phases observed at distances up to 370 km on the recordings of shots AD to the south and AA/SA from the Ligurian coast to the north, a velocity range between 6.2 and 7.4 kms^{-1} for the subducted lower European crust, shown as 6.5 kms^{-1} in Figure 3-12. Ye (1992) has shown that the same phases can also be interpreted differently (shown dashed in Figure 3-11a), which leaves open the question about the detailed sub-Moho velocity structure. Since we are dealing with a highly complex three-dimensional structure, the reader should be aware that, after proper projection of tectonic details and after a still more rigorous incorporation of the along-strike wide-angle profiles, some modifications of the crustal structure may still arise. Nevertheless, for the first time a continuous two-dimensional velocity–depth distribution derived independently from the EGT data can be combined with the properly migrated NFP-20 reflection data where the migration process itself is based on an independent determination of velocities from older, along-strike refraction profiles.

Po basin and Northern Apennines

The next section south of the central Alpine profile covers the tectonically complex region extending from the Insubric line across the Po basin and the northern Apennines to the Ligurian Sea (see Figures 3-1 and 3-12). It comprises the southernmost part of the seismic experiment EGT-S86, carried out simultaneously with EUGEMI, between shotpoints CE north of the Alps and AA/SA off the Ligurian coast. The data of the main profile are contained in the compilation by Aichroth *et al.* (1990) whereas the EGT data obtained in 1983 and 1986 on profiles and fans off the main line were compiled by Buness (1990) and Maistrello *et al.* (1991), respectively. These off-line data cover parts of the western and Southern Alps, the Po basin and the northern Apennines. It was only this large amount of data together with an already existing wealth of older profiles (see e.g. Giese *et al.* 1976) which allowed the derivation of the very detailed crustal structure shown in Figures 3-12 and 3-13.

The main crustal features can be described as follows (Figure 3-12): The crystalline basement with P-wave velocities between 6.0 kms^{-1} and 6.2 kms^{-1} along the profile is overlain in the north by the Neogene sedimentary cover of the Southern Alps, in the centre by the southward deepening sedimentary trough of the Po basin, and in the south by the sedimentary thrustbelt of the northern Apennines. The middle crust consists of a low-velocity zone with a decrease to 5.7 kms^{-1} (in places even to 5.4 kms^{-1}) which thickens southward and terminates under the northern part of the northern Apennines. The adjacent upper crust to the south apparently contains no such low-velocity zone but is even more differentiated than is shown in the section of Figure 3-12 derived by Buness and Giese (1990), as can be seen if compared to Figure 3-15a by Egger (1992). This layer is underlain by a 5 km thick transition zone with 6.3 kms^{-1} velocity, with the same north–south extension as the low-velocity zone above it. The deepest part of the crust shows the strongest variation along the profile both in thickness and velocity, with changes from 6.5 kms^{-1} in the north to 6.8 kms^{-1} in the south. Its thickness decreases from the Insubric line towards the Po basin as depicted in Figure 3-11 and pinches out beneath the northern margin of the northern Apennines.

In contrast to the model shown in Figure 3-11a by Ye (1992), where the thickness of the Adriatic crust decreases continuously from the Insubric line to the Po basin, Buness and Giese (1990) propose a stepwise shallowing of the Moho under the northern margin of the Po Basin with the indication of a northward oriented thrust combined with a reduction of the sub-Moho Pn velocity to 7.8 kms^{-1}. This transitional layer in the upper mantle thins out to the south. The Moho deepens again to 48 km under the front of the northern Apennines and to nearly 60 km under the northern coast of the Ligurian Sea. It is overlain by an undifferentiated subducted slab of the Adriatic crust with a low average velocity of only 6.2 kms^{-1}. Similar to the subduction structure under the Central Alps, the lithospheric unit of the northern Apennines overthrusts this subducted Adriatic slab. The crustal thickness of this unit decreases from 26 km at its northern margin to 22 km under the southernmost shotpoint AA/SA (see also Figure 6-30).

Giese and Buness have attempted to map regionally the different crustal units, which are presented as depth contours in Figure 3-13a and in a three-dimensional view in Figure 3-13b. The S- and E-dipping Moho of the European plate can be traced to a depth of 65 km. This is overlain in the north and northwest by the Adriatic microplate with Moho depths of up to 46 km under the Southern Alps, 30 km under the Po basin, 20 km at its western edge, and another downwarping to 50 km under the northern Apennines. The crustal base of the northern Apennines shallows southward from 24 km in Liguria to 19 km under the central Ligurian Sea (cf. Figure 3-15).

Ligurian Sea to Sardinia Channel

The Mediterranean part of the Southern Segment covers the area with the most recent and greatest large-scale tectonic mobility of the entire traverse. It crosses continental, transitional and partly oceanic realms, which also becomes obvious in the lithospheric structure.

Next, the EGT traverses an area of active rifting and oceanization. In 1981 a two-ship

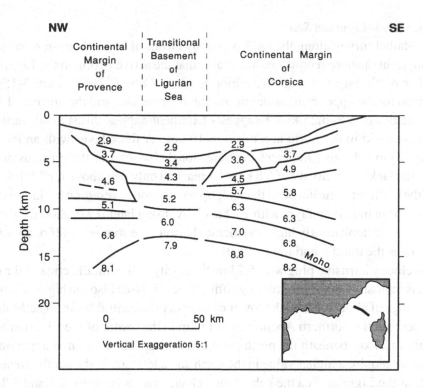

Figure 3-14. Seismic structure cross section of the Ligurian Sea derived from interpretation of offshore profiles west of the EGT line, simplified after Le Douran et al. (1984).

seismic survey along expanding spread profiles was carried out independently of EGT in the nothwestern Mediterranean Sea (Figure 3-14, Le Douaran *et al.* 1984). Complete crustal cross sections across the margins and the deep basins helped to elucidate the nature of the crust at the incipient and ongoing rifting in the Ligurian Sea and to clarify the amount of crustal thinning beneath the continental margins. Figure 3-14 shows a cross section of this survey, located to the west of the trace of EGT. Values of 4.5–4.9 kms^{-1} and 5.7–5.8 kms^{-1} are observed for the layers of the upper crust, whilst velocities of 6.8–7.0 kms^{-1} represent the lower crust on either side of the continental margins. Between the margins of Corsica and Provence a two-layered transitional crust was mapped that appears to be neither continental nor oceanic. P-wave velocities between 5.2 and 6.0 kms^{-1} for the 5 km thick crystalline crust and a velocity in the uppermost mantle of 7.9 kms^{-1} are considerably lower than on the flanks of the rift.

The next part of the EGT to the south (Figure 3-1) extends from the Northern Apennines across the Ligurian Sea and across the islands of Corsica and Sardinia to the Sardinian Channel. The cross section (Figure 3-15) shown here is based on a careful revision by Egger (1992) of various previous structural models derived from refraction seismic data along the entire section, with large shots fired on land at ALC and offshore at SA, SB, SC, SD and SE. In addition a limited amount of older shallow reflection data were incorporated. It also includes pre-EGT information about the shallow sediments and deep structure of the crust by Morelli *et al.* (1977), Nicolich (1985), Finetti and Del Ben (1986), and recently published structural models by Biella *et al.* (1987). For a complete list of references see Egger (1992). All refraction and wide-angle reflection data collected as part of the EGT project in 1982 (a small, test experiment), 1983 and 1985 are contained in the compilation by Egger (1990).

Northern Apennines–Ligurian Sea

The very detailed survey along the EGT proper by means of ocean-bottom seismometers (OBS) and adjacent onshore recordings has shown that the active rifting in the Ligurian Sea reaches further northeast to at least the region north of Cape Corse (Figure 3-15). The layering derived for the upper crust under the northern Apennines and the northern Ligurian Sea, with velocities of 5.8–6.25 kms^{-1} suggests basement nappes thrust northwards. The lower crust is reached in the north at a depth of 10 km. A thick layer with an increasing velocity from 6.3 to 6.4 kms^{-1} indicates the transitional character of the crust, as normally found under rift flanks. A velocity of 6.8 kms^{-1} is reached only just above the Moho. Below the Moho of the northern Ligurian Sea a thick high-velocity zone averaging 8.3 kms^{-1} is found beneath a rather thin transition layer with 7.7 kms^{-1}. A strong lateral variation from north to south, with layers of significantly higher velocities beneath the southern half of the Ligurian Sea, characterises the middle and lower crust.

The high-velocity intrusive plug with 6.7 kms^{-1} velocity in the middle crust of the central Ligurian Sea is indicative of the progressing rifting process. Note also the high-velocity area over a depth range of about 5 km in the lower crust under the central basin. The Moho rises from 24 km beneath the northern Apennines to 18 km in the centre of the Ligurian Sea and subsides again to 24 km beneath the north coast of Corsica. The measured apparent upper mantle velocities require a higher value in the north and a low one in the south. Beneath the southern half of the Ligurian Sea the sub-Moho velocity varies between 7.5 and 7.8 kms^{-1}.

Corsica–Sardinia

The islands of Corsica and Sardinia have a rather homogeneous Variscan-type continental crust (Figure 3-15) without spectacular features and with only minor indications of low-velocity zones. Since Corsica has no sedimentary cover the velocity in the upper crust increases in the centre of the island from the relatively high near-surface value of 6.1 –6.3 kms^{-1} at 12 km depth, followed by a small decrease to 6.1 kms^{-1}. The lower crust is reached at 19 km depth where the velocity changes abruptly to 6.6 kms^{-1}. The base of the crust deepens steadily from the Ligurian Sea to 32 km beneath the southern half of Corsica.

Relative to Corsica the actual seismic profile on Sardinia is offset to the west by about 50 km (Figure 3-1). The profile traverses Oligocene to Quaternary volcanics in the north and, in the south, first the Quaternary sediments of the Campidano graben and then Variscan, Caledonian and Cambrian units. This is expressed in a considerable lateral variation of the near-surface structure, with locally rather low seismic velocities. Below that the gross velocity structure is very similar to the one beneath Corsica, with a clear separation into an upper and lower crust based on the difference of velocities between, respectively, 6.1–6.25 kms^{-1} and 6.6 kms^{-1}. The structural results derived from cross lines and parallel profiles to the main N-S line (Egger 1992) lead to the conclusion that Corsica and Sardinia belong to the same lithospheric block. The uppermost mantle velocity increases southwards from 7.5 kms^{-1} beneath the Ligurian Sea to 7.7 kms^{-1} below Corsica and 8.0 kms^{-1} beneath Sardinia.

Sardinia Channel

The seismic profile across the Sardinia Channel connects the European Corsica-Sardinia block with the orogenic units of the northernmost part of the African plate. The relative motion between these two tectonic units has seriously affected the crustal structure (Figure 3-15). The shallow data have indicated a three-layered sequence of sediments which thin

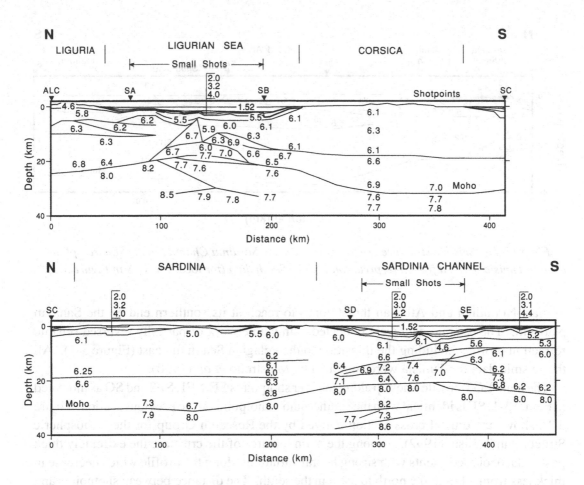

Figure 3-15. Seismic structure cross section from the Ligurian Sea across Corsica and Sardinia to the Sardinia Channel derived from interpretation of EGT-South data (for location see S in Figure 3-1).

above a number of basement highs or ridges. Associating the velocity of about 6.0 kms^{-1} with the seismic basement, this interface deepens from 4 km under shotpoint SD to 5.5 km under the central Sardinian Channel and rises again to 4 km near shotpoint SE with a sharp rise 40 km north of SE. In the northern third of the Sardinia Channel the velocity increases with depth in the upper crust from 6.05 kms^{-1} to 6.3 kms^{-1} at about 10 km. In the centre around shotpoint SE a high-velocity layer with 6.95 kms^{-1} had to be introduced into the model to fit the travel times observed on the OBS similar to the Ligurian Sea. At 13 km depth the velocity changes sharply from laterally variable values between 6.3 kms^{-1} and 6.4 kms^{-1} to values between 6.9 and 7.4 kms^{-1} immediately below this interface. This roughly 8 km thick transitional layer with an average velocity of 7.5 kms^{-1} probably has the same significance as the one found under the central Ligurian Sea. It wedges out both to the north and to the south to disappear beneath the Sardinian and Tunisian coast lines. The Moho has been modelled as a distinct boundary where the velocity increases to 8.0 kms^{-1}. Its depth decreases from 25 km under southern Sardinia to 20 km north of shotpoint SE from where it dips towards the Tunisian coast.

Tunisia

In order to construct a complete section of the European lithosphere the seismic profile of the EGT was extended across the Sardinia Channel into Tunisia where it traverses the

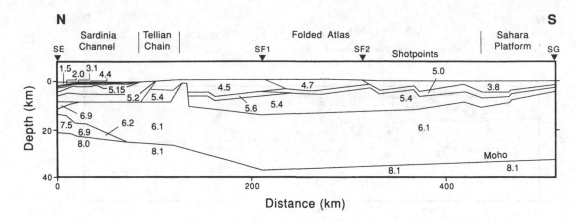

Figure 3-16. Seismic structure cross section from the Sardinia Channel to the Saharan platform in Tunisia derived from interpretation of EGT-South data (for location see S in Figure 3-1).

Tellian–Numidian and Atlasian thrust zones to reach at its southern end on the Saharan platform. The seismic data were collected within a network of partially reversed seismic refraction profiles including the transition to the Pelagian Sea in the east (Figure 3-1). All the seismic data for Tunisia were compiled by Maistrello *et al.* (1990).

The main EGT profile lines up along the four shotpoints SE, SF1, SF2 and SG as indicated in Figure 3-1. SE is identical with the southernmost shotpoint on the previous section. Figure 3-16 shows the crustal cross section derived by the Research Group for the Lithospheric Structure in Tunisia (1992). Among the main features of the crust are the extremely thick post-Palaeozoic sediments with strong lateral variations along the profile which decrease in thickness from 8 km in the north to 3 km in the south. The distance between shotpoints and the quality of the data did not allow the shallow seismic structure to be determined independently. Most of the information about the sediments was taken from results of hydrocarbon exploration (Bobier *et al.* 1991) and incorporated schematically in the seismic model (Figure 3-16) which includes the folded Tellian–Numidian and Atlasian terranes. A velocity of 6.0 kms^{-1} is found everywhere at unusually great depths of 8–14 km. No clear intracrustal layering could be detected on the main north–south profile. The striking feature beneath the sediments is its low average velocity of 6.05 kms^{-1} for the entire crust. The thickness of the crust increases steadily southwards from 21 km under the centre of the Sardinian Channel to 38 km in the central folded Atlas but then it decreases again slightly to 33 km under the southern end of the traverse. The upper mantle velocity has a normal value of 8.1 kms^{-1} beneath the African continent.

3.3 SEISMIC EXPLORATION OF THE UPPER MANTLE ALONG EGT

3.3.1 EVIDENCE FROM P-WAVES

The Moho is, by definition, the boundary at which P-wave velocity increases to values of 8.0 kms^{-1} or more and is generally a sharp transition. From the very large amount of P-wave refraction and wide-angle reflection data across Europe, supplemented by normal incidence reflection data, it is possible to map the Moho depth with an accuracy better than ±2 km.

Figure 3-17 shows a general map of Moho depth across Europe compiled from Meissner *et al.* (1987) and Ziegler (1990). A more detailed map of Moho depth contours along the EGT swathe is presented in EGT Atlas map 2. By linking the Moho with the crust/mantle boundary, this map can be regarded as showing the base of the crust and the top of the upper mantle. In considering its general features, the Moho map is usually viewed as the base of the crust and is compared with surface geology as expressed, for example, in Figure 2-3. General features worth noting are the crustal roots underlying the Pyrenees, Alps, Dinarides, Carpathians and their NW prolongation along the Teisseyre Tornquist zone (Figure 2-4). Equally striking are the contrasts between the thicker (45 km) continental crust of the Baltic Shield and east European platform, the thinner (30 km) continental crust of western Europe and the oceanic crust (20 km) of the Mediterranean basins.

Across the Baltic Shield, FENNOLORA and the POLAR Profile have defined the broad framework of crust and upper mantle structure. The crust is generally 45–50 km thick and the lithosphere reaches a maximum thickness of over 200 km, reducing to less than 100 km to the south of the Tornquist zone. Results from FENNOLORA (Guggisberg and Berthelsen 1987) presented in Figure 3-18 reveal a layered structure within the upper mantle with bands up to 50 km thick that have V_p reduced by 0.5 kms^{-1}. This profile provides an insight into the heterogeneity of the lithosphere beneath the crust, sometimes called 'the lower lithosphere', which is beyond the range of normal-incidence reflection profiling and beyond the limits of resolution of seismic tomography from earthquake sources. It is thus a key piece of evidence about the nature of continental lithosphere. There are glimmers of similar heterogeneity developed in the lower lithosphere beneath the Alps, their northern foreland and Sardinia, but elsewhere along EGT there are insufficient data to reveal the nature of any consistent heterogeneity that may be present.

A complementary approach to determining upper mantle heterogeneity on a broader scale has been taken by Spakman (1986, 1988, 1990a,b, 1991) from the inversion of over half a million observations of P-wave residuals covering western Europe and the Mediterranean. The residuals represent departures in the P-wave travel-times from a concentrically layered but otherwise uniform model of the Earth, such as the Jeffreys–Bullen model, which arise from local heterogeneities in P-wave velocity. Given a sufficiently dense network of pathways of the P-waves through a volume of the Earth, inversion can map the heterogeneity of this volume in three dimensions, which is known as seismic tomography. From his seismic tomography analysis of Europe, using earthquakes in southern and eastern Europe and the Mediterranean recorded at observatories spread all over Europe, Spakman (1990a,b) has produced a series of cross sections with contoured values of P-wave velocity to a spatial resolution of 50-100 km. Relatively high values of V_p have been mapped in three dimensions which coincide with lithosphere thickness variations, including subducted slabs from the African plate across the western Mediterranean, the Calabrian and the Hellenic arcs. From his most recent analysis (Spakman 1991) he has derived a cross section along EGT from central Scandinavia to Tunisia that shows P-wave velocity variations down to 1300 km depth. As can be seen in Figure 3-19, this gives a spectacular view of the variability within the upper mantle. Of specific interest is the region of relatively high velocity in the upper portion of the section which marks the lithosphere. This thickens to more than 200 km under the Baltic Shield, exactly as observed along FENNOLORA (Figure 3-18). It also thickens to 150–200 km beneath the Alps to form a lithospheric root. This lithospheric root beneath the Alps was discovered by Panza and Mueller (1978) from surface wave analysis and later confirmed by P-wave residuals, gravity modelling and geothermal considerations (Kissling *et al.* 1983). The asymmetrical form of this root, first noted by Panza *et al.* (1980a), is clearly evident in

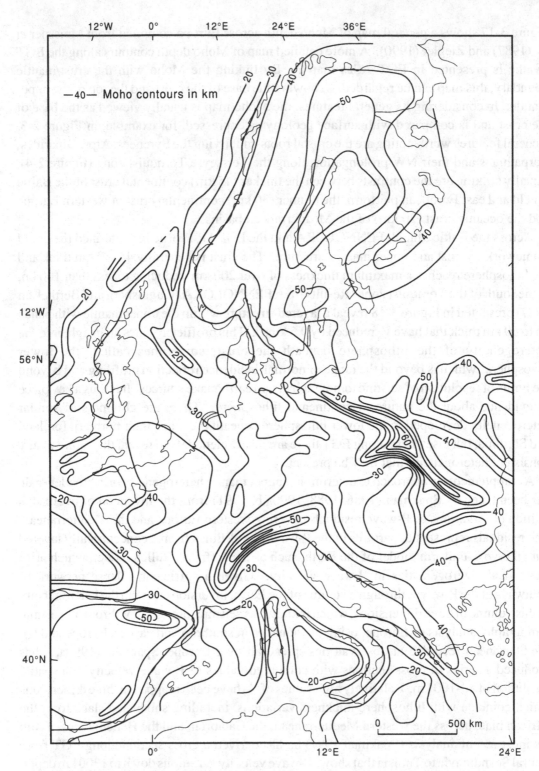

Figure 3-17. Contour map of Moho depth in km across Europe, after Meissner et al. (1987) and Ziegler (1990).

Spakman's cross section. The high-velocity lithosphere region remains around 100 km thick beneath Corsica and Sardinia but its position beneath North Africa is unresolved. In contrast, the central, Variscan part of EGT north of the Alps shows relatively low P-wave velocities within 100 km of the surface, indicative of thinned lithosphere. Exploring this in three dimensions, Mueller (1989) has been able to demonstrate that the thinned lithosphere with low velocities relates to the central European system. Although not clearly visible on the

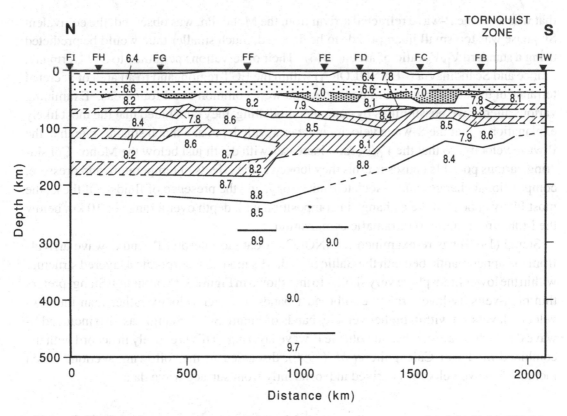

Figure 3-18. Seismic structure cross section of the crust and upper mantle of the Baltic Shield, based on P-wave velocities derived from interpretation of FENNOLORA.

cross section in Figure 3-19, further to the east, parallel cross sections prepared by Spakman (1990b) clearly indicate a dominant feature beneath the Hellenic arc that is continuous to the surface, which he has identified there as the down-going subducted lithosphere slab of the African plate. A similar feature is also visible to the west of the EGT profile beneath the Atlas mountains. The north-dipping high-velocity region beneath the EGT section may therefore represent subducted African lithosphere which has detached and is now sinking within the upper mantle. Further north, beneath the Alps, there is another region of relatively high P-wave velocity between depths of 300 and 600 km, clearly separated from the lithosphere, which could possibly represent the residual of former subducted lithosphere. At the northern end of the section the high-velocity thick lithosphere is underlain by a deeper region of relatively high velocity to about 500 km depth and there appears to be a distinct lateral change from this to the region immediately to the south.

3.3.2 EVIDENCE FROM S-WAVES

Although most of the major explosion experiments have deployed three-component seismometer stations with a view to recording S-waves (which should show better on the horizontal component seismometers) as well as P-waves, most attention has been given to the latter. This has been due in part to the rather weak S-wave signals generally obtained in such experiments. Gajewski *et al.* (1990) have made a point of examining the refraction of S-waves at the Moho and their propagation through the uppermost 10–15 km of the mantle in continental lithosphere. Despite observing good S-wave reflections off the Moho, they found that wherever the P-wave refracted arrival from the Moho, Pn, was observed, the equivalent Sn phase was too small in amplitude to be detected, much smaller than would be predicted using a standard V_p/V_s ratio of around R(3). Their observations pertained to SW Germany, France and Scandinavia (FENNOLORA) within the EGT region, and to an area of accreted terranes in the northeastern United States, and were common to all four areas. Examining various alternative possibilities by computer modelling, they concluded that the most likely explanation is that the S-wave velocity increases with depth at a much smaller rate than the P-wave velocity, so that the V_p/V_s ratio increases with depth just below the Moho. Considering various possible causes for this they looked at the effects of temperature and pressure, compositional changes and associated anisotropy, and the presence of fluids. Of these, the most likely appears to be a change in composition with depth over a range of 10 km below the Moho from mafic to ultramafic composition.

Stangl (1990) has re-examined FENNOLORA data to interpret P- and S-wave arrivals from the upper mantle beneath the Baltic Shield. His models incorporate a layered structure within the lower lithosphere very similar to that shown in Figure 3-18, although Stangl prefers thin high-velocity layers relative to broader bands of lower velocity rather than the low-velocity layers set within higher velocity bands of Figure 3-18. Stangl has also included S-wave data, which are consistent with the P-wave layering. This is entirely in accord with the combined model of Calcagnile *et al.* (1990), discussed in the following section, which includes S-wave velocities derived independently from surface wave data.

3.3.3 EVIDENCE FROM SURFACE WAVES

Panza *et al.* (1980b) carried out an inversion analysis of Rayleigh waves from group

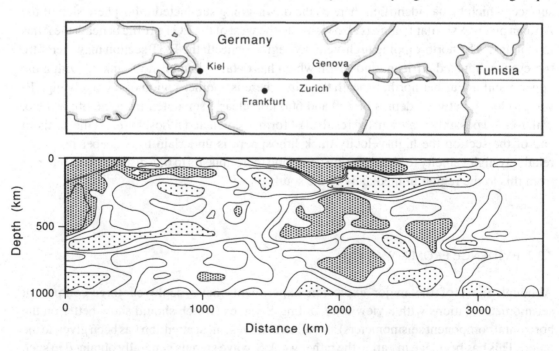

Figure 3-19. Seismic tomography cross section of the crust and upper mantle along the line of EGT. The darker shading refers to areas of higher P-wave velocity and the lightly shaded areas to lower P-wave velocity relative to a standard Earth model (Spakman 1991).

velocities and the phase velocities of the fundamental mode observed on long period seismometers at a relatively dense network of observatories spread across Europe. Their analysis also provided some measure of the variability of S-wave velocities in three dimensions beneath Europe as well as estimates of the V_p/V_s ratio from body waves. Their approach was to subdivide the Rayleigh wave pathways into segments within each of which the phase velocity could be determined, so that phase velocity curves could be obtained for each subdivision of their map. These curves were then inverted to produce models of S-wave distribution within the crust and upper mantle made up of a number of layers whose thickness and S-wave velocity values could be varied. The lower lithosphere was shown to be a region of relatively high S-wave velocity, of variable thickness, overlying the asthenosphere with relatively low S-wave velocity. Building upon the P-wave velocity model derived from FENNOLORA, Calcagnile *et al.* (1990) have analysed the dispersion of Rayleigh wave paths crossing Scandinavia in the same way, to produce a cross section of the crust and upper mantle that combines both P-wave and S-wave velocity variations. Their model has been combined with a cross section of the remainder of EGT derived from Panza (1985) and Calcagnile and Panza (1990) to give a complete cross section as shown in Figure 3-20. There is a clear correspondence with Spakman's P-wave velocity cross section, Figure 3-19, with thick lithosphere beneath the Baltic Shield and a lithosphere root beneath the Alps. Within the lower lithosphere of the Baltic Shield, the low-velocity bands are more pronounced in V_p, reduced by 0.5 km s^{-1}, than in V_s which is reduced by around 0.15 km s^{-1}. The distinctive layering and the reductions in V_p and V_s suggest compositional banding as the cause rather than the effects of any thermal variations. The cross section also provides an opportunity to check the V_p/V_s ratio within the lower lithosphere, where it equals R(3) within experimental error (cf. Panza *et al.* 1980b). A contour map showing the depth to the base of the lithosphere and the range of S-wave velocities, both in the lower lithosphere and the asthenosphere, is presented in Figure 3-21, based on the analysis by Panza (1985) and Calcagnile and Panza (1990). Uncertainty in determining its depth is in the order of 20–30 km. Values of S-wave velocity in the lower lithosphere range between 4.2 and 4.8 kms^{-1} and are generally around 4.4 to 4.5 kms^{-1} whilst those in the asthenosphere mostly lie between 4.2 and 4.4 kms^{-1}.

Lateral variability of S-wave velocity in the upper mantle has also been addressed by Nolet (1990) by means of new waveform inversion techniques using both shear and surface wave information recorded by the NARS array (see Figure 3-1). He has successfully fitted computed waveforms to those observed for S-wave groups and surface waves (fundamental mode and overtones) to map the variation of V_s to a resolution of 100 km vertically and 400 km horizontally along a cross section of the mantle beneath the NARS array to a depth of 500 km, presented in Figure 3-22. The most striking feature of this model cross section (known as WEPL3) is the upper mantle structure. It indicates the presence of a region of low S-wave velocity (V_s = 4.3 kms^{-1}) centred at 80 km depth beneath the southern half of the profile compared with higher values around 4.4 kms^{-1} for the northern half, and a region of relatively high velocity between 200 and 500 km depth beneath the European part of the section that contrasts with the Baltic Shield part in the north. The 'low' in V_s at 80 km is markedly lower than would be expected for peridotite at this depth (V_s ~4.5 kms^{-1}) but could be explained as a zone of partial melting where, given dV_s/dT = -4 x 10^{-4} kms^{-1} °C^{-1} at $T \simeq 1100$°C, temperatures are some 300° greater in the southern than in the northern half of the profile. However, the contrast around 300 km depth between 4.6 kms^{-1} and 4.8 kms^{-1} appears to be too great to be explained by temperature alone but indicates a compositional change at this depth in the region of the Trans-European fault and represents a significant difference in lithosphere structure between the European platform and the Baltic Shield. Relatively high

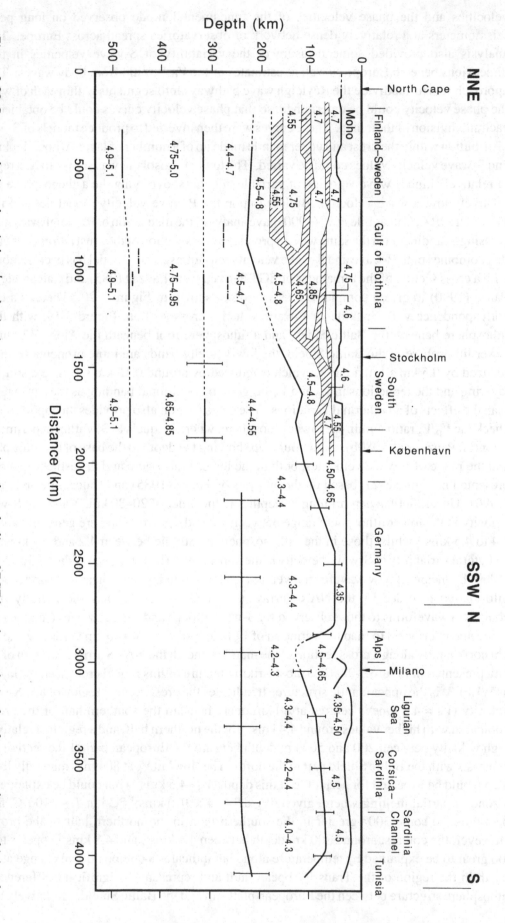

Figure 3-20. S-wave velocity values in kms[-1] and seismic structure cross section along the line of EGT based on analysis of surface wave data, combining the interpretation of Calcagnile et al.(1990) for the northern and Suhadloc et al.(1990) for the southern part of the geotraverse. Vertical bars indicate uncertainty.

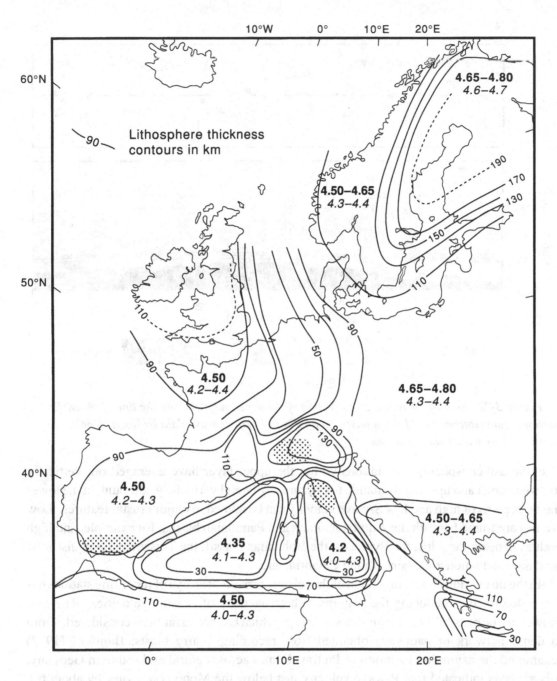

Figure 3-21. Contour map of the depth in km to the base of the lithosphere of Europe derived from an interpretation of the analysis of surface wave data, after Panza (1985). Numbers refer to S-wave velocities above and below (in italics) the base of the lithosphere. Shaded areas indicate where there are lithosphere roots.

values of V_S just below the Moho (also observed by Panza *et al.* 1980b) and again below 220 km beneath the Paris basin also coincide with locations of high P-wave velocity observed by Spakman (1986).

Lateral heterogeneity within the upper mantle below Europe has also been mapped by Snieder (1988) from large-scale waveform inversion of surface waves. He averaged in a vertical sense and produced maps of S-wave velocity variation for two fixed layers, one from 0 to 100 km depth, the other from 100 to 200 km. He found a greater and better defined variability within the lower of the two layers, containing the upper mantle low-velocity zone

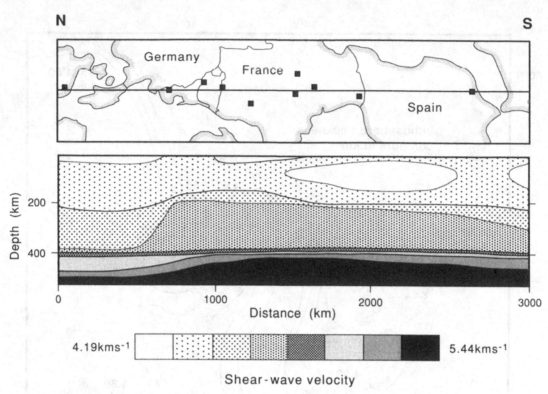

Figure 3-22. Seismic structure cross section of the upper mantle along the line of the NARS stations from Göteborg to Malaga based on analysis of surface wave data by Nolet (1990). S-wave velocity values shown in kms⁻¹.

(i.e. the asthenosphere). Variations within the upper layer have averaged out vertically between crust and uppermost mantle to obscure any lateral variability, although in the lower layer, regions of high and low S-wave velocity can be related to major crustal features. Low values are found beneath the Aquitaine, Paris and Pannonian basins, for example, and high values beneath the Alps, the Pyrenees, the Bohemian massif, the Teisseyre Tornquist zone and the subduction zone beneath the Calabrian arc.

In the discussion above, inversion of the seismic arrivals at arrays of recording stations has been described as deducing the patterns of heterogeneity of seismic velocities. There is, however, the possibility of azimuthal anisotropy which has not so far been considered. From a dense network of pathways obtained from recording quarry blasts, Bamford (1977) examined the azimuthal variation of Pn travel times across central and southern Germany. his analysis indicated that P-wave velocity just below the Moho could vary by about 6% around 8.05 kms⁻¹ with maximum values aligned approximately NE–SW. However, he acknowledged the difficulty in distinguishing anisotropy from the effects of lateral heterogeneity. Further studies by Fuchs (1983) and by Yanovskaya *et al.* (1990), who analysed Rayleigh wave phase velocites over eastern France, northern Italy southern Germany and Switzerland, have also revealed anisotropy of around 2.5 % in S-wave velocity just below the Moho with maximum values broadly NE–SW. They link this alignment with the direction of maximum horizontal tension across the region (see Figure 5-1). Although anisotropy may be present, it would appear not to have too significant an effect on velocity structure in the upper mantle, which is borne out by the way that the same features are found using analyses of both P-wave and S-wave velocity variations by such a variety of methods, based on independent datasets. It is therefore reasonable to draw geological inferences from the variations as mapping mantle heterogeneity. As a consequence of these findings it was decided to set up the ILIHA (Iberian LIthosphere Heterogeneity and Anisotropy) experiment as part of the EGT project, in order to investigate a region of the European lithosphere with a stable and relatively uniform crustal block, the Iberian Peninsula, of sufficient size to give a reasonable chance of investigating both the nature of any heterogeneity within the underlying mantle and, if possible to examine whether anisotropy might also be present. To this end, the NARS array was deployed across the region from April 1988 to March 1989 and a long-range refraction and wide-angle reflection experiment using explosive sources was conducted in October 1989. Analysis of the results from this difficult and complex set of experiments has yet to be completed.

The lower bounds to the upper mantle are marked by discontinuities at 400, 520 and 670 km depth, at which P-waves and S-waves increase sharply. Studies by Paulssen (1988) of converted P- to S-waves, which are sensitive to the nature of the transitions, have demonstrated that the 670 km transition forms a very sharp discontinuity less than 3 km in width, with a well-defined velocity increase downwards. This effectively defines the base of the upper mantle and can be seen on the P-wave tomography image of Figure 3-19. The 400 km transition is not as distinctive nor as sharp, being less than 10 km thick and laterally more variable in character, possibly due to lateral heterogeneity in the layer above. A further weak transition at around 520 km depth has also been recognised, and may feature beneath Europe. Although some controversy remains, the transitions are generally regarded as due to phase transitions rather than compositional boundaries. Their relevance to continental geodynamics will be examined further in Chapter 7

4 Europe's lithosphere – physical properties

4.1 PHYSICAL PROPERTIES OF THE LITHOSPHERE

E. Banda and S. Cloetingh

There are two ways of discovering the nature of the rocks in the Earth that are too deep to be directly accessible by drilling. The first is by means of geophysical observations made at the Earth's surface to determine *in situ* the physical properties in its interior. In the last Chapter we reviewed the evidence that can be obtained from measuring the passage of seismic waves through the ground. In this chapter we consider other geophysical methods. First, we use measurements of the outward flow of heat from the Earth's interior to try to find out the temperatures of rocks at depth. We then make use of observations of gravity mapped over the Earth's surface and corrected for the effects of elevation and latitude on the gravity field. The residual, Bouguer gravity anomalies are indicative of localised mass deficiencies or excesses within the Earth which can be interpreted in terms of density contrasts between various units. Because there is no unambiguous interpretation of gravity data, models are usually presented that are constrained by other evidence, such as the seismic structure cross sections shown in Chapter 3, and by reasonable values of density for rocks likely to be present.

Variations in the Earth's magnetic field mapped over its surface and suitably corrected for global effects, also record regions where rocks are magnetised sufficiently to create local perturbations of the main geomagnetic field. Magnetic anomaly maps can, like gravity, be interpreted in terms of the spatial distribution of magnetic rocks. Also, like gravity, there is an inherent ambiguity that cannot be resolved between the size of the anomalous magnetic body and the strength of its magnetisation. Interpretation is therefore achieved inductively by setting up models, generally constrained by other evidence, whose magnetic fields match those actually observed. Nearly all the magnetic bodies are within the upper crust.

Information from analysis of time variations of naturally occurring electromagnetic fields observed at the Earth's surface, magneto-telluric (M-T) measurements, relate to electrical currents induced in the Earth's interior and can be interpreted in terms of distributions of electrical conductivity in one- or two-dimensional models. There is an inherent ambiguity in this interpretation which has to be treated in a similar way to gravity and magnetic data.

The mechanical properties of the Earth's interior can be deduced from a mixture of observations: from temperature and pressure information at depth in the Earth, from laboratory experiments of the mechanical behaviour of rocks subjected to such temperatures and pressures, and from observations of the geometry of the bending of the Earth's surface when subjected to a load, which measure its flexural (elastic) properties, and its slow recovery from loading, which measure its viscoelastic properties.

A second means of investigating the nature of rocks at depth is to make use of samples of rock that have been brought to the surface from considerable depth by volcanic processes. Fragments of such rocks incorporated into igneous rocks at the surface, known as xenoliths, yield valuable information about rocks from deep in the crust or from the upper mantle, especially when this is combined with the geophysical information as presented in Section 4.3. Rocks that were once at considerable depth that have been brought to the surface by tectonic uplift and erosion provide the other line of direct evidence. But these give information about the nature of the rocks at depth in former times whereas the geophysical information provides evidence of the *in situ* properties that are present now. It is the synthesis of all these lines of evidence that taxes our imagination.

Intensive observation of the lithosphere using all these geophysical methods during the EGT field work period added a great deal of fundamental information to geophysical and geological data sets already in existence, allowing the main physical properties of the European lithosphere to be determined. Seismic aspects have already been discussed in Chapter 3. Here we continue by looking first at physical characteristics involving temperature, thermal and electrical conductivity and density distribution.

The precise determination of physical properties of the present day lithosphere is an essential prerequisite for understanding why and how the lithosphere evolves with time. Our results form the input data to be used in modelling lithospheric processes. One example of this is given in Section 4.2 where different data sets are used to study the distribution of strength with depth.

4.1.1 THERMAL STRUCTURE

Measuring the surface heat flow from the lithosphere requires reliable observations of both the temperature gradient and the thermal conductivity. Basically, there is no major technical difficulty in measuring this temperature gradient either on land or at sea. The problem, however, arises when extrapolating the measurements at depth. In addition, temperature gradients obtained near the Earth's surface are affected by ground-water circulation, long term climatic variations, sedimentation and erosion effects, uplift, subsidence and lateral variations of the thermal conductivity, and need to be corrected for these effects. Most of the effects are, in principle, identifiable but their quantification is not straightforward. A large amount of data can certainly help to improve the corrections. A second problem, however, comes from a scarcity of data, which forces undesired interpolation when studying a large area. This is because the most reliable data come from temperature measurements carried out in deep boreholes, the distribution of which is rather poor. At sea, temperature gradient observations are more readily obtained, although even these are not free from water circulation effects, nor from poor distribution.

Similar problems are faced when measuring thermal conductivity of rocks. This physical property is required to obtain surface heat flow density when conduction is the only mechanism assumed to be responsible for heat transfer. Thus

$$q = -k \, dT/dz$$

where q is heat flow density, k is thermal conductivity and dT/dz is the vertical temperature gradient.

Thermal conductivity of rocks, which typically ranges from 0.8 to 6.0 $Wm^{-1}°C^{-1}$, depends basically on mineralogical composition, size and orientation of crystals, porosity and physical conditions. Thermal conductivity is measured in samples recovered from boreholes. Here the difficulty comes from the small number of boreholes where both temperature gradient and thermal conductivity can be measured and also on how representative are the available samples.

The thermal structure of the lithosphere along EGT has been studied in fair detail, the large amount of data being sufficient to attempt its use for modelling purposes. In the Baltic Shield about 150 sites of thermal gradient and conductivity measurements are available as reported by Balling (1990b). Once the paleoclimatic correction had been applied and the effects of ground water movement removed, to match the assumption of heat transfer by conduction, the regional variation of heat flow density in the Baltic Shield could be mapped. Similarly, thermal data from the Variscan of Germany was critically revised and updated by Schulz (1990). A good part of the data set corresponds to measurements of bottom hole temperature (BHT) routinely carried out in well logging operations. In this case, correction of the original data is necessary due to perturbations during drilling and mud circulation. More than 7300 sites are presently available, although thermal conductivity measurements are available in only 113 sites. This data set was paleoclimatically corrected and a heat flow density map was obtained. In the Alpine–Mediterranean part of the EGT, several published and unpublished data sets were gathered by Della Vedova *et al.* (1990). The area includes part of the Mediterranean where a fair amount of marine heat flow measurements are available. They gathered over 1000 sites, with just over half of them corresponding to marine measurements. Their critical revision and analysis provided a reasonably accurate heat flow density map of the area. A general view of the heat flow map of the European Geotraverse area can be seen in Figure 4-1a. This is presented in more detail in EGT Atlas Map 13. In the Baltic Shield the surface heat flow distribution appears to increase from values around 40 mWm^{-2} in the northeast to 70 mWm^{-2} in the southwest. In the Variscan of central Europe the heat flow is fairly uniform around values of 70–80 mWm^{-2} except in the North German basin where it falls to around 60 mWm^{-2} and in the area north of Lake Constance (the Urach field) where heat flow density increases to values which can locally reach 100 mWm^{-2} or more. Values around 70 mWm^{-2} are found in the Alps, 45 mWm^{-2} in the Po Plain, 80–100 mWm^{-2} in the Ligurian Sea and northern Corsica, 60 mWm^{-2} in southern Corsica and northern Sardinia, 80–100 mWm^{-2} in the Sardinia and Sicily channels and about 75 mWm^{-2} in Tunisia (see Figure 5-16).

Once the surface heat flow distribution is known there are still two basic parameters which are needed before attempting to determine the temperature distribution in the lithosphere: the heat production and the thermal conductivity of lithospheric rocks. It is typically assumed that radioactive decay of U, Th and K is the main source of heat, making it essential to know the concentration of those elements in the rocks making up the lithosphere. This is, in the best case, solved by using results from heat production measurements of rocks representative of the whole lithosphere. The concentration of radioactive elements is proportional to the SiO_2 content of the rock. Therefore the major contribution to heat production comes from the upper crust. Several models to account for heat production in the crust have been proposed such as an exponential decrease with depth or a step-like function. Cêrmák and Bodri (1992) use two different exponential distributions, one for the upper crust and another one down to the Moho. Subcrustal lithosphere rocks appear generally to have a rather constant value

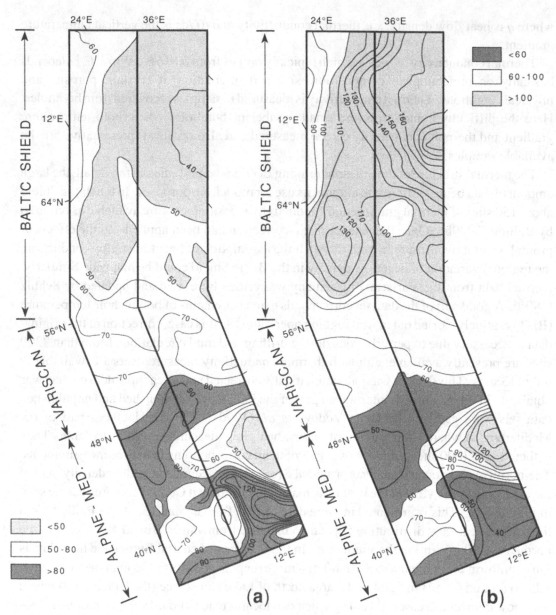

Figure 4-1. (a) Distribution of surface heat flow values along the EGT as compiled by Cêrmák and Bodri (1992). The variation is fairly smooth in the Baltic Shield and Variscan central Europe while the Alpine–Mediterranean part is characterised by sharp changes.
(b) Thickness of the thermal lithosphere according to Cêrmák and Bodri (1992).

around 0.01 μWm⁻³ which can be taken as fairly accurate. As far as thermal conductivity of the lithospheric rocks is concerned, typical values measured in a variety of representative rocks have generally been used in modelling the European lithosphere.

Several attempts to model the temperature field along different parts of the European Geotraverse have been published in the last few years, following the solution of the steady-state heat conduction equation. For instance, Cêrmák and Bodri (1992) have used a 3-D numerical solution of the steady state heat conduction equation

$$\text{div}\{k(dT/dz)\} + A = 0$$

where A is the radiogenic heat production function, which was taken to be $0.04\,qe^{-z/10}$, q being the measured surface heat flow, for depth z between 0 and 10 km, $2.5e^{-z/10}$ for z between 10 km and Moho depth, and 0.01 μWm⁻³ below the Moho. The adopted values for thermal

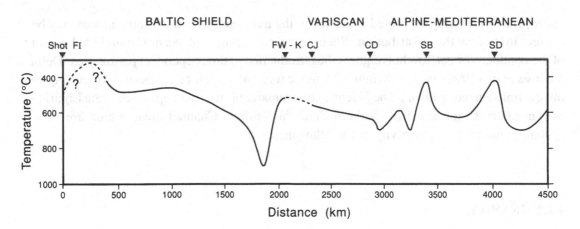

Figure 4-2. Estimated temperature of the Moho along EGT (after different authors, see text).

conductivity k were 2.5 and 3.5 $Wm^{-1}°C^{-1}$, respectively, for the crust and mantle. Other authors have adopted slightly different assumptions although their results are fairly consistent within about 100°C.

The depth to the Moho along the EGT is well known from the evidence given in Chapter 3. Therefore the temperature at the Moho can be estimated from the temperature field as shown in Figure 4-2. It can be seen that the Moho is far from an isotherm, with temperatures ranging from over 800°C in the Baltic Shield to 400°C in the Mediterranean. In fact the Baltic Shield Moho temperatures are fairly uniform with values around 400–500°C although an increase to over 800°C in the southern part of the shield can be attributed to a pronounced local thickening of the crust to more than 50 km. In the Alpine–Mediterranean area the Moho temperature is highly variable within the range 400-800°C. However, temperature estimates may in general be in error by as much as 100°C.

The depth to the base of the thermal lithosphere can be estimated from the temperature field data assuming that the base of the lithosphere–asthenosphere boundary corresponds to a given $T(x, z)$ curve. A constant value for the temperature T_m at the base of the lithosphere of around 1300°C is usually taken. Alternatively, Della Vedova *et al.* (1990) use the expression

$$T_m = 0.85(1100°C + 3z))$$

where z is in km, which does not correspond to an isotherm, resulting in the thermal lithospheric thickness map shown in Figure 4-1b. In this model the thickness varies from 100–170 km in the Baltic Shield. Pasquale *et al.* (1991) follow a slightly different approach and suggest that the thermal lithosphere in the central part of the Baltic Shield is more uniform and may be as thick as 200 km. This is reduced to 60–100 km in the Variscan of central Europe with values ranging between 50 and 70 km in the southern part of the EGT. The thinnest thermal lithosphere is found in the Mediterranean area where it barely reaches 50 km. The thickness of the thermal lithosphere can be compared with that obtained from analysis of surface wave dispersion data shown in Figure 3-20. Although the main trends are reasonably similar, locally they may differ significantly as, for instance, in the Southern Alps and Po Plain. The main question arises about the resolution of both methods as well as the comparability of the concepts of thermal and seismic lithosphere.

The Alpine–Mediterranean part of the EGT is the one showing the most noticeable lateral heat flow variations. This pattern has been studied in detail by Della Vedova *et al.* (1990) who argue that inversion of surface heat flow values to determine temperatures at depth is not appropriate in recently deformed areas due to the transient thermal component that may be present in the heat flow distribution. Such a transient component can not be included as part of the conductive heat flow but requires the introduction of time dependent parametres. Della Vedova *et al.* (1990) use a 2-D finite difference forward modelling technique which singles out the transient component. They identify two remarkable transient episodes in the Ligurian Sea–northern Corsica and in southern Sardinia–Sardinia Channel areas which are most probably due to modern activity of the lithosphere.

4.1.2 GRAVITY

Gravity data are generally used to deduce the distribution of density with depth in the lithosphere provided that either the geometries of the main discontinuities or the lateral changes of density are known. Seismic data are the best constraint for gravity modelling since they provide fairly accurate depths to likely density discontinuities and also because the lateral density variations can be obtained from a number of empirical velocity–density relationships. Gravity data are of particular value in inferring deep crustal structure in areas where seismic constraints are available. However, a recent compilation of laboratory measurements (Mengel and Kern 1992) shows that, although a broad correlation exists

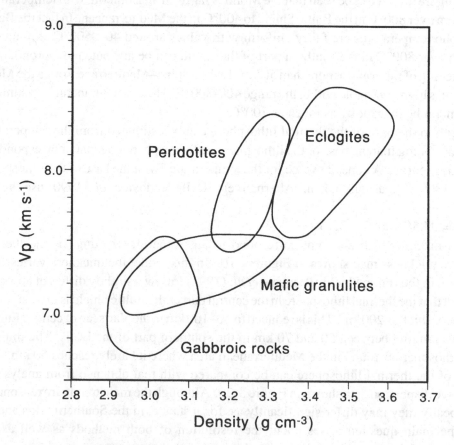

Figure 4-3. Experimentally determined P-wave velocities and densities of gabros, mafic granulites, eclogites, and peridotites for 600 MPa confining pressure (after Mengel and Kern 1992). Note that P-wave velocities of eclogites and peridotites overlap and that densities of eclogites are higher than those of peridotites.

between seismic P-wave velocity and density, the range of posibilities is large (Figure 4-3). They suggest that densities can only be inferred from V_p values with an accuracy of about 0.2 gcm^{-3}.

Along the EGT a gravity data compilation by Klingelé *et al.* (1990a) (EGT Atlas Map 9) is available. Gravity modelling using this data set has already been attempted by Henkel *et al.* (1990) and Marker *et al.* (1990) in the Baltic Shield and by Klingelé *et al.* (1990b) along a profile from the Alps to northern Africa. In the Baltic Shield, detailed gravity studies of outcropping upper crustal rocks have been used to account for their effect on the gravity field and allowed them to present a model for the lower crust and upper mantle shown in Figure 4-4a. The lower crust displays a fairly uniform density of 3.05 gcm^{-3} with anomalous bodies of higher density at its base. These anomalous bodies were introduced to reduce the density contrast with the upper mantle and meet seismic wide-angle reflection results. However, seismic data in the areas where the high density bodies are localised are not of high resolution and it may well be that a less pronounced geometry could also fit the gravity data. In Figure 4-4b we also show the density depth profiles at several sites where the seismic P-wave velocity has been converted into densities according to the empirical relationship of Woollard (1975) and Nafe and Drake (1963). Along the POLAR profile, combined gravity and magnetic interpretation has resulted in a much better definition at depth of structures such as the Lapland Granulite belt. It has been shown that it dips gently to the north beneath the Inari

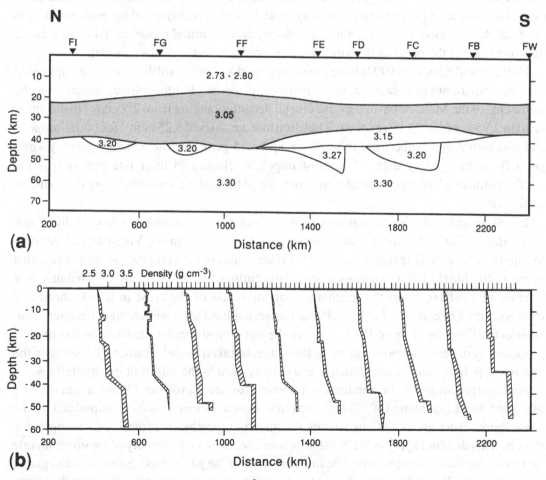

Figure 4-4. Average density model in gcm^{-3} for a profile in the Baltic Shield (a) according to Henkel et al. (1990) and (b) from simple conversion of P-wave velocities into densities according to Woollard (1975) and Nafe and Drake (1963). Shotpoints locate profile in Figure 3-1.

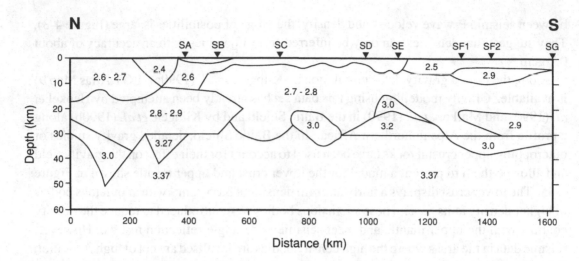

Figure 4-5. Density model along a profile from the Alps to Tunisia (simplified from Klingelé et al. 1990b).

terrain to depths of 15–20 km, in accord with the seismic evidence shown in Figure 3-3b.

In their 2-D interpretation of the southern segment of EGT shown in Figure 4-5, Klingelé *et al.* (1990b) find a good fit between the observed data and computed gravity using seismic constraints which they incorporate into the starting density model. A trial-and-error procedure was used to optimise the density model. A remarkably good agreement between the final density model converted into P-velocity and the initial model taken from seismic interpretation of the order of 0.2 kms[-1] gives confidence to the density model.

Holliger and Kissling (1992) have also interpreted a gravity profile across the Alps using seismic constraints to obtain the gross features of the whole lithosphere. Except for the sediments of the Molasse basin they use crustal densities ranging from 2.7 gcm[-3] in the upper crust to 2.95 gcm[-3] in the lower crust. They find that densities of 3.25 gcm[-3] for both European and southern Alpine subcrustal lithosphere, and 3.15 gcm[-3] for the asthenosphere, gives a good fit to the gravity data. The most important finding in their interpretation is the confirmation of a SW-dipping European continental lithosphere underthrusting the southern Alpine one.

In spite of unavoidable uncertainties in density values, various interpretations indicate that in the Baltic Shield the lower crust is on average denser than in the Variscan and Alpine–Mediterranean areas (Figures 4-4 and 4-5). In fact, what could be taken as normal lower crust in the Baltic Shield has anomalous high density further south. This may be attributed to a fundamental difference in the mechanism of formation of the crust in the Archean and Proterozoic when compared with the Phanerozoic, a subject to be addressed in Chapter 7. An apparent difference of up to 0.12 gcm[-3] in the uppermost mantle densities in the models proposed by different authors cannot, at this stage, be taken as real. Rather it depends on the relationship between P-wave velocity and density used in the different interpretations.

An interpretation of the medium wavelength of the European Geoid anomalies by Marquart and Lelgemann (1992) indicates that anomalies can mostly be explained by the lithosphere structure as obtained from seismic tomography, together with topographic effects. This does not apply to the Po basin where a decrease in the density of the upper mantle portion of the lithosphere seems to be one way to meet the geoid data. Similarly, a negative anomaly in the Baltic Shield is thought to be due to post-glacial rebound, to be discussed further in Chapter 5.4. Geoid anomalies also require a small density contrast between the

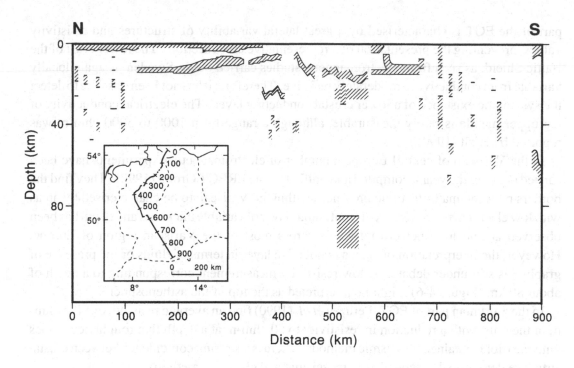

Figure 4-6. Conductive areas (shadowed) identified in the crust of the EGT Central Segment (after EREGT Group 1990).

lithosphere and the asthenosphere in the Baltic Shield, which is more pronounced in the Variscan and Alpine parts, generally confirming the seismic and gravity data.

4.1.3 ELECTRIC CONDUCTIVITY

An important parameter contributing to our knowledge of the physical properties and state of the lithosphere is the electric conductivity distribution at depth. This can be obtained from electromagnetic experiments, yielding information on the presence of fluids, fractures, composition and temperature. Electromagnetic studies along the EGT have been performed in parts of the Baltic Shield, in the Variscan of central Europe, in Switzerland and Sardinia. These are presented in EGT Atlas Maps 11 and 12. Results mainly relate to the crust and upper mantle, but occasionally they reach the asthenosphere (Figure 4-6).

In the Baltic Shield a number of experiments carried out with the same instrumentation, data processing and modelling give different results, although generally the upper crust has a high resistivity (10^3–10^4 ohm-m) whilst at lower crust depths there are fairly low resistivities. In the POLAR Profile, which lies on the Archean and Early Proterozoic of the Finnish part of the Baltic Shield, Korja *et al.* (1989) have reported a variety of results from electromagnetic studies. The uppermost part of the crust displays a variable resistivity of between 200 and 1000 ohm-m. A highly conductive layer (i.e. very low resistivity, 0.1 ohm-m), some 10 km thick on average and coincident with a region of high seismic reflectivity, is located at various depths in the middle and upper crust, occasionally reaching the surface. This high conductivity layer, however, cannot be taken as a general feature of the Baltic Shield since results from the Swedish part generally do not confirm it. This may be due to the local presence of fluids and major fractures in the upper and middle crust or, more probably, to graphite-bearing structures (Korja 1992). In general, however, the Archean and Proterozoic part of the EGT is characterised by a great lateral variability of structures and resistivity values, preventing the presentation of a representative cross section. The lower crust of the Baltic Shield, as seen from electromagnetic studies carried out in Sweden, is again locally variable from conductive to moderately resistive. Therefore, it has not been possible to detect the systematic existence of a lower crustal conducting layer. The electrical conductivity of the upper mantle is rarely measurable, although a range from 1000 to 5000 ohm-m was reported by Hjelt (1990).

In the Variscan of central Europe a number of electromagnetic experiments have been carried out over the years, compiled and unified by the EREGT Group (1990). They find the borders between major tectonic provinces within the Variscan to be characterised by areas with low electrical resistivity. A discontinuous layer of variable thickness and depth has been observed at mid to upper crustal levels across most of the Variscan region of Europe. However, the interpretation of such a conductive layer in terms of fluids or the presence of graphite is still under debate. A low resistivity measurement corresponding to a depth of about 80 km (Figure 4-6) might be interpreted as the top of the asthenosphere.

In the Sardinian part of EGT, Peruzza *et al.* (1990) find an average resistivity of 10^4 ohm-m in the crust with a reduction in resistivity to 50 ohm-m at a depth that roughly coincides with the Moho obtained by seismic methods. There is a striking coincidence between crustal structure determined independently by seismic and electrical methods.

4.2 MECHANICAL STRUCTURE

S. Cloetingh and E. Banda

It has long been recognised that the Earth's response to applied surface loads is both elastic and viscous. The elastic response is indicated by its ability to support stresses on the level of hundreds of MPa for times of millions of years (Figure 4-7). This mechanical behaviour is analogous to a uniform elastic plate. The viscous response is required by the observation that, following the removal of long wavelength loads such as ice sheets, the Earth responds in a time dependent fashion, rather than instantaneously. The two responses are traditionally treated by assuming that a near surface elastic layer is underlain by viscous material with negligible strength on a geological time scale. Initially the mechanical behaviour of the upper part of the lithosphere was described in terms of a uniform elastic plate with a thickness defined by an isotherm of 300–600°C based on studies of the flexure and intraplate seismicity in oceans. Hence the elastic part of the lithosphere was found to correspond roughly to the upper half of the total thermally defined lithosphere, displaying an increase in the effective elastic thickness (EET) with thermal age.

For continental lithosphere, which has undergone repeated deformation with often complicated loading scenarios, flexural studies are not as straightforward as the studies that have been carried out for oceanic lithosphere. All these investigations have shown, however, that the lithosphere is capable of supporting significant tectonic stresses over long geological time scales over a temperature range that is lower than half the melting point (which corresponds to temperatures at the base of the lithosphere estimated in the range 1200–1400 °C). The concept of a uniform elastic plate has been quite useful as a first order description of the thickness of that part of the lithosphere which has the potential to accumulate tectonic stress. For an actual comparison of stresses in the lithosphere (Zoback *et al.* 1989) with

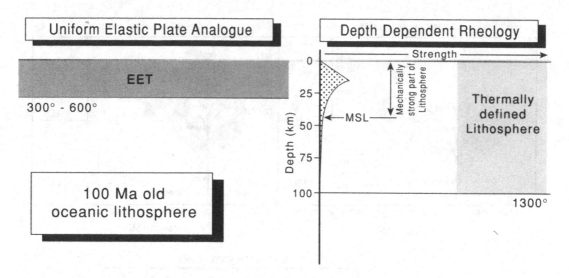

Figure 4-7. Panel illustrating for 100 Ma old oceanic lithosphere the relationship between thermal lithosphere (defined by the 1300°C isotherm), the effective elastic thickness, EET, and the mechanically strong part of the lithosphere, MSL, with a brittle/ductile rheology and a finite depth-dependent strength, rapidly decreasing at temperatures around 750–800°C.

strength levels, however, it is essential to account for the finite strength of the lithosphere inferred by laboratory experiments on lithospheric rocks.

A large number of geological and geophysical observations demonstrate that the upper part of the lithosphere must itself be rheologically zoned (e.g. Meissner and Kuznir 1987). Laboratory experiments on rocks show that, since their properties depend dramatically on temperature and pressure, significant variation must occur with depth in the lithosphere. It is possible to define the mechanically strong part of the whole lithosphere (MSL) and its crustal segment (mechanically strong part of the crust, MSC) for which lower boundaries can be defined by the depths at which ductile strength is lower than a threshold level of 50 MPa (Figure 4-8). The thickness of the MSL, reflecting the depth dependent and finite strength of the lithosphere, is by its nature in excess of the EET, since it represents the integrated response of the lithosphere for a plate with an infinite strength (Figure 4-7).

4.2.1 DEPTH-DEPENDENT RHEOLOGY OF THE LITHOSPHERE.

In the previous Section 4.1 we concentrated on the thermal and other physical properties of the lithosphere along the EGT. Here we concentrate on the consequences of the thermal structure for the mechanical properties of the lithosphere.

Despite the complexity of rock deformation, very simple approximations can be powerful in lithospheric studies. These approximations provide functional representations of the dependence of lithospheric properties on the most geologically significant factors. These formulations are from continuum mechanics, in which the material is treated as an aggregate with certain macroscopic features. These properties reflect, but do not explicitly deal with, the controlling physical properties. The lithosphere is modelled as two regions with different mechanical properties. In the upper region, the differential strength, which is the maximum differential stress that can be sustained, is limited by brittle fracture. For applied stresses less

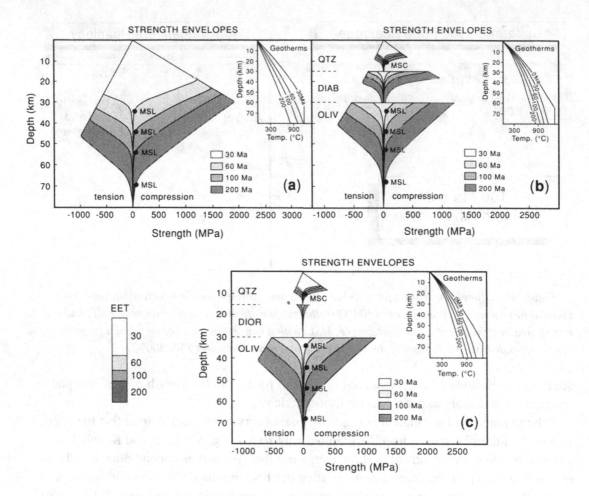

Figure 4-8. Depth-dependent rheological models for the lithosphere.
(a) Oceanic lithosphere;
(b) Continental lithosphere with a rheology based on quartz/diabase/olivine mineralogy.
(c) Continental lithosphere with a rheology based on quartz/diorite/olivine mineralogy.
Different curves show the effects of various geotherms. Minima in crustal strength develop only for steep thermal gradients associated with the presence of thinned lithosphere. The mechanically strong part of the crust and lithosphere are defined by the parameters MSC and MSL indicating, respectively, the depth in the crust and the upper mantle where ductile strength becomes smaller than 50 MPa. Lithospheric strength is controlled primarily by the petrological layering and the temperature profile in the lithosphere while an order of magnitude variation in strain rate introduces a shift in the depth of MSL and MSC of only a few km (modified from Stephenson and Cloetingh 1991). The inset in (a) shows the relationship between MSL and the concept of EET inferred from flexural studies. As these loading studies adopt a uniform elastic plate with infinite strength, EET is smaller than MSL.

than the brittle strength, the material deforms elastically such that stress and strain are linearly related. If the applied stress reaches the brittle strength, fracture occurs.

Laboratory experiments indicate that the brittle strength is a linear function of the applied normal stress and is largely insensitive to variations in temperature, strain rate, and rock type. Thus, in the brittle regime, lithospheric strength increases linearly with pressure. In the lithosphere, within the brittle regime, the lithostatic pressure σ is a function of density ρ

multiplied by gravity acceleration g and depth z, so that $\sigma = -\rho gz$ and is assumed to be one of the principal stresses. If the rock contains a pore fluid pressure $P_f(z)$, the relations between the stresses for sliding are given in terms of the principal effective stress, σ_1 and σ_3. Common assumptions are that the rock is dry, $P_f(z) = 0$ or that the pore pressure is hydrostatic (equivalent to assuming that pores are connected up to the surface) so that $P_f(z) = -\rho gz$ where ρ is the fluid density. Alternatively, the pore pressure can be assumed to be a fixed fraction of the lithostatic pressure. The strength of the lithosphere is defined by the maximum difference between the maximum (σ_1) and minimum (σ_3) effective stress that the rock can support.

Laboratory studies of rocks subject to differential stress demonstrate that, at temperatures in excess of approximately half their melting point, stresses relax by creep. Thus, with increasing depth, the effects of temperature become dominant and ductile deformation occurs in the lower portion of the lithosphere. Ductile flow is described by different creep equations for various differential stresses (Goetze and Evans 1979).

A very important creep process takes place by thermally activated climb of dislocations obeying a non-linear relation between stress and strain rate

$$\dot{e} = \dot{e_0}\sigma^N \exp(-Q/RT)$$

where Q is the activation energy, R the universal gas constant, σ the differential, or deviatoric, stress ($\sigma_1 - \sigma_3$), $\dot{e}$ the strain rate, T the temperature and N the exponent in the creep law (typically in the range of 2–5). For olivine, Q is 510 kJmole^{-1}, for sigma measured in bars the value of the pre-exponential constant is 70 bar^{-3}s^{-1}. Power law creep holds for conditions of high T and relatively low deviatoric stresses. For stresses in excess of 200 MPa, dislocation glide is the dominant creep process in olivine.

The basic equation for this process (Dorn creep) is

$$\dot{e} = \dot{e_0}\exp[-Q/RT(1 - \sigma/\sigma_p)^2]$$

where σ_p is 8500 MPa (85 kbar) and Q is 535 kJmole^{-1}. The value of the pre-exponential constant is 5.7×10^{11} s^{-1}.

Flow laws can be combined with failure criteria to estimate the strength of the lithosphere as a function of depth. For this purpose a strain rate and a geothermal gradient giving temperature as a function of depth are assumed. At each depth the strengths in both brittle fracture and ductile flow are calculated, and the smaller in magnitude is the relevant strength. At shallow depths deformation occurs in the brittle regime with a brittle strength, linearly increasing with pressure and therefore with depth, that differs for compression and tension. In general at larger depths, due to the increase in temperature, flow is the dominant effect and the limiting strength is given by the flow law for ductile deformation.

Figure 4-8 gives strength envelopes for oceanic and continental lithosphere. The oceanic lithosphere, reflecting the olivine rheology of a cooling plate, has a much simpler structure than layered continental lithosphere where the strength depends primarily on petrological layering and thermal structure. In addition, wetness, and to a lesser extent, strain-rate also affect the strength levels that can be supported within the lithosphere. We have calculated strength profiles for continental lithosphere for several positions along EGT, constructed from extrapolation of rock mechanics data using the petrological information and the thermal structure discussed in Sections 4.1 and 4.3 respectively. The profiles are given for different locations along the EGT, corresponding to three sites in the Baltic Shield (Figure 4-9), three locations in Variscan part of central Europe (Figure 4-10) and three sites in the northern part of the Alpine belt (Figure 4-11). The rheology of the lithosphere is for a quartz/ diorite/ diabase/ pyroxenite/ olivine petrological layering (Carter and Tsenn 1987), see Table 4-1, depending on the available petrological and lithospheric structure data available for the actual location. We adopt a strain rate of 10^{-16} s^{-1} which is characteristic for long-term geodynamic processes operating on the lithosphere.

Table 4-1. Rock mechanics data: values for creep constants (after Carter and Tsenn 1987) adopted in ductile flow laws used for the construction of strength envelopes.

	Baltic	Variscan	Alpine	Q (kJmole^{-1}) dry	Q (kJmole^{-1}) wet	e_0 (Pa^{-N} s^{-1}) dry	e_0 (Pa^{-N} s^{-1}) wet	N dry	N wet
layer 1	Quartzite	Quartzite		134	172.6	6.03×10^{-24}	1.26×10^{-13}	2.72	1.9
layer 2		Granite	Granite	186.5	140.6	3.16×10^{-26}	7.94×10^{-16}	3.3	1.9
layer 3	Diabase	Diabase	Diabase	276	212	6.31×10^{-20}	1.26×10^{-16}	3.05	2.4
layer 4		Ortho Pyroxenite		293	271	1.26×10^{-15}	1.00×10^{-19}	2.4	2.8
layer 5	Olivine/ Dunite Power law	Olivine/ Dunite Power law	Olivine/ Dunite Power law	510	498	7.00×10^{-14}	3.98×10^{-25}	3.0	4.5
	Olivine Dorn creep	Olivine Dorn creep	Olivine Dorn creep	535		5.7×10^{11}		σ_p 8500 MPa	

Inspection of the strength profiles shows pronounced changes in the distribution of lithospheric strength with depth between northern and central Europe. The lithosphere in the northern and central parts of the Baltic Shield is, according to the model predictions, quite strong with characteristic values of the mechanically strong lithosphere (MSL) in the range 80–95 km. The large values of MSL in the northern and central part of the Baltic Shield reflect the temperature distribution with low gradients in the mantle part of the lithosphere, which also leads to the presence of significant strength at subcrustal levels. In contrast, the predicted strength profiles in the southernmost part of the Baltic Shield and in the Variscan and Alpine parts of Europe show relatively low values of MSL (30–50 km) and a strong reduction in upper mantle strength. As a result of higher temperature gradients, pronounced minima in crustal strength occur, leading to discrete cores of strength at upper crustal levels at depths of 5–15 km in some parts of the profiles. A striking feature predicted by the rheological profiles is the occurrence of a minimum in strength at the base of the crust in Europe underlain, in most cases, by a strong subcrustal lithosphere. This leads to a large difference between MSL and MSC values, corresponding to isotherms of 750–800°C and 300–400°C, reflecting the creep strength of olivine and crustal rocks respectively. This feature has interesting tectonic implications as such minima in crustal strength are often the sites for crustal decollements.

4.2.2 RHEOLOGY AND INTRAPLATE SEISMICITY

The depths of continental earthquakes provide an interesting source of information to test the predictions of a depth dependent rheology for the lithosphere. Chen and Molnar (1983)

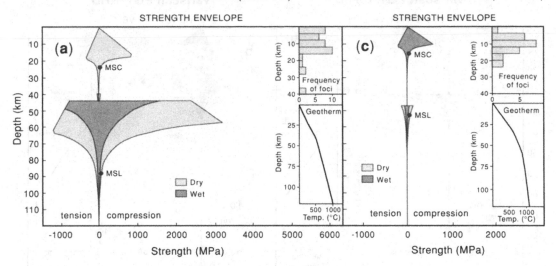

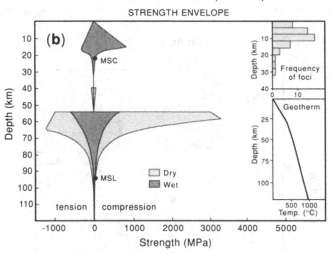

Figure 4-9. Strength profiles and distribution of intraplate seismicity for three different positions along the Baltic Shield. The strength profiles (left hand panel) are constructed from extrapolation of rock mechanics data and from geotherms (right hand panel) discussed in the text;

(a) Northernmost Baltic Shield,

(b) location in the central part of the Baltic Shield,

(c) Southernmost Baltic Shield.

Locations FG, FE and FC are given in Figure 4-12. The rheology of the lithosphere is for a quartz/ diorite/ diabase/ pyroxenite/ olivine petrological layering (Carter and Tsenn 1987, see Table 4-1).

proposed that the maximum depth of continental intraplate seismicity increases with thermo-tectonic age. Therefore, information on the rheological zonation of the lithosphere can also be inferred from the depth of seismicity, which occurs at depths where the lithosphere has sufficient strength to allow the build up and relaxation of tectonic stress (Stein *et al.* 1989, Govers *et al.* 1992).

It is interesting to compare the depth distribution and cut off depths of intraplate seismicity in the European lithosphere with the predictions of the rheological models. An interesting correlation exists, as shown in Figures 4-9, 4-10 and 4-11, between zones of deep intraplate

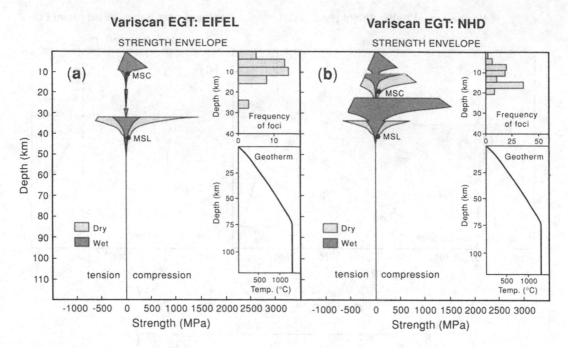

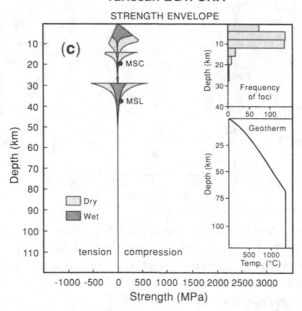

Figure 4-10. Strength profiles and intraplate seismicity distribution for three locations along the Variscan of central Europe
  (a) Eifel
  (b) North Hessian Depression
  (c) Urach.
Locations are given in Figure 4-13. Figure conventions as in Figure 4-9.

seismicity and areas where the strength envelopes indicate the ability of the lithosphere to support tectonic stresses at large crustal depths. In the northern and central part of the Baltic Shield both the depth of intraplate seismicity and the strength envelopes predict the presence of strong lithosphere with MSC values of about 20–25 km (Figure 4-9). To the south the seismicity cut off and the strength envelopes point to a crust with a mechanically strong part of 15–20 km thickness (Figures 4-9 and 4-11).

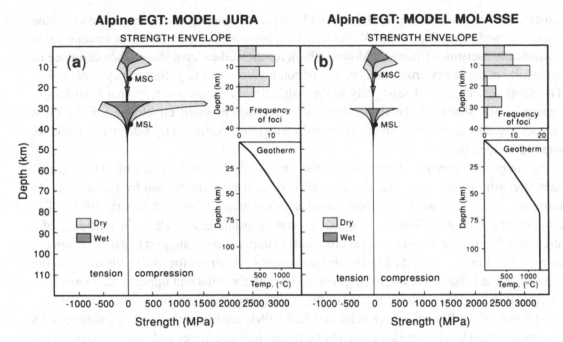

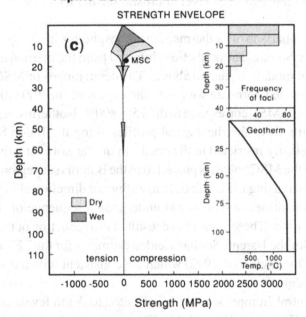

Figure 4-11. Strength profiles and seismicity distribution for three locations in the northern part of the Alpine belt
(a) Jura
(b) Molasse basin
(c) Aar-Gotthard.
Locations are given in Figure 4-13. Figure conventions as in Figure 4-9.

Correlations within individual tectonic provinces occur on a smaller spatial scale. For example, in areas of the Baltic Shield where the yield profiles predict a reduced strength with an MSC of 15–20 km (southernmost Baltic Shield) intraplate seismicity does not occur below depths of 25 km whereas in areas of large MSC values (central and northern Baltic Shield) the cut off level of intraplate seismicity is at larger depths, which is consistent with the predictions of the rheological models for the crust. It appears that both the upper and middle crust in the northern and central Baltic Shield can accumulate tectonic stress as shown by the seismicity and strength profiles. While initially it has been suggested that the lower crust is intrinsically aseismic (Chen and Molnar 1983), recent studies have shown the occurence of seismicity in the lower crust for a number of continental areas (e.g. Shudofsky *et al.* 1987). The depth distribution of seismicity in the Baltic Shield shows a pattern that is strikingly similar to the shape of the depth dependent rheological yield envelopes for the crust characterised by an increase in brittle strength with depth followed by a decrease in ductile strength in the crust.

Although the subcrustal lithosphere in the northern and central Baltic Shield has significant strength, according to the rheological models, it is characterised by the absence of intraplate seismicity, as is also found in other shield areas (Chen and Molnar 1983). The absence of intraplate seismicity in the subcrustal lithosphere, as well as the existence of a strength minimum between the upper crust and the upper mantle, suggest that the lithosphere responds in a two-layered fashion to applied stresses. Whereas for low levels of stress the reponse of the lithosphere might be governed by both the crust and upper mantle parts, it is likely that with an increase in the level of stress with time, deformation is concentrated in the crustal part of the lithosphere (Kuznir and Park 1984, see Figure 7-2). The existence of a present day high level of stress induced by both plate tectonic forces and post-glacial rebound is also supported by the relatively high level of crustal seismicity in the Baltic Shield (see Chapter 5.1).

Figure 4-12 gives a comparison of isotherms of the lithosphere with the spatial distribution of seismicity in the Baltic Shield. The predictions inferred from the thermal models compare well with the rheological models discussed above. The depth ranges of MSC where crustal strength minima are predicted correlate well with the depths of the 300–400°C isotherms, whereas the depth ranges of MSL correspond to the 750–800°C isotherms. There is a certain continuity in the main trends of the rheological profiles along the Baltic Shield, with the intraplate seismicity basically restricted to the crust. In the far north at latitudes of around 70°N a rapid decrease in the MSC values is predicted in the Barents sea region which deviates for the overall trend of increasing MSC values in a northward direction along this part of the EGT. The Barents Sea continental margin has undergone a sequence of rifting events in Mesozoic and Cenozoic time. These events have resulted in a reduction of the strength of the crust. Basin modelling for the Barents Sea has yielded estimates for the EET of these regions of the order of 20 km (Cloetingh *et al.* 1992) which are consistent with the values predicted by the EGT thermal models.

In the Variscan of central Europe, seismicity is limited to depth levels corresponding to isotherms of 450–600°C (Figures 4-10 and 4-13). The maximum depth of seismicity is in the range of 20–30 km, being essentialy restricted to crustal levels. The strength profiles display minimum values both within the crust and between the crust and the upper mantle. A comparison of strength profiles and seismicity shows a minimum in seismicity coinciding with a minimum in crustal strength in the Eifel and Urach regions. The complex rheology of the North Hessian Depression, reflecting the petrological layering, is also expressed in the distribution of seismicity which shows a peak at mid-crustal depth.

Seismicity in the northern Alpine segment of the EGT is essentially limited to depth levels coinciding with isotherms of 450–600°C (Figures 4-11 and 4-13). In the Jura mountains a minimum in seismicity occurs in association with a minimum in strength, seismicity being restricted to the upper and lower crust. An interesting pattern of seismicity occurs in the Molasse basin with a peak in seismicity in the upper crust and a cluster at Moho depth. Both the seismicity and the strength profiles point to a weak lower crust in the area. The cut off

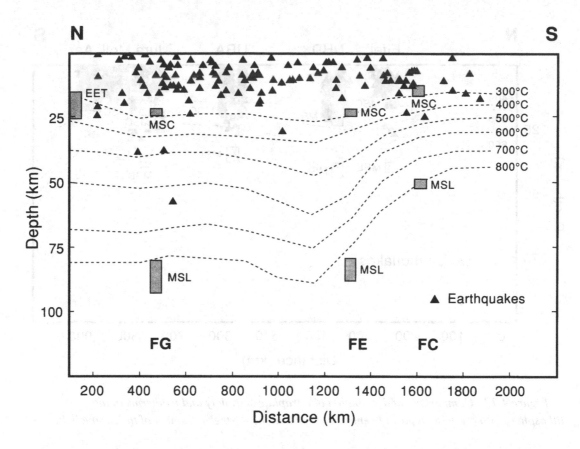

Figure 4-12. Comparison of distribution of intraplate seismicity in the Baltic Shield (Ahjos 1990) with depth of isotherms in the Baltic lithosphere (after Pasquale et al. 1991). Boxes indicate the depth ranges of MSC and MSL corresponding to the strength envelopes discussed for the three sites in the Baltic Shield (see Figure 4-9). The box in the top left hand corner gives an EET estimate obtained from flexural analysis of the Barents sea continental margin (Cloetingh et al. 1992).

depths of intraplate seismicity in the Jura mountains and Molasse basin are in the range 25–30 km. In contrast, in the area of the Alpine nappes an increase in the depth of the 400°C isotherm occurs to levels of 40 km accompanied by a decrease in the level of intraplate seismicity to depths of less than 20 km (Figure 4-14). As noted also by Deichmann (1992), this observation seems at first sight a paradox. However, due to the thickening of the crust, the associated decrease in strength is largely in excess of the effect of the temperature on strength. Also noticeable in the Aar–Gotthard region is the prediction of a zero strength for the subcrustal lithosphere.

Due to high levels of tectonic stress associated with bending of the lithosphere under the influence of Alpine thrust sheets, intracrustal decoupling has probably occurred, as discussed in Chapter 6.4, leading to a concentration of deformation in a zone restricted to upper crustal levels. The rheological profiles and the distribution of seismicity point to important lateral variations of the mechanical properties of the lithosphere underlying the Alpine foreland and provide useful constraints for the study of foreland flexure in the Molasse basin. The EGT data support the presence of a weak lithosphere suggesting that the flexure of the Alpine foreland is governed by the bending of a plate with low EET values. In this context, it is interesting to note that studies of flexural downbending under mountain loads in other parts of the European Alpine system (e.g. Zoetemeijer *et al.* 1990, Van der Beek and Cloetingh 1992) have also revealed the existence of mechanically weak lithosphere with characteristic

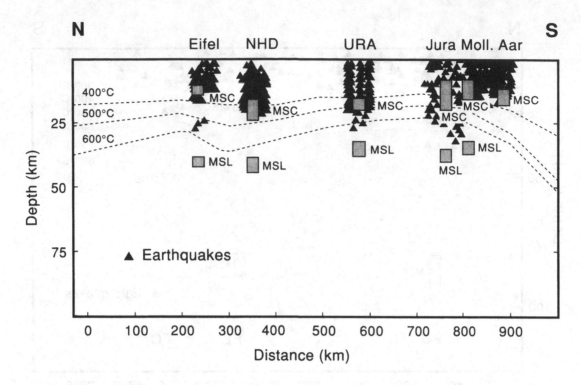

Figure 4-13. Comparison of distribution of intraplate seismicty and isotherms in the lithosphere of the Variscan part of central Europe and the northern segment of the Alpine belt.

EET values of 10–15 km. These values seem, therefore, to be characteristic for the mechanical response to Alpine tectonics of continental lithosphere affected by Mesozoic extension on pre-existing rifted margins. These points are discussed further in Chapter 7.2.

Figure 4-14 presents the values of MSL and MSC (roughly corresponding to 750–800 and 300–400°C isotherms respectively) inferred from the analysis of the EGT profile as a function of thermo-tectonic age. Previous studies for other continental areas (e.g. McNutt *et al.* 1988, Bechtel *et al.* 1991) have demonstrated the existence of a relationship between the time span elapsed since the last phase of thermo-tectonic activity and the thickness of the mechanically strong part of the lithosphere. Inspection of Figure 4-14 demonstrates that the mechanical thickness of the European lithosphere is characterised by an increase with age. MSL values are in the range of 80–90 km for Archaean and Early Proterozoic lithosphere in the Baltic Shield with a strong reduction for younger thermo-tectonic ages to values of 40–50 km.

The inferred spatial variation in the mechanical properties of the lithosphere is important in view of its control on the magnitude of the stresses supported within the lithosphere. The presence of a thick strong lithosphere will decrease the stress level, while stress magnitudes will be amplified by the presence of a thin weak lithosphere. The mechanical properties of the lithosphere are also of crucial importance for understanding the formation and evolution of sedimentary basins located on the European lithosphere (Ziegler, 1988). For example, the evolution of rifted basins depends critically on lithospheric flexure (Cloetingh *et al.* 1989) which is in turn determined by the MSL and MSC values discussed here. The same is true for compressional foreland basins, since their geometry and wavelengths are controlled by the thermo-tectonic age of the underlying lithosphere.

The European lithosphere provides a wide spectrum of thermo-tectonic ages and associated temporal and spatial of thermo-mechanical properties and thus offers a unique natural

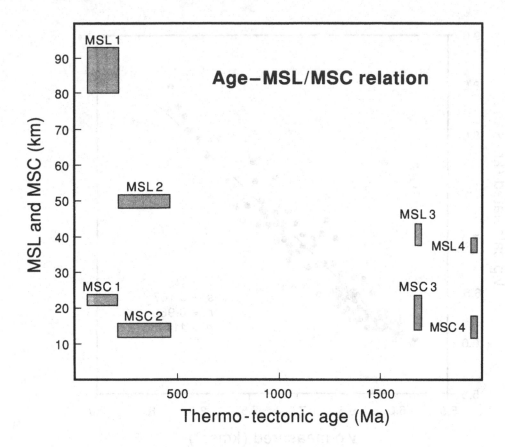

Figure 4-14. Mechanical properties of the lithosphere as a function of thermotectonic age along EGT. Ranges corresponding to MSC and MSL roughly coincide with the 300 and 800°C isotherms, respectively. Indexes 1, 2, 3 and 4 indicate age ranges corresponding to lithosphere in the northern and central part of the Baltic Shield, the southernmost Baltic Shield, the Variscan of central Europe and the northern part of the Alpine belt.

laboratory to study the temporal evolution of continental lithosphere and sedimentary basins, including the natural and energy resources they contain.

4.3 EVIDENCE FROM XENOLITHS FOR THE COMPOSITION OF THE LITHOSPHERE

K. Mengel

Refraction seismic experiments provide information about the distribution of elastic properties within a crustal segment but the nature of rock types occurring at depths remains largely unknown. On the other hand, petrological data derived from xenolith populations cannot be used to construct unequivocal lithological profiles through the upper lithosphere because their depth of origin is poorly constrained by thermobarometry. An alternative approach is to combine refraction seismic profiling with information on elastic properties of xenoliths which is potentially useful to assign rock types found as xenoliths to observed P-wave velocities at various depths.

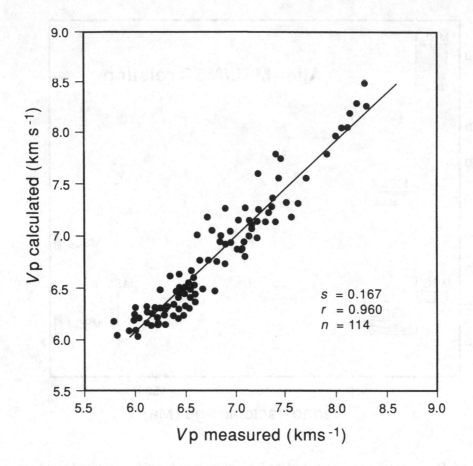

Figure 4-15. Calculated (Vp$_{calc}$) versus experimentally determined (Vp$_{meas}$) compressional wave velocities of 114 samples at 400 MPa confining pressures (literature data). Rock types include granitoids, felsic and mafic gneisses and schists, granulites, eclogites and peridotites. For source of data see Mengel (l990b): r = linear regression coefficient, s = standard deviation for n = 114 values (Vp$_{calc}$ - Vp$_{meas}$).

A drawback of xenolith-based lithospheric models is that they are one-dimensional. Even if we allow geophysicists and xenolith petrologists to extrapolate their evidence some 50 or 100 km beyond their actual sites of observation, it seems inappropriate to construct large scale 2-dimensional models. Although high-grade rock types found as xenoliths may also occur at surface outcrops it is generally not feasible to interpolate between outcrops and xenolith localities in order to arrive at a consistent model for the lithosphere along the trace of a refraction seismic experiment. Thus the evidence from xenoliths for the structure of a crustal segment remains fragmental, but it still represents a unique possibility to look deep into the continental lithosphere. This section concentrates on the interpretation of petrological and chemical data of xenoliths entrained in young volcanic rocks along the EGT central segment.

4.3.1 CALCULATION OF P-WAVE VELOCITIES IN XENOLITHS

In many cases, the size of xenoliths is too small to allow experimental determination of elastic properties at elevated pressures and temperatures. The methodology of calculating P-wave velocities for xenoliths is based on the simple assumption that the isotropic bulk rock velocity (V_p*) is the sum of mineral proportions (X_i) multiplied by P-wave velocities (V_i) of mineral species by the simple equation

$$V_p^* = \sum_{i=1}^{n} X_i V_i$$

The validity of this approach can be tested by comparing V_p data calculated in this way (V_{pcalc}) with experimentally determined P-wave velocities (V_{pmeas}) of rocks for which modal data are available. Figure 4-15 shows a plot of V_{pcalc} versus V_{pmeas} for a total of 114 samples of common rock types compiled from the literature, with rock types ranging from granites and meta-pelitic rocks to amphibolites, granulites, eclogites, and peridotites (for source of data see Mengel 1990b).

Elastic properties of deep crustal rocks are also a function of *in situ* pressures and temperatures and of crack volumes. Mengel *et al.* (1991) have demonstrated that the effect of crack volumes is not relevant to xenoliths and that the derivative dV_p/dP (4.5x10^{-4} kms^{-1} MPa^{-1}) is compensated by the derivative dV_p/dT (-5x10^{-4} kms^{-1}°C^{-1} compiled from data of Kern 1982, Kern and Schenk 1985) for *P–T* regimes corresponding to surface heat flow values between 50 and 90 mWm^{-2} (Haenel 1983). This approach to calculating P-wave velocities is therefore also assumed to be valid for elevated pressures and temperatures in stable platform areas and relatively young orogens. The uncertainty of the method is estimated to be ± 0.17 kms^{-1}.

4.3.2 XENOLITH POPULATIONS ALONG THE EGT

Along the northern segment there are four major xenolith-containing volcanic provinces; Seiland (northern Norway), Kalix and Alnö (central Sweden) and Scania (southern Sweden). All four xenolith localities are spatially close to the EGT trace but their age of emplacement may be too old for direct comparison with present lithosphere features (1.1–1.7 Ga at Scania to 1.15 Ga at Kalix, Griffin and Kreston 1987). Along the southern segment, the east Sardinian and west Sardinian post-Miocene volcanism has brought up numerous mantle xenoliths, with spinel-lherzolites and harzburgites being the most common rock types (Morten 1987). When plotted in an olivine-pyroxene triangle diagram, these xenoliths cover much of the modal compositions observed within mantle xenolith suites of continental Europe (e.g. Eifel, North Hessian Depression, Massif Central). Crustal xenoliths have also been mentioned but altogether there is very little information on the petrology and composition of these xenoliths so that construction of crustal cross sections is not possible at present. Detailed xenolith studies are also available for the Variscan crust of the French Massif Central (e.g. Downes *et al.* 1990) and for east German Tertiary volcanics (e.g. Kramer 1988). These areas are, however, located some 200–400 km off the EGT trace and are therefore not reviewed here.

The area between the Alps and the North German lowlands includes several thousands of individual volcanic outcrops of Cenozoic age. With respect to the last orogenic processes in Variscan times, the ages of xenolith host rocks are young enough to allow comparison with refraction seismic data. Fortunately, xenoliths are found within the three major tectonic parts of the central Variscan orogen; the Hegau and Urach volcanics in the Moldanubian zone, the Heldburg Gangschar in the Saxo-Thuringian zone and the North Hessian Depression (NHD)

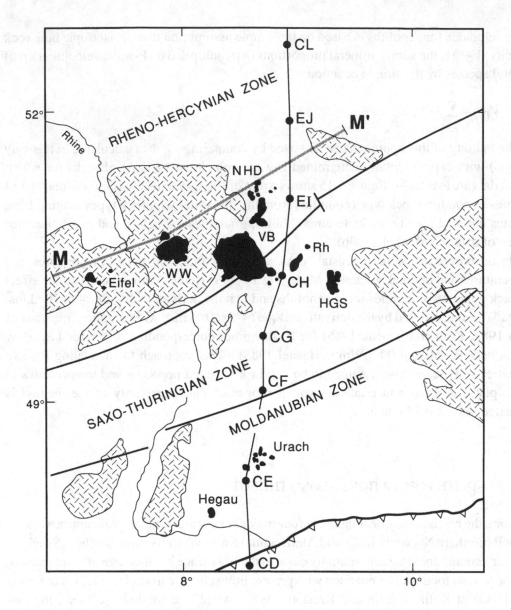

Figure 4-16. Distribution of Neogene volcanic fields (black areas) in Germany. Hatched areas are exposed Paleozoic units. Solid lines: positions of refraction seismic profiles, CL to CD Aichroth and Prodehl (1990), M-M´ Mechie et al. (1983). NHD, WW, VB, HGS and Rh refer to xenolith localities of the North Hessian Depression, Westerwald, Vogelsberg, Heldburger Gangschar and Rhön, respectively.

and Eifel volcanic fields in the Rheno-Hercynian zone (see locations in Figure 4-16). The latter two localities have been subject to extensive petrological and chemical investigation (Wörner *et al.* 1982, Stosch *et al.* 1986, Mengel 1990a). Sachs (1988) has investigated the Urach/Hegau xenolith suite and the Heldburg Gangschar samples are currently being investigated by Huckenholz. Owing to the relatively close spatial association of xenolith localities to the EGT trace, V_p depth profiles can be compared directly with the respective xenolith information (bulk rock V_p). Xenolith data from the Eifel are correlated with the refraction seismic profile described by Mechie *et al.* (1983).

This section mainly summarises published xenolith data presented in a review paper by Mengel *et al.* (1991) and the reader is referred to the original publications of these authors for further details. Huckenholz has provided a wealth of unpublished modal data of xenoliths from the Heldburg Gangschar.

4.3.3 PETROLOGY OF XENOLITHS

At Hegau and Urach, mafic rock types are rare. The mafic pyroxene and garnet granulites found at other sites are absent in that collection. The suite is dominated by medium- to high-grade meta-sedimentary and meta-igneous gneisses, with Moldanubian granitoids occurring in the uppermost part of the profile. The few mafic varieties include meta-pyroxenites and hornblendites, and clinopyroxene-biotite schists (Sachs 1988).

The data base for lower crustal xenoliths from the Heldburg Gangschar consists of an unpublished set of modal data, which was kindly provided by Huckenholz. The vast majority of samples from this region consist of mafic granulites of basaltic or gabbroic (cumulate) composition. A few felsic rocks have very recently been collected. Beneath the NHD and the Eifel, mafic granulites and felsic (granitoid) gneisses and granulites are common rock types, as well as some meta-sedimentary gneisses and granulites. Whereas all mafic granulites from the North Hessian Depression lack garnet, both pyroxene and garnet granulites are observed in the Eifel.

Thermo-barometric estimates for crustal and mantle xenoliths from the North Hessian Depression, the Eifel and Urach/Hegau localities have been summarised by Mengel *et al.* (1991, their table 3). If we take the maximum pressure estimates as a measure for their depths of origin, the corresponding temperature values would indicate extremely steep thermal gradients for the three regions. Using the temperature–depth relations of Pollack and Chapman (1977) based on regional heat flow, thermo-barometric estimates of 900–1000 MPa and 800°C for mafic granulites and 700 MPa, 600°C values for medium- to high-grade metasedimentary rocks would equal a surface heat flow of more than 120 mWm^{-2}. The compilation of regional heat flow patterns for Germany (Haenel 1983) reveals 60–70 mWm^{-2} for the North Hessian Depression and Heldburg Gangschar area, 70–80 mWm^{-2} for the Eifel and 80–100 mWm^{-2} for the south German Urach/Hegau provinces. Thus, it is regarded as quite unlikely that the reported thermo-barometric estimates have any significance for the present thermal conditions in the middle or lower crust. Instead, they are believed to reflect orogenic processes which had affected these crustal segments, presumably during Variscan orogenic cycles.

4.3.4 CONSTRUCTION OF LITHOLOGY–DEPTH PROFILES

Voll (1983) presented a crustal profile for the Eifel which is based solely on the abundances of rock types observed within his xenolith collection. There is, however, no convincing evidence that the sampling process by uprising magmas is representative with respect to actual abundances of crustal rock types within a crustal segment. As an alternative it is assumed here that all major crustal rock types have actually been sampled but that the relative abundances of individual rock types have no significance.

The comparison of P-wave velocities of xenoliths with velocity–depth profiles along the EGT suggests the model crustal structures as presented in Figure 4-17. P-wave velocities for xenoliths from the Eifel, North Hessian Depression and Urach/Hegau are reported in Mengel *et al.* (1991). V_p data for the Heldburg Gangschar xenoliths have also been calculated from their modal mineralogy according to the procedure described in Section 4.3.1. The pressures and temperatures for which the P-wave velocities are calculated are those expected in the middle or lower crust and in the upper mantle.

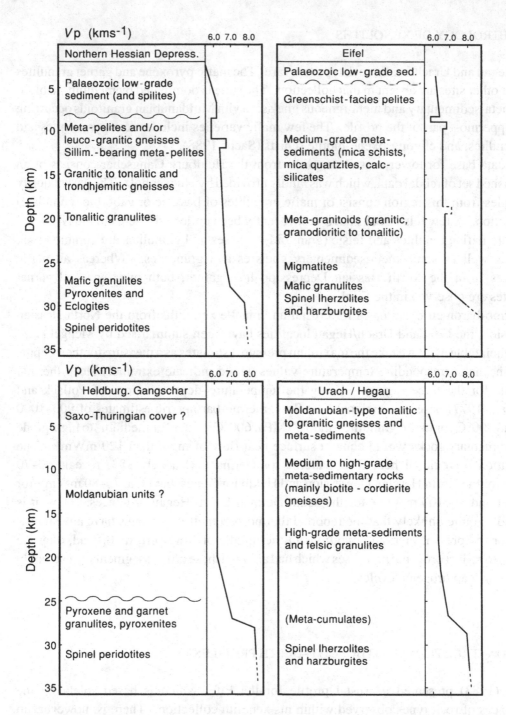

Figure 4-17. Xenolith-based crustal profiles for the Hessian Depression, Eifel, Heldburg Gangschar and Urach-Hegau localities. The Vp–depth relations are from Aichroth and Prodehl (1990), except for the Eifel which is from Mechie et al. (1983). The relative positions of rock types are arranged according to calculated P-wave velocities of xenoliths.

North Hessian Depression

Beneath an 8–10 km thick Phanerozoic cover which includes about 2 km of Mesozoic sediments, there is a low velocity layer between about 9 and 11 km depth with a small high velocity layer immediately beneath. The P-wave velocities of leuco-granitic gneisses and low-grade meta-sediments match the lower velocities whereas the high velocity layer may be explained by sillimanite bearing meta-sediments.

P-wave velocities measured in the middle crust between 13 and 20 km depth are conformable with felsic rock types found as xenoliths. The NHD granitic gneisses have somewhat lower velocities than the NHD tonalitic granulites and it appears likely that the latter originate from depths of 20–25 km. The increase in V_p between depths of 25 and 32 km of about 6.6–7.3 kms^{-1} suggests that the mafic granulites constitute the lowermost section of the NHD crust. The fact that the seismic Moho is not a sharp discontinuity implies that the mafic granulites may grade into ultramafic rocks below 30 km depth. Compressional wave velocities calculated for websterites and spinel peridotites also fall in this range of high velocity values (Mengel *et al.* 1991). The crust–mantle transition zone is believed to consist of a complex mafic/ultramafic layer.

Eifel

The crustal profile for the Eifel has much in common with that of the NHD, with the exception that calcsilicate rocks, meta-quartzites, and mica-schists occur in the middle part of the Eifel crust. The depth of origin of meta-igneous gneisses is not well constrained because the observed increase in V_p with depth is quite continuous and a further subdivision is beyond the resolution of the approach employed.

Like the NHD, observed P-wave velocities increase from values near 6.7 kms^{-1} around 28 km to >8 kms^{-1} below 33 km. The only metamorphic rock types that match such values are mafic pyroxene and garnet granulites. It has been suggested by Stosch (1987) that these rocks grade into spinel peridotite upper mantle, thereby also forming a complex mafic/ultramafic crust–mantle transition zone.

Heldburg Gangschar

Compared with the Rheno-Hercynian crustal sections of the NHD and the Eifel, the distribution of P-wave velocities in the central part of the Saxo-Thuringian is more simple: between 8 and 19 km depth, observed V_p values range between 5.9 and 6.2 kms^{-1}. The lower crust (20–27 km) is separated from the middle crust and from the upper mantle by sharp discontinuities. Within the xenolith suite from the Heldburg Gangschar, there is currently no information about the upper and middle crust. P-wave velocities calculated for the pyroxene granulites range from 6.8–7.5 kms^{-1} with an average value of 7.2 kms^{-1} which is slightly too high for the Heldburg Gangschar lower crustal segment between 20 and 27 km depth. It is therefore suggested that the lowermost part of the Saxo-Thuringian lower crust is composed of mafic granulites which are compositionally similar to those of the Rheno-Hercynian lower crust as exemplified by the Eifel and NHD mafic xenoliths. P-wave velocities calculated for the websterite xenoliths from Heldburg Gangschar vary between 7.8 and 8.3 kms^{-1} and it seems likely that these rocks are present in a thin crust–mantle transition zone between 27 and 29 km depth and in the uppermost Saxo-Thuringian mantle.

Urach/Hegau

In southern Germany near shotpoint CE (Figure 3-1), the crustal P-wave velocity distribution is characterized by a steady increase from values of 6 kms^{-1} at about 7 km depth to 6.8 kms^{-1} near 28 km. The Urach/Hegau xenolith suite is dominated by medium to high-grade felsic meta-sedimentary and meta-igneous rocks and it is quite problematic to construct a fine scaled lithology profile. However, Sachs (1988) and Glahn *et al.* (1992) suggested that garnet–sillimanite bearing samples with P-wave velocities of 7.0 to 7.6 kms^{-1} originate from the lowermost part of the crust; the few meta-mafic xenoliths are expected to originate from the same zone. The other felsic xenoliths cannot be assigned to crustal depths more specifically. The distribution of rock types shown in the Urach/Hegau column of Figure 4-17 is drawn tentatively by assuming that the slightly faster high-grade rocks originated from greater depth than the medium grade cordierite gneisses. Moldanubian type tonalitic and granitic gneisses probably occur in the uppermost part of the section.

Summarizing the main aspects of the crustal sections presented it is important to note that each xenolith locality comprises a distinct lithology suite. The differences between the Rheno-Hercynian xenolith profiles (NHD and Eifel) and the Moldanubian Urach/Hegau zone seem to reflect contrasting tectonic styles as well as substantial compositional differences, probably derived by pre-Variscan crustal evolution in independent terranes which were welded together during Variscan collision tectonism (see Chapter 6.3).

4.3.5 CHEMICAL COMPOSITION OF DEEP CRUSTAL XENOLITHS

As well as the information from petrological and P-wave velocity data, the chemical and isotopic composition of bulk rock samples can provide further clues for understanding evolutionary processes in the deep crust. Whereas extensive discussions of chemical and isotope signatures for the North Hessian Depression and the Eifel localities have been published (e.g. Stosch 1987, Mengel 1990a and references therein), the investigation of xenoliths from Urach/Hegau and from the Heldburg Gangschar is in a less advanced state. Therefore, only the interpretations for the two former regions in the Rheno-Hercynian zone are summarized here. We have to await more detailed and complete analyses of the other two xenolith localities before discussing aspects of the chemical composition and evolution of the deeper crust beneath these xenolith suites.

North Hessian Depression

Among the NHD granitoid gneisses, two different precursor types are identified:
(a) granitic gneisses which are very similar in their mineralogy and their major and trace element composition to Variscan S-type granites,
(b) tonalitic and trondhjemitic gneisses which have many major and trace element characteristics in common with tonalite–trondhjemite suites of the North Atlantic region (Mengel 1990a).

Among the few meta-sedimentary fragments there are some distinctly depleted samples. The strongly restitic nature of these rocks implies that episodes of intra-crustal partial melting of sedimentary protoliths have depleted the deeper crust in granitic components. However, trace element modelling fails to explain the latter as partial melting products of these meta-sedimentary rocks so that the actual restite material related to the S-type meta-granites remains unknown.

The NHD tonalitic granulites are characterized by generally low contents of K-feldspar and the presence of clinopyroxene. They are strongly (light rare-earth element) LREE-enriched and do not show Eu-anomalies. The range of oxygen isotope ratios in the tonalitic granulites (9–12 ‰ $\delta^{18}O$ SMOW) overlaps with that of the majority of basaltic-type mafic granulites. Mengel (1990a) has discussed in detail how the tonalitic and the mafic granulites are related by an intra-crustal partial melting event, where the tonalitic rocks are regarded as partial melting products and the mafic granulites represent the corresponding restite material. In a conventional Sm–Nd isochron diagram, four tonalitic granulites and three highly depleted mafic granulites plot close to a 1.4 Ga reference line (Mengel and van Calsteren 1989). This 'date' could represent an early stage of intra-crustal melting in the present NHD lower crust or a mixing line between upper mantle and lower crustal units.

Among the suite of mafic xenoliths, there are two distinctly different types of mafic granulites:

(a) mafic granulites of basaltic composition,

(b) those which formed from fractionation of gabbroic intrusion and from cumulates.

The basaltic-type granulites have formed from slightly evolved tholeiitic protoliths. They are generally depleted in K, Rb, and Th and the most depleted samples have lower LREE abundances than the rest. This may reflect the extraction of felsic LREE-enriched melts which probably are seen in the tonalitic granulites.

The average SiO_2 content (44.1 wt%) of the basaltic-type mafic granulites is much lower and their Li concentrations and $\delta^{18}O$ ratios are significantly higher (5–32 ppm Li; $\delta^{18}O = 6$– 11 % SMOW) relative to modern tholeiitic rocks including MORB. The enrichment in Li and ^{18}O has been explained by Mengel and Hoefs (1990) as a result of seawater interaction with ocean floor basalts. Normalised incompatible trace element patterns of NHD mafic granulites are identical to those of enriched-type MORB (Mengel 1990a, his Figure 7A). This observation together with the evidence from $Li–\delta^{18}O–SiO_2$ relations favours an origin of these mafic xenoliths from an oceanic crust-type protolith which was transformed from an early greenschist-facies stage via amphibolite-facies regional metamorphism into granulites.

The noritic to anorthositic granulites are high grade metamorphic equivalents of gabbroic cumulates derived from a subduction-related parental magma. Their enrichment in Al_2O_3 and their distinct positive Eu-anomalies reflect accumulation of calcic plagioclase. REE modelling indicates that the parental magma may be represented by an island arc tholeiitic (IAT) precursor. Two xenoliths with distinct IAT signatures have been reported by Mengel (1990a) which could represent such parental magmas. The association of ocean floor and island arc rocks together with the presence of former tonalitic rocks suggests that relics of an island arc association exist in the lower crust of the Rheno-Hercynian zone.

High Mg-numbers and low K, Ti and P contents of the garnet websterite and pyroxenite whole rock samples indicate that they have most probably formed from mafic (picritic to olivine tholeiitic) magmas which intruded at near Moho depths to form pyroxenites. During a subsequent pressure increase, they were transformed into eclogites. Major elements and normalized REE patterns suggest that their parental magma was different in composition from MORB (Mengel 1990b). This, and the high-T, medium-P thermo-barometric data, exclude an origin from a subducted oceanic slab and instead favour an origin from episodes of magmatic underplating.

Eifel

The amphibolite-facies xenoliths (mica-schists, gneisses) can be divided into meta-sedimentary and meta-igneous rocks (Reys 1988, Schmucker 1989). Among the criteria used to distinguish between meta-sedimentary and meta-igneous rocks are normative corundum, K/Na, Cs/K (higher in meta-sediments at given SiO_2 contents) and the presence of Eu-anomalies in REE patterns (strongly negative in meta-sediments: Eu/Eu* 0.58–0.75). The felsic meta-igneous amphibolites range in SiO_2 from 58 to 71 wt%. LREE over HREE fractionation of this suite is about twice as high as that of the meta-sediments. Also, their high Sr contents (450–650 ppm) and their fairly low Rb/Sr and K/Na ratios indicate that the protoliths of these samples had I-type granodiorite and tonalite precursors.

The major element and compatible trace element composition of the mafic granulites has been interpreted in terms of magma fractionation processes (Stosch *et al.* 1986, Loock *et al.* 1990). Samples with high Mg-numbers, relatively low REE abundances, and LREE over HREE enrichment and positive Eu-anomalies may represent early cumulates of a mafic parental magma; samples with lower Mg-numbers, high LREE abundances and high LREE/ HREE ratios and negative Eu-anomalies may represent fractionated liquid compositions.

On the basis of Sm–Nd isotope studies, Stosch and Lugmair (1984) suggested that the igneous age of at least some granulite protoliths might be 1.5 Ga. However, Pb isotope compositions of some granulite xenoliths seem to indicate a much younger formation of their magmatic protoliths, possibly only 400 Ma (Rudnick and Goldstein 1990).

4.3.6 IMPLICATIONS FOR THE EGT CENTRAL SEGMENT – CONCEPTS AND QUESTIONS

Unlike the exposed granulite-facies rocks of the Variscan orogenic belt north of the Alps, the majority of granulites from the xenolith localities reported here are mafic in composition. Such an observation is not new: Bohlen and Mezger (1988) have reviewed thermo-barometric and chemical data from exposed granulite terrains compared with those from crustal xenoliths. The fact that, on a world wide basis, P-estimates for xenoliths are significantly higher than those for deeply eroded granulite terrains and that xenoliths are on average more mafic in composition implies that in exposed terrains, combined uplift and erosion has not reached crustal levels as deep as are represented by xenoliths.

In this context it is therefore not surprising that the deep crust beneath the North Hessian Depression, Eifel and Heldburg Gangschar localities also has a quite mafic composition. Although from a petrological point of view the crustal profiles of the Eifel and NHD do look similar to each other, it has to be noted that detailed chemical investigations have revealed some significant differences.

The NHD xenolith suite contains relics of an island arc system which includes former oceanic crust, island arc volcanics and mafic cumulates as well as tonalitic rocks. For the Eifel lower crust, chemical and isotope data indicate the presence of depleted mafic granulite which probably formed by intra-crustal intrusions and subsequent granulite facies meta-morphism. Unlike the NHD lower crust, a typical island arc assemblage is not indicated beneath the Eifel. Although depleted mafic granulites also reveal Middle Proterozoic ages of formation (see Stosch 1987) it seems unlikely that lower crustal processes beneath the Eifel are genetically connected to those observed within the NHD lower crust.

This example of subtle but important differences in the chemical evolution of two xenolith suites from one orogenic belt demonstrates the conceptional limits of xenolith investigation and it must be emphasized that the one-dimensional xenolith evidence should not be extrapolated too far, even along strike within the same orogenic unit (Rheno-Hercynian).

The maximum pressure estimates for mafic xenoliths from the crust–mantle boundary are around 900–1000 MPa. These estimates are very low for the deep parts of an orogenic crustal segment and imply that the base of the crust in the Rheno-Hercynian zone of the Variscan belt has not been thicker than about 28–30 km. This is also the depth of the seismic Moho along

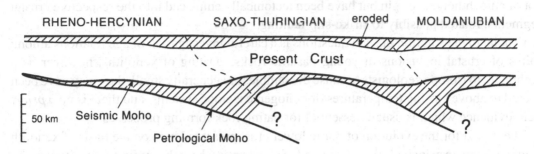

EGT CENTRAL SEGMENT

RHENO-HERCYNIAN   SAXO-THURINGIAN   eroded   MOLDANUBIAN

Present Crust

50 km   Seismic Moho

Petrological Moho

Figure 4-18. Conceptual model to explain crust–mantle lithology relations beneath the Variscan belt along EGT central segment. The Rheno-Hercynian is overthrust by the Saxo-Thuringian zone which is in turn overthrust by the Moldanubian zone. The missing roots might be represented by eclogite facies rocks (hatched field in the lower part) implying that the petrological Moho is not identical with the seismic Moho currently observed at around 30 km depth throughout the central segment.

the entire central segment. The tectonic framework of this part of the Variscan belt reflects an assemblage of terranes and ocean basins (see Chapter 6.3). Collision tectonics have involved all three major zones and it is surprising that there is no seismic evidence for the existence of orogenic crustal roots (see Figure 3-9). If it is correct that the deep crust in the region has an overall mafic composition as indicated by xenoliths, basaltic or gabbroic precursors may have been transformed into garnet granulites or eclogites upon orogenic crustal thickening.

Mengel and Kern (1992) have outlined some aspects of the petrological and physical evolution of mafic lower crust, emphasizing that former mafic crustal rocks now occurring in eclogite facies might not be detected by refraction seismics as part of the continental crust. The reason is that P-wave velocities of eclogites overlap with those of peridotites (Figure 4-3). Thus, the flat seismic Moho at 28–30 km depth (Figure 3-9) is not necessarily identical with the petrological base of the crust. Petrologists would define eclogite-facies mafic rocks derived from former mafic crustal precursors as belonging to the deepest parts of the continental crust. Therefore, the lithology below the seismic Moho might at least in some areas, contain eclogite facies rocks which constituted the roots of Variscan thickened crust. This idea is shown in Figure 4-18 which represents a pre-erosional model for Variscan collision and overthrusting involving the Rheno-Hercynian, Saxo-Thuringian, and Moldanubian zones. The hatched area beneath the 'petrological Moho' might contain dense eclogite-facies assemblages which possibly represent deep orogenic material subducted in a southward direction which now appears seismically as mantle.

The fact that densities of eclogites are significantly higher than those of spinel peridotites allows us to suggest that delamination of eclogite-facies rocks into lithospheric mantle is a viable concept for the disappearance of orogenic roots in refraction seismic profiles. It is, however, unclear whether this idea is also applicable to the lack of mafic lower crustal rocks beneath the Urach/Hegau region.

The xenolith-based lithology profiles presented here leave other important questions unanswered. Significant proportions of the NHD and Eifel crust are suggested to contain meta-granites of S-type and I-type origin. If these rocks had an overall thickness of only 5 km we have to expect 10–25 km thick layers of depleted meta-sedimentary or meta-igneous restites. Such rock types are either not observed or their chemical composition is incompatible with an origin as restites. It might thus be speculated that the meta-granitoid gneisses are not of autochthonous origin but have been tectonically emplaced into the respective crustal segments, related possibly to strike-slip faulting.

Crucial to the solution of such questions is a careful investigation of age relations among suites of crustal inclusions in young basaltic rocks. Dating of xenoliths, however, is a nightmare to isotope geologists because the samples are generally small, they may have been heated to above closing temperatures of radiogenic isotope systems, and there is no *a priori* field evidence which is usually essential for dating rock forming processes.

The model for the evolution of mafic lower crust presented here on the basis of xenolith evidence may, in principal, also be applied to other areas where the mafic composition of the crust/mantle transition zone is suspected to be mafic in composition. This may be particularly important where the seismic Moho has approached depths of more than 35–40 km, and the temperature at the seismic Moho is below 600°C, e.g. in the Precambrian of Scandinavia (see Chapters 6.1 and 7).

4.4 INTEGRATED LITHOSPHERIC CROSS SECTION

D. Blundell

The total view of a complete lithospheric cross section of Europe on a continental scale can now be given by drawing together the information contained in Chapters 2, 3, and 4. The starting point is to take surface geology to depth by utilising the seismic structure models derived in detail for the crust from deep reflection profiling combined with the wide-angle reflection and refraction events and, with less resolution, for the upper mantle from P-wave refaction, delay time tomography and surface wave inversion studies. A crustal structure cross section is produced along the entire length of EGT by combining the cross sections derived, from north to south, from the POLAR Profile (Figure 3-3), FENNOLORA (Figure 3-4), EUGENO-S (Figure 3-6) and BABEL (Figure 3-7) over the northern segment, from EUGEMI (Figure 3-9) and DEKORP-2 (Figures 3-10) across the central segment, and from EUGEMI, EGT-S86 and other profiles across the Alps and northern Appenines (Figures 3-11 and 3-12), including deep reflection profiles NFP-20, linked with EGT-South (Figures 3-15 and 3-16) to complete the southern segment to Tunisia. The crustal sections have been unified into one profile along the entire length of EGT and are presented as a composite lithospheric section in Figure 4-19. This section shows crustal boundaries which form discontinuities, including the Moho, and divides the crust into four areas on the basis of P-wave velocities of less than $6.0 \, \mathrm{kms^{-1}}$, 6.0-$6.5 \, \mathrm{kms^{-1}}$, 6.5-$7.0 \, \mathrm{kms^{-1}}$ and above $7.0 \, \mathrm{kms^{-1}}$. Since the experiments are all comparable in resolution, it is reasonable to interpret the extent of lateral and vertical heterogeneity as having geological significance. Data for the upper mantle come from Figures 3-18, 3-19 and 3-21.

The major features that are clearly brought out in Figure 4-19 are:

(a) the relative uniformity of the crust of the Baltic Shield with thickness in excess of 45 km and evidence of sharp, vertical displacements of the Moho.

(b) the heterogeneity within the crust, yet uniform 30 km thickness, beneath the Variscan region of the central segment of Germany, and the virtual absence of crustal material with Vp above $6.5 \, \mathrm{kms^{-1}}$.

(c) crustal thickening beneath the Alps and evidence of imbrication, extending beneath

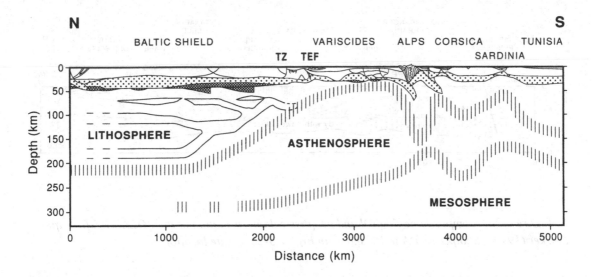

Figure 4-19. Composite lithospheric cross section of Europe along EGT from seismic data (for sources see text). TZ Tornquist zone; TEF Trans-European fault.

the northern Apennines, directed towards the south.

(d) crust of 30 km thickness and lithosphere of 70 km thickness beneath Corsica and Sardinia with a P-wave crustal velocity structure indicative of continental crust typical of western Europe, separated from neighbouring continental blocks by anomalously thin crust beneath the Ligurian Sea to the north and the Sardinia Channel to the south.

(e) the thick continental crust of North Africa beneath the Atlas Mountains and central Tunisia.

(f) thick lithosphere beneath the Baltic Shield, matching thick crust, with distinctive banding within the upper mantle portion.

(g) thin lithosphere beneath central Europe, particularly near the Rhine graben but a thick lithospheric root is developed beneath the Alps and northern Apennines, displaced southeast from the thick crustal root of the Alps. Lack of structure within the upper mantle portion of the lithosphere may simply be due to the lack of resolution of the seismic measurements.

(h) thin asthenosphere beneath the Baltic Shield, thicker beneath central Europe and highly variable in the Alpine/Mediterranean region.

The Baltic Shield

At the extreme north the POLAR Profile crosses Archaean crustal units interleaved with tectonically disturbed Early Proterozoic units. An integrated geophysical and geological interpretation of the POLAR Profile by Marker (1990) is presented in Figure 4.20. In this crustal cross section he has brought together refraction seismic, reflection seismic and electromagnetic models along with gravity and magnetic interpretation. Overall Moho topography and seismic velocity structure are derived from the seismic refraction model. The crust is thinner and does not include a high velocity ($Vp > 7$ kms[-1]) region within the lower crust beneath the Archaean age Inari terrain and Early Proterozoic Lapland granulite belt, in contrast with the Sörvaranger terrain and Karelian province to either side. The Moho here gives strong seismic reflections at normal incidence. The Lapland granulite belt is shown to be allochthonous above the highly sheared Tanaelv belt which forms a gently dipping basal

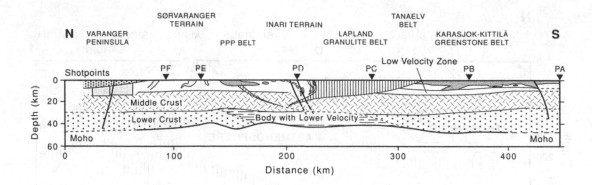

Figure 4-20. Composite geological and geophysical interpretation of the POLAR Profile, after Marker (1990). Shotpoints PA to PF shown in Figure 3-1 indicate location.

thrust zone around 10–15 km depth rooting northeastwards down into the crust, picked out by strong seismic reflectors and a zone of high electrical conductivity. The NE boundary of the Lapland granulite belt is less well imaged geophysically but appears to be a shear zone steeply dipping to the NE (Marker *et al.* 1990). North of the Inari terrain, the Early Proterozoic Polmak–Pasvik–Pechenga (PPP) belt forms a supracrustal sequence preserved in a NE directed thrust belt. On the cross section (Figure 4-20) this thrust is continued down to the mid-crust and may extend down to the Moho, as shown, although this cannot be resolved geophysically. The surface geology and the crustal-scale structural features shown on the cross section provide the basis for geodynamic interpretation of this region in terms of plate tectonic concepts as the Lapland–Kola orogen, as discussed in Chapter 6.1. This is interpreted as having been created from the closure of an Early Proterozoic Kola ocean around 1.95 Ga which brought together various Archaean continental or island arc fragments.

Strong evidence of plate tectonic processes acting during the Early Proterozoic comes from the BABEL deep seismic reflection profiles across the Gulf of Bothnia (BABEL Working Group 1990). The reflection patterns of the migrated seismic section shown in Figure 3-5 closely resemble those observed across Phanerozoic orogens such as the Alps (Figure 3-11) and the Pyrenees (Choukroune and ECORS Team 1989) where plate tectonic processes are better understood, and interpreted as collision zones. The BABEL Working Group conclude that the Moho offset and crustal reflection patterns can be attributed to collisional tectonics, with NE directed subduction, of Early Proterozoic (1.9 Ga) age. They suggest that the kinematics of plate collision have not changed significantly from that time to the present even though the crust is substantially thicker than is generally preserved in Phanerozoic orogens. Furthermore, the geometry of the structures of this ancient orogen has been preserved undeformed and has survived to the present day as indicated by the seismic images displayed by the BABEL profile.

An integrated interpretation by Henkel *et al.* (1990) has brought together gravity, magnetic, electrical and seismic information along the FENNOLORA profile to give the composite cross section shown in Figure 4-21 The northern end of FENNOLORA crosses Caledonian crust that has been thrust to the SW over Archaean terrain of around 45 km crustal thickness with a relatively thin or even absent high velocity ($Vp > 7.0$ kms^{-1}) lower crustal zone. To the south, the Early Proterozoic Sveco-Fennian Province crust is generally thicker, with Moho undulating and reaching depths of 50 km in places, and contains a significant thickness of high velocity lower crust. Henkel *et al.* (1990) used the seismic profile as the main framework for their interpretation of the other geophysical data, all of which, being

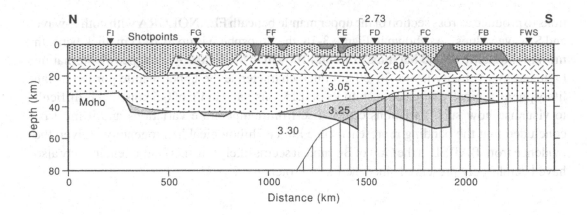

Figure 4-21. Composite interpretation of FENNOLORA profile (Henkel et al. 1990).
Shotpoints FH to FW shown in Figure 3-1 indicate location.

potential field methods, required some form of inverse modelling. As explained in Section 4.1, they discovered that the Moho topography of the Sveco-Fennian province was not reflected in the gravity profile and concluded in consequence that the Moho could not represent a significant density contrast. They were able to match gravity and magnetic anomalies extremely well by placing most density and magnetic contrasts within the upper part of the crust, using values consistent with surface geology. The longer wavelength gravity variation was modelled with a mid-crustal density contrast of 0.29 gcm^{-3} placed where Vp increases from around 6.4 to 6.6 kms^{-1} (see Figure 3-5). The lower crust high velocity region required a density in the region of 3.2 gcm^{-3}, not far short of the density of 3.3 gcm^{-3} used for the upper mantle. The Sveco-Fennian province also has an associated regional magnetic low that extends across a large area to the east (Wonik and Hahn 1989). The scale of this anomaly is such that it is difficult to assign it to contrasts in the upper crust but would allow it to be placed in the lower crust or upper mantle. Henkel *et al.* (1990) have outlined a tentative source region in their model as shown in Figure 4-21, covering the southern half of the FENNOLORA profile. They set the susceptibility to be 0.1 SI, corresponding to about 2.5% magnetite. Temperatures are such that this region down to 80 km depth could lie above the Curie isotherm. A region of high electrical conductivity in the crust beneath central Norrland in Sweden, observed by Jones (1981) and by Rasmussen *et al.* (1987) along FENNOLORA, has been modelled as a north-dipping mid-crustal feature situated to the south of the Skellefte belt (see Chapter 6.1) and extending for 100 km or so.

Henkel *et al.* (1990) attempted to use the geophysical properties, particularly the combination of density and seismic velocity, to pin down the geological composition of the crust. Those appropriate for the upper and middle sections of the crust could be linked with surface geology and extrapolated to depth by having an increasingly mafic composition. They compared the lower crustal zone, with its combination of a density of 3.2 gcm^{-3} and Vp of 7.1 kms^{-1}, with data from Carmichael (1989) to argue that it must be composed of eclogite. In support, they pointed out that 'eclogites are reported to occur as tectonically emplaced fragments in west Norway and in tectonic lenses in the Caledonian thrust sheets'. The upper mantle, they suggest, has a peridotite composition, which has been found in xenoliths from upper mantle depths beneath the Baltic Shield (Griffin and Krestan 1987).

The structure of the upper mantle beneath the Baltic Shield has been investigated on various scales from surface wave inversion and from the P-wave refraction data of FENNOLORA (see Chapter 3.3). Calcagnile *et al.* (1990) have combined these investigations to produce a cross section of the upper mantle beneath FENNOLORA with both P-wave and S-wave values. As shown in Figure 3-18, the lithosphere is up to 200 km thick beneath most of the Baltic Shield but thins to around 100 km in the south. A sequence of alternating high and low velocity layers 30–50 km thick characterise the upper mantle portion of the lithosphere, typically with contrasts in Vs of 0.15 kms^{-1} and in Vp of 0.5 kms^{-1}. It is difficult to visualise how such variations could arise from temperature variations and it must be concluded that the banding represents some kind of lithological heterogeneity. Given the evidence from BABEL in the Gulf of Bothnia, it seems likely that this heterogeneity may also have been largely preserved since the Early Proterozoic.

Tornquist zone and Variscan region

The region of transition from the Baltic Shield to the North German basin was examined in detail through the EUGENO-S seismic experiment, which provided the main crustal model cross section (see Figures 3-6 and 6-7). Thybo (1990) has put together the EUGENO-S and EUGEMI seismic data and linked them with surface geology and gravity and magnetic data. The crust thins from 45 km beneath the Shield in a series of steps as the Sorgenfrei Tornquist zone is crossed to about 30 km beneath the Danish Basin and changes in character, both in terms of seismic reflection patterns (Figure 3-7) and in seismic velocity structure. The Danish basin is filled with 4 km of Mesozoic–Cenozoic sediments and possibly up to 6 km of Palaeozoic sediments, leaving a crystalline crust below of about 20 km thickness. To the south, a basement ridge, the Ringkøbing-Fyn high, intervenes between the Danish and North German basins. Crustal thickness increases to 34 km beneath the high, with a sharp step on its northern flank that is well constrained geophysically. Its southern boundary is coincident with the location of the Trans-European fault (see Chapter 6.2) which may extend to lower crustal depths beneath the North German basin where, coincident with the Elbe line, the character of gravity and magnetic anomalies changes markedly (Dohr *et al.* 1983, Wonik and Hahn 1989) as is evident on EGT Atlas Maps 9 and 10. This may well mark the true boundary between Baltic Shield and European Phanerozoic crust. It is also in the general location of the deep lithosphere/upper mantle changes noted by Nolet (1990) from his 2-D model derived from waveform inversion analysis (Figure 3-22).

The Variscan crust of Germany has been intensively studied along the central segment of EGT by a wide variety of geophysical methods. The interpretation of these has been integrated with surface geology by Franke *et al.* (1990b) and in Chapter 6.3 to create the cross section shown in Figure 4-22. South of the Elbe line the crystalline crust is uniform in composition with the Moho flat at 30 km depth. This continues southwards, with nearby DEKORP-2N deep seismic reflection profile (Figure 3-10) indicating largely transparent middle and lower crust beneath the Rhenish massif, although, further south within the Rheno-Hercynian zone, listric faults from surface dipping south can be traced to mid-crust and sole out at the top of the transparent zone where they coincide with a thin layer of high electrical conductivity. In the deep crust, beneath the North Hessian Depression, the Moho is locally observed as a transitional region rather than a sharp boundary. Here, evidence from xenoliths not only provides an opportunity to link seismic velocities with lithologies, as explained by Mengel in Section 4.3, it allows an interpretation to be made that links the presence of eclogite at depths regarded as uppermost mantle on the seismic-based cross section with a Variscan history of island arc/continent collision. The evidence from this particular section has significance for other parts of the EGT section where the presence of eclogite is suspected. Comparable lithologies are found for xenoliths from the Heldburger Gangschar beneath the

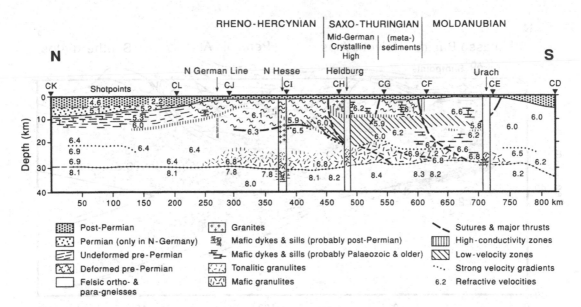

Figure 4-22. Composite cross section of the crust of the EGT central segment (Franke et al. 1990b, Prodehl and Aichroth 1992), reproduced with permission. Shotpoints CD to CK shown in Figure 3-1 indicate location.

northern edge of the Saxo-Thuringian zone, but contrasting xenolith lithologies from Urach in the Moldanubian zone further to the south suggest that the structure and dynamic evolution of this zone may be fundamentally different from the others. Boundaries between these zones, whilst steep near the surface, appear to dip south at angles decreasing to 30° near to the Moho, and appear to have associated high electrical conductivity. Whilst each zone displays individual geophysical characteristics with noticeable lateral as well as vertical heterogeneity and P-wave velocities within the crust that are relatively low, the Moho remains flat and close to 30 km in depth, expressed for the most part as a sharp discontinuity.

The Alps and northern Apennines

As the Alps are approached, and the Molasse basin thickens to the south, the Moho starts to deepen. The combination of seismic refraction information with coincident NFP-20 eastern and southern deep reflection profiles along EGT provides a high resolution structural framework through the crust upon which to develop geological interpretation, taken up in detail in the next chapter. Within the crust, seismic P-wave velocities are similar to those of the Variscan zones to the north, displaying considerable lateral variability, but with no evidence for any significant thickness of material with Vp exceeding 7.0 kms^{-1}. However, where the Moho is deepest it becomes less sharp as a boundary. Though this may be due to a loss in resolution, it could also be geologically less well defined. Perhaps the most dramatic effect is the duplication, even triplication of lower crustal units in imbricate stacks, each with a Moho at their base, interleaved with material having upper mantle properties. Gravity modelling is consistent with the seismic evidence, but only when the variability of the whole lithosphere is taken into account. A lithospheric root to the Alps was discovered by Panza and Mueller (1978) on the combined evidence of P-wave residuals, surface wave inversion modelling and gravity, as already explained in Chapter 3.3, and further confirmed from the P-wave seismic tomography of Spakman (1986, 1990a,b). Significantly, the lithosphere root is displaced to the south of the crustal root, a point that will be discussed again more fully in

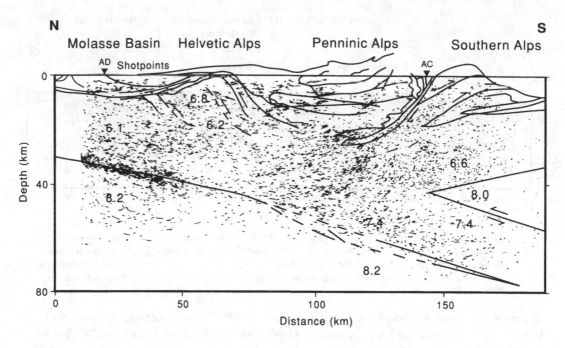

Figure 4-23. Geological cross section of the Alps derived from combining surface geology (after Pfiffner et al. 1990) with seismic information from Figure 3-11. Shotpoints CC to CD shown in Figure 3-1 indicate location.

Chapter 7. The geophysical information from the Alps has been combined with surface geology by Pfiffner *et al.* (1990) to produce a geological cross section of the crust as shown in Figure 4-23. This and other similar interpretations have provided the starting point for various authors attempting to reconstruct crustal geometries of the Alps at various times during the past, a subject taken up in Chapter 6.4.

Mediterranean region

Although the three-dimensionality of crustal and lithosphere structures should always be stressed, it is particularly important to know the 3-dimensional geometry of crustal scale structures along EGT south of the Alps if any understanding of their evolution is to be gained. This point is abundantly clear from the structure contour map of the various Moho units beneath the Alps shown in Figure 3-13. The islands of Corsica and Sardinia form a single block of continental crust and lithosphere with Moho at about 30 km and lithosphere base around 80 km depth, and P-wave velocity structure consistent with European lithosphere to the north. This block is surrounded by thin oceanic lithosphere of the Mediterranean basins and its evolution to its present location demands both lateral movement and rotation, as discussed in Chapters 6.6 and 7.2. The block is separated from Europe by the Ligurian Sea, which is oceanic in character (Figure 3-14) and is tectonically active at present, as evidenced from the high heat flow anomalies that Della Vedova *et al.* (1990) show can only be explained as transient thermal phenomena (see Section 4.1). To the south the block is separated from Africa by the Sardinia Channel, which also has thin crust and lithosphere of likely oceanic affinity which is tectonically active, and where strong heat flow anomalies are also due to transient effects.

The southern end of the EGT is firmly planted in the continent of Africa where across Tunisia both crust and lithosphere are of thicknesses comparable with those of the Baltic

Shield. No high P-wave velocity lower crustal zone has been found, however. There is evidence from P-wave seismic tomography (Figure 3-19) of the African lithosphere continuing north under the Mediterranean as a subduction slab. Although this is observed both to the west, beneath Algeria, and to the east, associated with the Hellenic arc, along EGT from Tunisia to Sardinia the high velocity region in the upper mantle, which can be interpreted as representing a downgoing slab, is separated from the lithosphere, as shown in Figure 3-19.

Despite its limitations as a single transect, EGT provides a comprehensive coverage of the geology of the European continent on a lithospheric scale, which makes possible the interpretation of the tectonic evolution of Europe in a consistent fashion through space and time, the subject of Chapter 6. It also gives, for the first time, the continuity of geophysical information along the lithospheric profile to support and test quantitative geodynamic models of continental lithosphere, the subject of the final chapter of this book.

5 Europe's lithosphere – recent activity

N. BALLING AND E. BANDA

A variety of studies and data compilations to obtain information about the state of geodynamic activity of the lithosphere both now and in the recent past have been carried out along EGT. Knowledge of the present state of activity is of critical importance for understanding modern tectonic development, and for geodynamic processes that have been operating in the past.

In this chapter we discuss observational evidence on recent activity in the lithosphere of Europe along and adjacent to the EGT swathe as revealed from seismicity information (given in EGT Atlas Maps 3 and 4), including focal depths and focal mechanisms (EGT Atlas Map 5), state of stress in the lithosphere (EGT Atlas Map 6), recent crustal movements, active volcanism and transient heat flow. We discuss dynamic activity in relation to tectonic units, the state of stress both regionally and locally, and seek explanations in terms of present-day and recent dynamic processes within the lithosphere and in the lithosphere–asthenosphere system.

Since EGT encompasses a wide variety of tectonic provinces from the old Archean and Proterozoic Baltic Shield to the young active Alpine–Mediterranean system, present-day dynamics has to be viewed and interpreted in terms of different processes acting on a variety of different structures. In the southern part of Europe the present state of dynamic activity clearly reflects active tectonic processes including lithospheric continental collision, extension and subduction. In the north, old, mature lithosphere has been subject to glacial loading and unloading and is influenced by plate tectonic processes in the north Atlantic.

5.1 SEISMICITY

Present-day activity in the lithosphere is generally reflected in the seismicity of an area. For most tectonic provinces along EGT, macroseismic observations, including historical seismicity and detailed instrumental observations covering the past 20–30 years, are available. The combined use of information from regional and local networks often gives very detailed information about the activity of an area. Epicentres, hypocentral depths and focal mechanisms (fault plane solutions) can be determined accurately and these allow seismotectonic information to be extracted. It is generally accepted that earthquakes are the result of dynamic slip on faults. The fact that small to intermediate size intraplate earthquakes

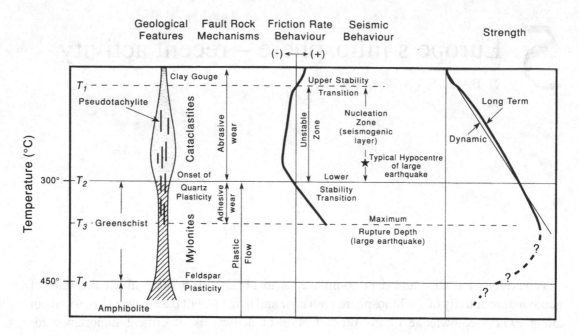

Figure 5-1. Model of a shear zone and the seismogenic layer for continental quartz–feldspar rheology (after Scholz 1990). With increasing temperature (depth), deformational behaviour changes from brittle above semi-brittle to plastic.

are quite common in the uppermost continental lithosphere shows that a significant part of the continental lithosphere is subject to stress at levels near to failure, at least in zones of weakness.

Laboratory studies have shown that pure brittle behaviour in crystalline materials gives way at elevated temperature and pressure to crystalline plasticity. Combined laboratory and field observations clearly indicate that brittle faulting occurs at shallow depth. This is seen to give way over a transition zone, which may be quite broad, in which deformation is semi-brittle, to plastic shearing at greater depth. The depth range over which these changes occur in nature depends largely on lithology and temperature, as discussed in Chapter 4.2.

A synoptic model of a continental shear zone and the seismogenic layer is shown in Figure 5-1 where a quartz-feldspathic rheology is assumed. Quartz begins to flow at about 300°C (Voll 1976, Kerrich *et al.* 1977) and feldspar at about 450–500°C (White 1975, Voll 1976). Fully plastic flow does not occur by gliding alone, but is associated with recovery and recrystallization. Water content is a very critical variable in determining the brittle to plastic transition in silicates. Studies by Tullies and Yund (1980) have shown that the addition of small amounts of water to dry granite induces a significant weakening and enhances ductility. The temperature for the brittle–plastic transition was, for 0.2% wt. water, found to be reduced by about 150–200°C for both quartz and feldspar, compared with 300–400°C and 550–650°C, respectively, in 'dry' granite. These results emphasise the importance of fluid content in understanding the activity of the lithosphere.

Figure 5-2. Regional distribution of large and intermediate size earthquakes in Europe (adapted from Simkin et al. 1989). The database used includes large events (magnitude >7) from 1897 and onwards and instrumentally recorded earthquakes after 1960 with magnitude >4. The Alpine and Mediterranean areas of present active tectonics show markedly higher seismic energy release than the older tectonic units in northern Europe. Tectonic maps for comparison are shown in Figures 2-3 and 6-2.

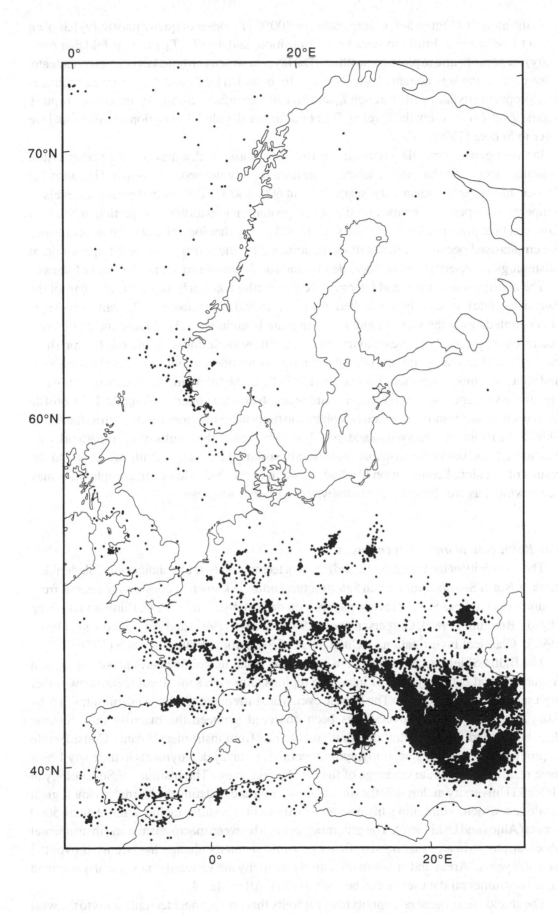

In the model of Figure 5-1, a temperature of 300°C (T_2: onset of quartz plasticity) has been taken to indicate the brittle to semi-brittle transition, and 450°C (T_4: onset of feldspar plasticity) the semi-brittle to plastic transition. The layer in which earthquakes normally nucleate, referred to as the seismogenic layer, is generally bounded by T_1 and T_2. Large earthquakes may propagate dynamically through T_1 and breach the surface. Similarly, maximum rupture depth (level T_3) is below the level of T_2. For a further detailed description of this model we refer to Scholz (1990).

In the uppermost mantle where olivine rheology is assumed, a new brittle or semi-brittle zone may occur. The base of the seismogenic layer for olivine rheology, as found in particular for oceanic intraplate seismicity, seems to be at 600–800°C. Depths to the various levels of temperature depend, of course, on the actual geothermal conditions, in particular the heat flow and heat production profile (see Chapter 4.1). The rheological and thermal conditions are emphasised because of their critical importance for the seismotectonic interpretation of seismological observations on focal depths and focal mechanisms, and the state of stress.

The distribution of large and intermediate size earthquakes in Europe (western part of the Eurasian plate) is clearly controlled by plate tectonic processes. Present activity is concentrated along the western and southern plate boundaries. The Alpine and Mediterranean areas of presently active tectonics including lithospheric collision and subduction show markedly higher seismic energy release compared to the older tectonic units, the Baltic Shield and adjacent units, in northern Europe (Figure 5-2). In the following sections we summarise and discuss recent results on seismicity and seismotectonics obtained along the EGT profile from north to south and with special emphasis on the principal tectonic units such as the Baltic Shield, the Rhine graben system and the Alpine arc. When the results of local networks are considered, including the distribution of small earthquakes, many details are added to the results of a regional compilation like that shown in Figure 5-2. However, complexities may occur which, as we shall see, are not always so easily interpreted.

The Baltic Shield and adjacent areas

The seismicity of the Baltic Shield and adjacent tectonic units, the Scandinavian Caledonides, parts of North Sea and Norwegian Sea structural units, is known in considerable detail from a number of studies. Recent review papers include Husebye *et al.* (1978), Bungum and Fyen (1980), Bungum (1989), Gregersen *et al.* (1991), Slunga (1991) and Ahjos and Uski (1991, 1992). Classical investigations include those of Sahlström (1930) and Båth (1956).

The Baltic Shield is characterised in general by small magnitude earthquakes of modest frequency, and is thus typical of intraplate regions, although some local areas show rather high seismicity (Figure 5-3). The largest events are infrequent and have magnitudes of 5.5–6.0. A good agreement overall has been observed between the macroseismic historic distribution and the epicentral distribution obtained from instrumental data. Considerable improvements in regional seismographic coverage and array deployments in the early 1980s have resulted in accurate coverage of the Nordic countries. The Institute of Seismology at Helsinki University undertakes the work of compiling and updating the Nordic seismological catalogue which, combining historic and instrumental results, includes more than 5000 events (Ahjos and Uski 1992). The general agreement between macroseismic and instrumental observations indicates that no significant general seismicity change has occurred over the past 500 years. Areas and zones of increased seismicity are generally more clearly outlined in the instrumental data set as can be seen in EGT Atlas Map 4.

The shield area in general exhibits lower activity than the younger tectonic units to the west

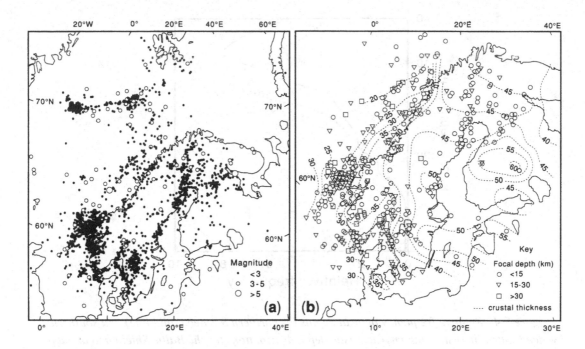

Figure 5-3. Seismicity of Northern Europe. Earthquake epicentres, 1965-1989, after Ahjos and Uski (1992) displayed: (a) according to magnitude; (b) according to focal depth. Tectonic maps are shown for comparison in Figures 2-3 and 6-2. Dashed lines show Moho depth contours in km.

and southwest (compare Figure 5-3a with Figures 2-4 and 6-2). Within the shield and the adjacent tectonic units, the Norwegian Caledonides and the Sorgenfrei Tornquist zone, we find the following areas with seismic activity markedly above general background: the southwest coastal area of Norway and adjacent offshore North Sea areas with an exceptionally high seismicity, southwest Sweden and the Oslo region, the western Gulf of Bothnia, northern Sweden and northern Finland (Lapland), and the west coast of central and northern Norway. Offshore western Norway, zones of seismicity are observed along the continental margin from the northern tip of Norway south to the North Sea. The seismicity associated with the mid-Atlantic ridge (Mohn's Ridge and Knipovich Ridge) is also apparent from Figure 5-3a.

Most events are localised in the upper part of the crust (Figure 5-3b). More than 80% of the Fennoscandian earthquakes occur at depths less than 20 km (Ahjos and Uski 1992). The deeper events are mainly observed offshore western Norway and northwest of Denmark, which is also the area with the larger magnitude events. The Baltic Shield exhibits only few events with magnitudes >4, although a number of historic events have been recorded (Gregersen *et al.* 1991).

A detailed analysis of the relative frequency of events at various focal depths from the shield of northern and southern Sweden (Slunga 1991) shows interesting characteristics (Figure 5-4). The highest frequency occurs in the upper 20–25 km with peak frequencies around 8 km in the north and 15 km in the south. In the middle part of the crust (from 20–25 km to about 35 km depth) the frequency of events is significantly reduced and the deepest part of the shield crust (between about 35 km and the Moho at 45–55 km, see Figures 3-4 and 5-3b) is almost aseismic.

Temperatures at 20–25 km depth in the shield are modelled at 250–350°C and at 35 km at 400–500°C (Balling 1992, Chapter 4.1). The depth distribution of the Baltic Shield earthquakes seems to be in good agreement with the general crustal shear and strength model

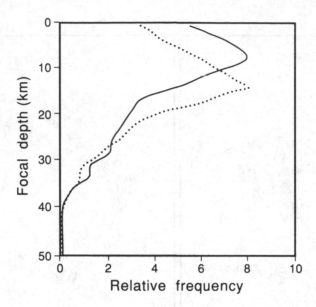

Figure 5-4. Relative frequency of focal depths for northern Sweden (solid line) and southern Sweden (dotted line) showing characteristic depth distributions for the Baltic Shield and about 8 km depth difference in earthquake peak occurrence between the northern and southern areas, after Slunga (1991).

of Figure 5-1. The differences in depth distribution between northern and southern shield areas seem, however, not to be explained by temperature. Heat flow is lower in the north than in the south. Events deeper than 30 km (down to 40–50 km) offshore northwest Jutland and southwest Norway seem localised in the uppermost mantle (Figure 5-3b). The uppermost mantle temperatures beneath the Norwegian–Danish basin at depths of 30–45 km are modelled at 700–850°C. Here, olivine-rich lithology apparently reacts in a brittle to semi-brittle mode.

Slunga (1991) has argued that aseismic slip along faults in the Baltic Shield is generally much more common than seismic slip. Based upon geodetic observations indicating fault movements, vertically as well as laterally, of about 1 mma^{-1} (Mörner 1977), he found aseismic slip to be about four orders of magnitude more extensive than seismic fault movements.

Some information is available on potential correlation between zones of increased seismicity and tectonic structures. In southwest Sweden a number of Precambrian shear zones are known, within and associated with the generation of Sveco-Norwegian crustal units (see Chapter 6.1.1). Recent investigations indicate that the southwest Swedish zone of highest seismicity is bounded towards the east by the so-called Protogine zone (Slunga 1991), a major shear zone in this area. Recent studies (Bungum 1989) show a zone of seismic activity along the Oslo graben. The western Bothnian zone clearly follows the trend of the east coast of Sweden and may be related to fault zones bounding a shallow Jothnian basin. The areas of elevated seismicity in northern Finland are reported to be in areas of intersecting fault zones.

A close correlation between zones of seismicity and tectonic units is found in the younger structures offshore Norway and Denmark. Zones of seismicity are clearly associated with the North Sea Viking graben (e.g. its western boundary fault, Bungum 1989), and most probably also with the central graben and a number of local zones offshore Norway are related to tectonic units.

A narrow zone of seismicity aligned NW–SE along the Norwegian–Danish basin (Gregersen 1979, Bungum 1989), seems to include uppermost mantle events (Figure 5-3b)

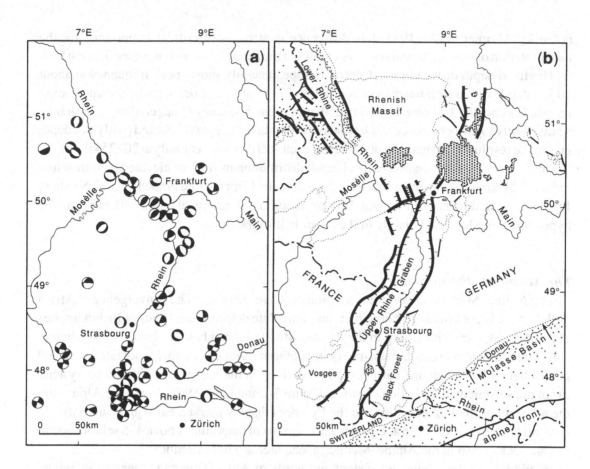

Figure 5-5. (a) Seismicity and earthquake focal mechanisms for the Rhine graben and adjacent areas (see EGT Atlas Map 5 for details)

(b) Geological units, after Franke, Figure 6-17).

and trends along a northern part of the Sorgenfrei Tornquist zone (see Figures 2-4 and 6-6). However, the seismicity zone is offset to the SW by 50–100 km from the surface expression of the main fault (the Fjerritslev fault) in this area.

An apparent correlation between zones of seismicity and marked gravity anomalies trending NE–SW and NW–SE over Fennoscandia has been noted by Balling (1980, 1984). Kinck *et al*. (1991) argue that earthquake zones in southern Scandinavia are in general related to areas of thinned crust.

Central Europe

The north German lowlands with Caledonian basement and deep sedimentary basins are almost aseismic. Increased activity is observed in central Variscan Europe towards the Alps and in the Rhine graben (Figure 5-2). Like northern Europe, seismicity is moderate; historical events show maximum magnitudes of 5–6. Relatively high activity is found in the Swabian Jura and upper Rhine graben, and above normal seismicity is also observed in the lower Rhine embayment, the Rhenish massif, the Cologne basin and in the northern Alpine foreland (Ahorner *et al*. 1983, Kunze *et al*. 1986, Leydecker 1980, Langer 1990). Detailed comparisons can best be made by comparing EGT Atlas Maps 1 and 4 .

The Rhine graben system forms the present day most active tectonic element north of the Alps. It is subdivided into two segments, the Upper Rhine graben in the south (see Figure 5-5) trending NNE–SSW, and the Lower Rhine graben in the north trending NW–SE. As noted by Ahorner (1970, 1975) this difference in strike is of critical importance for the interpretation of seismicity and seismotectonics of the Rhine graben system (see Section 5.3).

The focal depth distribution in Central Europe generally shows peak frequency at about 10 km depth with a significant decrease below 12–15 km. The decrease in seismic energy release is generally less pronounced than that in event frequency (Langer 1990) which may be due partly to the occurrence of a few larger events at the deeper levels and partly to a depth bias in the resolution of small events. The lower cut-off depth is generally at 20–25 km. There is a tendency towards a shallower focal depth distribution in areas of elevated heat flow and reduced thickness of the upper brittle crust such as the Upper Rhine graben and the Swabian Jura. Crustal thickness in central Europe is almost constant at about 30 km and the hypocentres are localised mainly in the upper half of the crust.

The Alpine–Mediterranean area

The Alpine–Mediterranean area is a region of active tectonics. The convergence of Africa with Europe is continuing to cause inter- and intraplate deformation. Plate collision implies marked lithospheric shortening in the Alpine–Mediterranean belt, as discussed in Chapter 2.4. The regional present-day seismicity of southern Europe is mainly associated with and controlled by the recent and current tectonic activity in the Alpine–Mediterranean system. From the seismicity distribution shown in Figure 5-2 and EGT Atlas Map 4, the Alpine arc, the adjacent northern Apennines and the Pyrenees display marked earthquake activity. In contrast to northern and central Europe, where events of magnitude above 4–5 seldom occur, magnitudes of 5–6 in the Alpine–Mediterranean area are not infrequent.

In addition to the Alpine arc system and northern Alpine foreland, zones of increased seismicity are outlined along the northern Apennines and in parts of central and southern Italy. Activity is observed in some areas of the Ligurian Sea. Major parts of the Po Plain show exceptionally low seismicity, and only scattered activity is observed along the main EGT profile from Corsica across Sardinia and the Sardinian Channel to the northern coast of Tunisia (Suhadolc 1990). Significant activity is reported from Tunisia and eastern Algeria with an E–W trend along the Tell mountains and a local N–S trend southwards from the Gulf of Tunis (Hfaiedh et al. 1985, Kamoun and Hfaiedh 1985).

For the Alpine arc, accurate seismicity information (epicentral maps, focal depth and focal mechanism solutions) is available from Switzerland and adjacent areas (e.g. Pavoni 1990, Deichmann and Baer 1990) and northwestern Italy (e.g. Eva et al. 1990) covering most of the central, southern and Western Alps, the northern Alpine foreland, the western Po Plain, and the northern Apennines. A generally good agreement has been found between maps of historic seismicity and recent instrumentally observed seismicity. Instrumental information is more homogeneous and generally more specifically outlines the areas and zones of seismic activity, but because this is limited to the past 20–30 years, not all local areas of known historic activity may be sampled. Marked seismic activity is observed in the eastern and southeastern part of Switzerland and in a broad region from the Rhine graben and Black Forest (see Figure 6-13) southwestwards to Neuchâtel (see EGT Atlas Map 4). The central parts of Switzerland appear to have been less active in recent years than historically. In the Western Alps, the epicentral distribution clearly indicates activity associated with the Briançonnais and Piedmont zones (see Figure 5-9). There is low activity in the external massifs, in contrast to that of the overthrust of the Embrunais folds. The inner and external crystalline massifs are nearly aseismic except for the Dora Maira massif where seismic activity occurs beneath the structure (Eva et al. 1990, see also Figure 5-10).

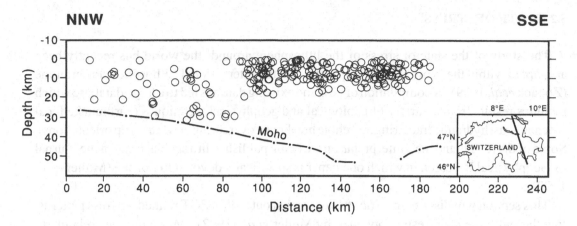

Figure 5-6. Focal depth distribution for the Central Alps and northern Alpine foreland, after Deichmann and Baer (1990) and Roth et al. (1992). Reliably located events recorded by the Swiss Seismological Service, 1972–1989, and by a temporary network, 1986–1988, in the area shown by the lower right inset map have been projected on to a cross section perpendicular to the line of the Alps. A remarkably sharp cut-off depth at around 13 km is observed within the Alpine arc.

Low magnitude activity is observed along the western Ligurian coast and in the adjacent Ligurian Sea. Seismicity south of the Argentera–Mercantour massifs trends along a neotectonic strike-slip line known as the 'Saorge-Taggia' line which has a history of high intensity events. The eastern Ligurian Sea (with continental crust) is practically aseismic, which contrasts with the marked neotectonic extension of this area. The northwestern Apennines are seismically active along a rather narrow NW–SE zone, generally following the main topographic divide. Areas of the folded structure of the buried Apenninic front show scattered activity (Eva *et al.* 1990).

The focal depth cross section along EGT for the northern Alpine foreland and the Central Alps (Figure 5-6) reveals remarkable lateral variation. In the Jura Mountains and the Molasse basin the whole crust is seismically active. The base of the seismogenic layer closely follows the Moho at a depth of 25–35 km. Lower crustal seismicity ends abruptly at the transition from the Molasse basin to the Helvetic nappes. No reliably located earthquakes are observed beneath 15–20 km, but marked activity is observed in the whole of the upper crust from near the surface to a depth of about 13 km beneath the Central Alps (Deichmann and Baer 1990, Roth *et al.* 1992). There are clear indications of lower crustal earthquakes again beneath the Southern Alps. The lower boundary of seismic activity beneath the Alps at 15–20 km is in agreement with the predicted level of brittle to plastic transition. However, the apparent abrupt depth cut-off may be taken to support the implication from seismic data of an active detachment between the upper and the lower crust (see discussion in Chapter 6.4). The relatively deep seismicity below the northern foreland is difficult to explain in view of heat flow and crustal temperatures above normal (Deichmann and Rybach 1989) but may be a result of the high level of stress there (see discussion in Chapter 7.2). Focal depths beneath the Western Alps indicate that the maximum depth of seismicity increases from west to east from about 10 km to about 20 km (Eva *et al.* 1990). We note that the deep Alpine lithospheric root is aseismic.

Deep earthquakes (200–300 km) are observed in SW Italy beneath the southwestern Tyrrhenian Sea and western Calabria and are intrpreted as associated with westerly dipping subduction of Ionian lithosphere (see Figure 6-40 and Chapter 6.7 which considers the recent tectonic development of the Mediterranean area).

5.2 STATE OF STRESS

The study of the state of stress of the lithosphere around the world has recently been attempted within the World Stress Map project of the International Lithosphere Programme (Zoback *et al.* 1989). A compilation of new and existing data has led to a large database which includes results from a variety of geological and geophysical techniques (earthquake focal mechanisms, hydraulic fracturing, borehole breakouts, overcoring and fault-slip orientations). Some of the results and their interpretation are being published in a special issue of the Journal of Geophysical Research in which one paper is specifically devoted to Europe (Müller *et al.* 1992).

This section will discuss specific studies carried out within EGT related to stress patterns, together with the main results obtained by Müller *et al.* (1992). Measurements relating to tectonic stress may reflect both regional stress patterns and more local perturbations. The only way to discriminate between these two possibilities is to define a large-scale stress (long wavelength) pattern on the basis of a large number of consistent observations distributed regionally. This pattern can then be interpreted to result from broad-scale tectonic forces from plate boundaries, large-scale flexural loading or unloading effects, inhomogeneous density contrasts within the lithosphere–asthenosphere system or other large scale phenomena. Perturbation of the regional pattern can then be attributed to local effects such as topography, erosion, and other local or induced stresses of small wavelength.

Figure 5-7 shows the results of the compilation by Müller *et al.* (1992) who consistently find that much of western Europe is subject to NW–SE to NNW–SSE compression and NE–SW to ENE–WSW tension. The majority of earthquakes show strike-slip mechanisms, suggesting that the intermediate principal stress is dominantly vertical. Extension prevails in the Aegean area and in the Appennines. Müller *et al.* (1992) found no correlation between stress pattern and small-scale geological structures. They also report no variation of the maximum horizontal stress (SH$_{max}$) with depth.

The maximum horizontal stress in northern Europe shows a scattered orientation around WNW–ESE. Fault plane solutions are available for more than 200 events from the Baltic Shield (Figure 5-8, Slunga 1991) and for an increasing number of events around its periphery, including some from the North Sea basins (Bungum 1989). The dominant type of faulting in the shield is transpressional on subvertical fault planes. Events from the Tornquist zone show both normal and strike-slip faulting.

One of the recent significant results from focal mechanism analysis is a regionally consistent NW–SE orientation of maximum horizontal stress release by earthquakes. This orientation is consistent with that expected for ridge push from the thermally elevated young oceanic lithosphere of the North Atlantic ridge. Results from earthquakes, *in situ* stress measurements at depths below 300 m (Stephansson *et al.* 1991), including breakout orientation analysis in the >6 km deep Siljan borehole (Qian and Pedersen 1992), appear consistent. This consistency clearly implies that most seismic faulting results from a systematic horizontal compression originating from the North Atlantic ridge push.

Secondary NE–SW oriented maximum horizontal compression observed in some areas of Scandinavia may be associated with postglacial rebound. It is now known that rebound induced stress played an important part during and just after deglaciation about 10000 years ago. Clear evidence for significant postglacial faulting in the northern Baltic Shield is available (e.g. Lundqvist and Lagerbäck 1976, Muir Wood 1989). The Pävie fault in northern Sweden has a length of 150 km and shows a fault displacement of 10–12 m. Several other faults have recently been mapped which show vertical fault displacements of up to 30 m. The

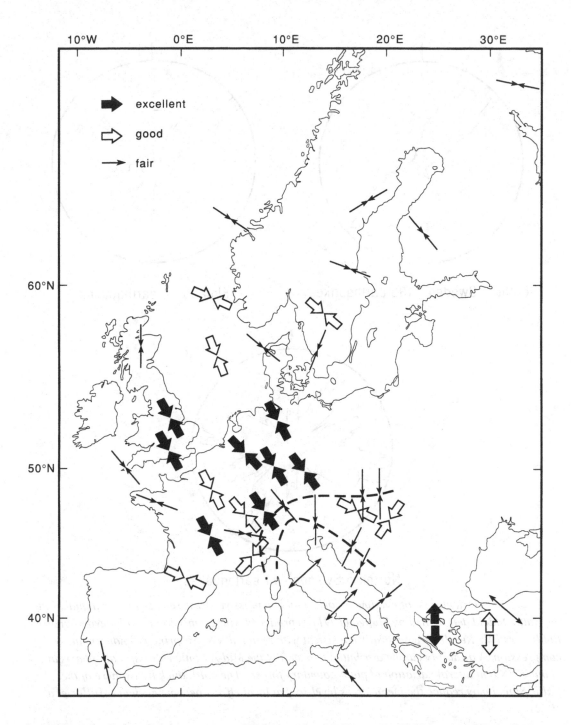

Figure 5-7. Generalised stress map of Europe, after Müller et al. (1992) and Mueller (1989). Inward directed arrows indicate the maximum horizontal compression direction. Outward directed arrows indicate the least horizontal stress in regions of extension. Thick solid, thick open and thin arrows indicate data quality in terms of number of observations. Dashed lines indicate the Alpine arc.

generation of the Pävie fault is estimated to be associated with an earthquake of magnitude 7.9 (Muir Wood 1989). These faults do not seem to be seismically active today.

In northern Sweden, the variability in direction of principal horizontal stress (Figure 5-8) may result from several sources. An offset of the Mid-Atlantic spreading axis at about 70°N may influence the lateral variation of plate boundary forces. Equally, the physical properties of the Baltic Shield, with low heat flow and thick lithosphere, may reduce the mean stress level and allow local forces to have a significant influence.

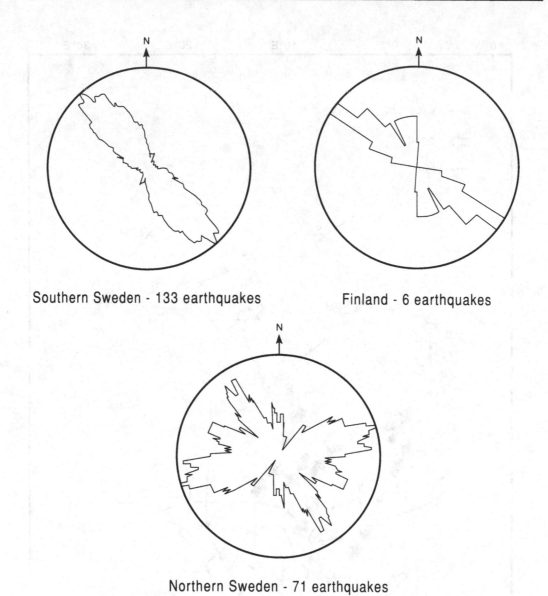

Southern Sweden - 133 earthquakes   **Finland - 6 earthquakes**

Northern Sweden - 71 earthquakes

Figure 5-8. Orientation of maximum horizontal compressive stress released in earthquakes in the Baltic Shield determined by Slunga (1991), reproduced with permission from Elsevier Science Publishers BV. Results are generally consistent with regional NW–SE principal horizontal compression. Variability can be attributed to a shift of the Mid-Atlantic ridge spreading axis at about 70°N and lateral variation of plate boundary forces. The cold, thick lithosphere of the Baltic Shield may reduce the mean stress level so that local effects become evident (Müller et al. 1992).

In western Europe, a well constrained mean SH_{max} orientation trending NW–SE to NNW–SSE has been determined. Fault plane solutions and other seismotectonic studies show the dominance of transpressional motion (Kunze *et al.* 1986). Seismic faulting seems to be controlled largely by the NW–SE compressional stress and associated NE–SW tension. Current seismic activity in the central area of the EGT (Figure 5-5, EGT Atlas Maps 4 and 5), including the Rhine graben system, shows predominantly normal faulting in the lower Rhine graben, the Rhenish Massif and the Cologne basin (NW part of the area), whereas further south in the Swabian Jura and adjacent areas almost all fault plane solutions show sinistral strike-slip movements along N to NNE stiking planes. In the upper Rhine graben a complex pattern of movements is observed, including horizontal strike-slip, together with

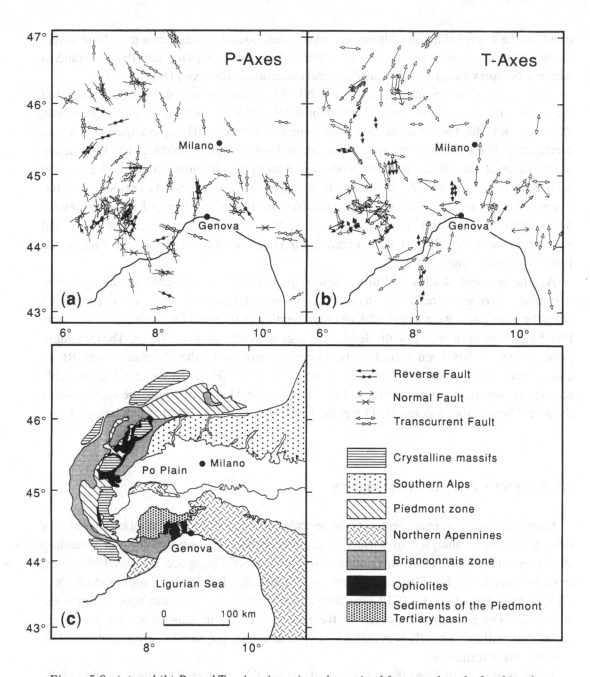

Figure 5-9. (a) and (b) P- and T-axis orientations determined from earthquake focal mechanisms for the central and western Alpine arc region. Lengths of arrows are proportional to the sine of the axis of inclination; after Eva et al. (1990): (c) Geological units after Buness and Giese (1990).

normal faulting (Figure 5-5), which seems to be due to the interaction of the regional stress field with graben features. Although details of the seismotectonic pattern may be complex and variations occur across the central EGT area, nearly all the fault plane solutions are consistent with a regional NW–SE orientation of maximum horizontal stress.

Focal mechanisms determined for nearly 200 events from the northern and western Alpine arc in general show P-axis orientation radially perpendicular to its trend (Figure 5-9), and thus T-axis generally parallel to strike (Pavoni 1990, Eva *et al.* 1990). Pavoni (1987, 1990) found that P-trajectories are aligned with the maximum horizontal crustal shortening derived from kinematic analysis of neotectonic structural features, indicating that the stress field and the mode of deformation have not changed much over the past few millon years. The regular trends of the T-axes along the Alpine arc indicate that extension may occur parallel to strike. Local variations occur in the southwestern part of the Western Alps towards the Ligurian Sea where a complex pattern of compression and extension is observed (Figure 5-9).

In Italy, along the Apennines, a consistent NE–SW extension roughly perpendicular to the main trend of the mountain range has been found (Figure 5-7). Some earthquakes, however, show a strike-slip focal mechanism with a highly variable SH_{max} orientation. A few earthquakes in the Adriatic sea tend to indicate an E–W compressional regime. Present day deformation of the Adriatic margins is not easy to interpret, although anticlockwise rotation relative to Eurasia and E–W continental collision have been suggested as possible mechanisms to explain the observations. These are discussed further in Chapter 7.2. In the western Mediterranean and North Africa the stress field is simpler, with its maximum horizontal stress directed NNW–SSE, roughly parallel to the relative motion between the African and Eurasian plates (Philip 1987).

Available stress studies in Europe demonstrate a fairly consistent orientation of the maximum horizontal stress. Such a consistency, coinciding with the direction of ridge push from the North Atlantic and the relative motion of the African and Eurasian plates, suggests that the stresses are actually controlled by forces generated by plate tectonics. The maximum principal stress has been found to be horizontal except in the Aegean, lower Rhine embayment and Apennines. The NW–SE to NNW–SSE SH_{max} in western Europe is affected only locally by major geological structures as the Alps, confirming the suggestion of the stress being largely controlled by plate driving forces acting on its boundaries.

5.3 RECENT CRUSTAL MOVEMENTS

Using the concepts of plate tectonics, large-scale horizontal movements that are tectonically driven have been analysed by means of geological markers such as sea floor spreading directions and rates, transform faults and the like. Recently, space geodesy techniques, namely Very Long Baseline Interferometry (VLBI), Satellite Laser Ranging (SLR) and Global Positioning Systems (GPS), are beginning to provide independent information to test the plate tectonic motion models. In the near future, the increasing accuracy of these techniques will undoubtedly provide fundamental data covering both interplate and intraplate motion and deformation.

In terms of plate motions the area covered by the EGT is affected by the spreading of the North Atlantic and by the interaction of the African and Eurasian plates. The present rate of spreading in the North Atlantic has been confirmed by space geodetic techniques (Smith *et al.* 1990) although at a somewhat slower rate than that proposed on geological grounds (De Mets *et al.* 1990). The Eurasia–Africa interaction covers an extended area where geodetic data are still scarcely available. The results to date, however, are consistent with convergence of Africa and Europe as inferred from geological models.

Intraplate motions have also been analysed with geodetic techniques. These indicate that no significant tectonic motion can be inferred within central Europe, suggesting that it behaves as a single block. Ongoing observations in the Mediterranean area have not yet been able to offer details on microplate movements. It will not be long, however, before the accumulation of results from the various geodetic techniques will provide fundamental information to give a better understanding of this complex area.

The results from geodesy available so far are consistent with the horizontal movements discussed in Section 5.2 on the basis of earthquake data and *in situ* stress measurements.

We now consider observational evidence on recent vertical crustal movements along EGT, and discuss their causes in terms of departures from isostasy and in relation to geodynamic processes. Repeated high precision levelling carried out over periods of more than 75 years, combined with tide gauge records in coastal areas covering even longer periods, provides quite detailed information on recent vertical crustal movements from most countries along EGT (Gubler 1990). Although some problems are encountered in comparing details from one region to another due to the use of different local reference levels, the main trends in recent vertical movements can be outlined. Variations along EGT are substantial (see EGT Atlas Map 6). The Fennoscandian area (Scandinavia and Finland) is subject to marked uplift with a maximum rate of uplift relative to mean sea level, reaching 8–9 mma^{-1} in northern Sweden and the northern Gulf of Bothnia. A land uplift of up to almost 1 m per 100 years is easily recognised. At the other end of the EGT, the Po Plain shows subsidence of up to 5 mma^{-1}. Between these extremes relatively small variations are encountered in most areas. Most of central Europe shows variations generally less than ± 1 mma^{-1} (Mälzer 1986). The Alps exhibit a small but significant uplift, up to 2 mma^{-1}. The uplift of the Alps and the subsidence of the Po Plain to the south are believed to be related to deep tectonic processes of the Alpine system (see Chapter 7.2). The very high uplift rates observed in Scandinavia are due to a different process; glacial loading and unloading. Isostatic responses are probably in action in both cases but in different ways. We shall discuss both these phenomena and consider isostatic problems in the light of information obtained from regional gravity and geoid anomalies.

5.3.1 GEOID ANOMALIES

Geoid and gravity data provide information on the lateral variation of subsurface density (see also Chapter 4) and thus yield essential information on the local and regional state of isostasy, and the extent to which crustal and lithospheric masses may be isostatically compensated or maintained by deep geodynamic processes.

A number of geoid maps exist for the European area. The one shown here (Figure 5-10) is from Marquart and Lelgemann (1992) and is based on the global geoid deduced from SEASAT and gravity data expanded in spherical harmonics to degree 180 (Rapp 1981). In order to emphasise components related to regional crustal and lithospheric structures, a low pass filter with spherical harmonic coefficients between degree 10 and 180 (wavelengths between about 200 and 3800 km) has been applied. We see a strong negative anomaly (<-10 m) in northern Europe centred over Fennoscandia. Central and southern Europe show a positive anomaly of 5 to 7 m amplitude. A number of minor regional highs correlate with high topography (e.g. southern Norway, Western Alps, Italy, and Yugoslavia), and negative anomalies may be associated with sedimentary basins (e.g. North Sea and Po basin).

Marquart and Lelgemann calculated the geoid effect of isostatically compensated topography and crustal thickness variations for the area shown in Figure 5-10 and modelled geoid variations along the main EGT seismic profiles, considering also the shorter wavelength variations. They concluded that the main features of the residual geoid field can generally be explained by topography and crustal thickness variations that are isostatically compensated at depth. In modelling regional as well as local variations, the combined effect of the crust and the lithosphere, and in some areas lateral density variations within the crust such

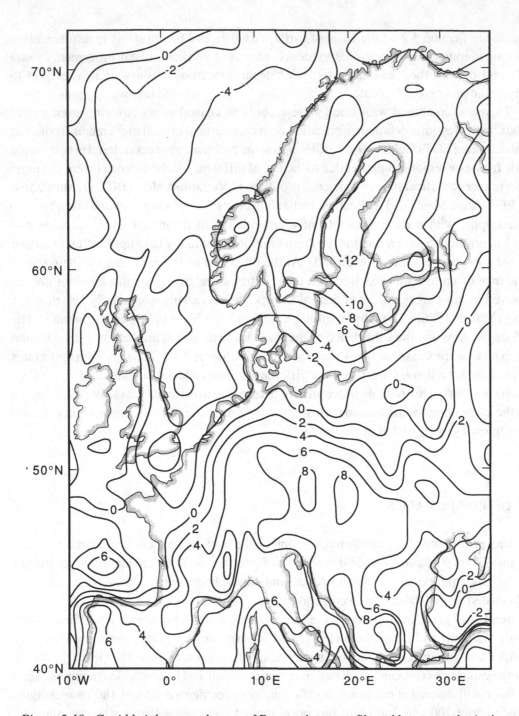

Figure 5-10. Geoid height anomaly map of Europe, low pass filtered between spherical coefficients 10 and 180, after Marquart and Lelgemann (1992). Contour values in m.

as those associated with sedimentary basins, were emphasised.

The particular features of the Fennoscandian and the Alpine areas of significant but very different recent activity are discussed separately in the following sections. The complex structures and recent activity along the southernmost parts of the EGT and adjacent areas are treated in subsequent sections of this chapter in relation to recent volcanic activity (Section 5.4) and transient heat flow (Section 5.5) and again in Chapter 6.6 and 6.7.

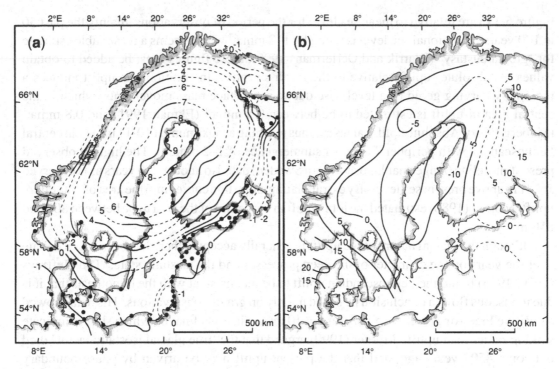

Figure 5-11. (a) Present rates of vertical crustal movement in Fennoscandia obtained from repeated precise levelling surveys and tide gauge records, after Balling (1980). Dots show locations of tide gauge stations. Dotted lines indicate areas lacking levelling profiles or sufficient tide gauge data. Rates of movement in mma^{-1} are relative to mean sea level.

(b) Regional Free Air gravity anomalies over Fennoscandia corrected for effects of topography, after Balling (1980).

5.3.2 FENNOSCANDIAN UPLIFT

Less than 20000 years ago great parts of northern Europe were covered with ice sheets probably reaching thicknesses in central parts of about 3000 m. The glacial maximum seems to have occurred between 20000 and 25000 years ago and deglaciation between 18000 and 10000 years ago (Denton and Huges 1981, Fjeldskaar and Cathles 1991). Through the glaciation and deglaciation cycles, nature is here providing a fascinating natural laboratory in which to study the response of the Earth to loading and unloading. Due to gravitational disequilibrium during and after deglaciation (unloading), masses within the Earth are redistributed, resulting in rapid vertical crustal movements and changes in sea level. The response of the Earth reflects a complex interaction between ice loads, water loads and deep masses, and is controlled by its deep rheological properties and the loading history. Observational data constrain mantle viscosity and may provide information on the extent to which the mantle density profile is near adiabatic (see Peltier 1982, 1989 for recent reviews).

The present rate of uplift in Fennoscandia and its uplift history are known in considerable detail (see Mörner 1980, 1990, for reviews). Classical studies include those of De Geer (1888/90), Haskell (1935) and Lidén (1938). An essentially elliptical region of emergence about 1800 km in length and 1200 km in width (Figure 5-11) is well established from repeated precise levelling over large areas of Sweden and Finland and for profiles in Norway and Denmark, supplemented for the whole area by tide gauge station records. Generally tide gauge and interpolated levelling results agree within ±0.5 mma^{-1}. Uplift rates shown in

Figure 5-11a are relative to mean sea level for the period of observations, mainly the past 50 to 100 years. If regional sea level is rising by 1–2 mma^{-1} which seems a reasonable estimate for 'global eustasy' (Warrik and Oerlermans 1990), these figures must be added to obtain values of 'absolute' uplift relative to the Earth's centre. In addition, the uplift induces a regionally varying geoid (sea level) rise due to subsurface mass movements, which in the central area of uplift is estimated to be between 0.5 mma^{-1} (Ekman 1991) and 0.8 mma^{-1} (Sjöberg 1989). Absolute uplift values are thus reaching a maximum of 10-11 mma^{-1} in central Fennoscandia, with peripheral areas of submergence of 1–2 mma^{-1}. The highest observed present shoreline (Ångermanland, northern Sweden) dated to about 9250 years BP is observed at 281 m above present sea level. By extrapolating shore line curves into the area of maximum uplift, Mörner (1980) estimated the total uplift in the centre of the area to be between 800 and 850 m.

Although the glacial nature of the uplift is generally accepted there has been much debate over the years on the character of the deep processes and the isostatic state. Thus Jeffreys (1970, 1975) found the Fennoscandian uplift to be inconsistent with the hypothesis that it is due to viscous flow, a conclusion he based mainly on gravity observations. His data showed a positive Free Air gravity anomaly (about 10 mGal) in Scandinavia, which should indicate sinking rather than uplift. Mörner (1990) argued that the 'true glacial isostatic factor' died out some 4500 years ago, and that the present uplift may be driven by phase boundary displacements, and/or readjustments within the lower lithosphere. Much of the debate is related to the problem of gravity and isostasy and to what extent a mass deficiency exists. Balling (1980) demonstrated that there is a regional gravity low over Fennoscandia (Figure 5-11b). Free Air gravity anomalies corrected for a regional positive correlation with topography due to the combined effects of masses above sea level and compensating masses at depth (long term regional isostatic compensation of topography along the Scandinavian peninsula) show a longer wavelength negative anomaly of 15–20 mGal covering the region of land uplift. Due to a least squares definition of residual gravity, there is in Figure 5-11a a 'balance' between positive and negative anomaly areas, and the zero contour does not represent isostatic equilibrium.

The geoid also shows a residual low over Fennoscandia (Bjerhammar 1980, Marquart 1989, Peltier 1989, Marquart and Lelgemann 1992, see Figure 5-10). The regional geoid shows a marked NW gradient in this area, and the geoid residual low is found after filtering. The low covering the area of uplift seems to be between 5 and 10 m. The centre of the low and to some extent its amplitude, depends upon the 'harmonic window' applied (cf. Bjerhammar 1980). Peltier (1989) showed geoid lows over the northernmost part of the Earth including Fennoscandia and Laurentia. He applied harmonic coefficients in the range 10–22 which, compared with the map in Figure 5-10, contain less effect from the crust. His map shows a geoid low of about 6 m covering that part of Fennoscandia which is in a state of uplift.

Mörner (1990) argued for the possibility that major parts of the gravity and geoid lows are related to crustal and lithospheric thickness variations and not to a residual glacial isostatic uplift. Marquart (1989) found a clear correlation between the geoid low and crustal thickness and only small geoid variations are left after she had corrected for isostatically compensated crustal thickness variations and topography. Marquart used existing regional Moho maps, which indicated a close correlation. However, recent compilations of crustal thickness variations in Fennoscandia including new results from seismic refraction and reflection investigations (see Figures 3-3, 3-4 and 3-17) show a broad region of deep crust not only beneath the central part of Fennoscandia but also in central and eastern Finland with Moho depths to 50–60 km. Thus we observe no clear general correlation between the regional

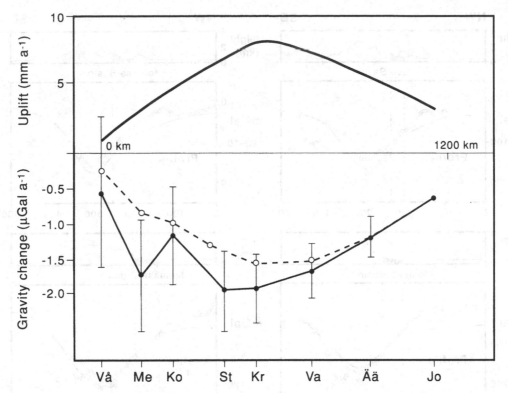

Figure 5-12. Present rate of uplift, and observed and calculated secular change of gravity across Fennoscandia along a line at about 63 °N, after Sjöberg (1989). Filled circles are observed values, with standard error bars; open circles are calculated values. Observations are from an 18 year period. A mantle flow model is used for the calculations.

gravity and geoid lows and crustal thickness variations, in particular not for the central and eastern part of the area. A certain contribution from the combined effect of crustal and lithospheric thickness variations should not be excluded, but modelling is difficult.

Due to the observations of generally small free air gravity anomalies we may conclude that crustal thickness variations are largely isostatically compensated, but the depth and nature of compensation are not known. Modelling a residual gravity, corrected for the effect of the crust and the compensating masses, including lithospheric thickness variations, involves the calculation of differences between very large numbers and is associated with significant uncertainty. Anomalies of the order of 10–20 mGal are not easily resolved.

Repeated precise measurements of gravity along lines crossing the Fennoscandian region are being carried out. A period of about 20 years has been covered and observations are beginning to yield significant results on the secular change in gravity (Mäkinen *et al.* 1986, Sjöberg 1989). Sjöberg's analysis for a line crossing Fennoscandia at about 63°N (the best line) shows in Figure 5-12 a reasonable agreement between observations and theoretical values using a viscous flow model. He obtained theoretical values of secular gravity variation typically about -0.18 mGalmm[-1] (absolute uplift) to comparewith -0.16±0.04 mGalmm[-1] calculated from observations. We recall that the theoretical Free Air gravity gradient (no mass flow) is -0.31 mGalmm[-1]. Estimates of the maximum remaining uplift range from 40–50 m (Fjeldskaar and Cathles 1991, Ekman 1991) to about 130 m (Balling 1980).

In a recent study of the Fennoscandian uplift, Fjeldskaar and Cathles (1991) investigated a number of viscosity models including the two 'extremes', a uniform viscosity mantle (deep flow) and a rigid mantle overlain by a low-viscosity asthenosphere (channel flow). They found that the best fit to the observed present uplift pattern came from a model with mantle

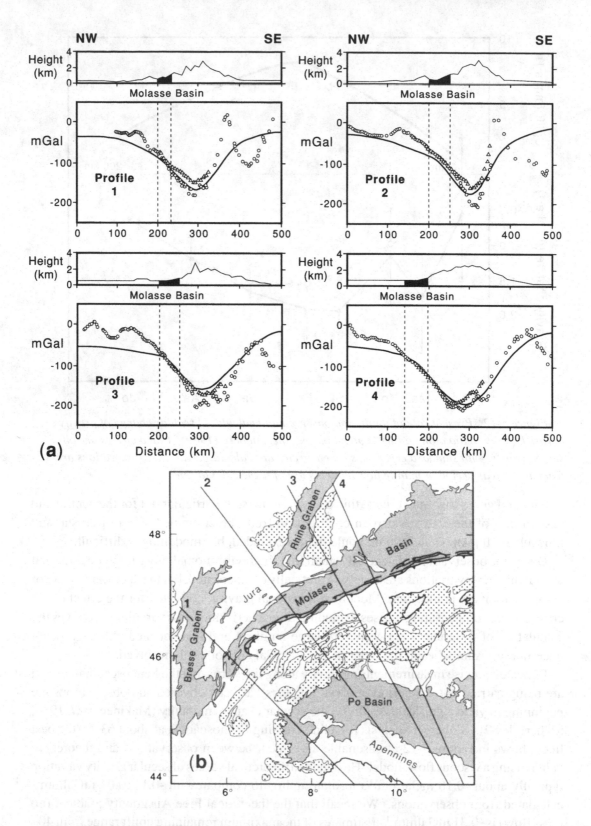

Figure 5-13. (a) Profiles across the Alps and surrounding regions showing average elevations and observed and calculated Bouguer gravity anomalies. Observed Bouguer anomaly values, indicated by triangles (with terrain corrections applied) or by dots (with no terrain corrections), are compared with Bouguer anomalies calculated assuming local isostatic equilibrium, rafter Lyon-Caen and Molnar (1989), eproducedwith permission from the Royal Astronomical Society.
(b) Location map of the four profiles. Basin areas are shown in dark shading.

of viscosity 1.2×10^{21} Pas overlain by a 75 km thick asthenosphere of viscosity 2.0×10^{19} Pas (both channel and deep flow). They emphasise the need for a low viscosity asthenosphere to explain the present uplift data. There are some trade-offs between asthenosphere viscosity and thickness and mantle viscosity and ice thickness that were not specifically addressed. They noted that changes of the lithospheric rigidity (within the range of $1-100 \times 10^{23}$ Nm) result in only minor changes in modelling the present rate of uplift. In another study, Peltier (1989) applyied viscoelastic theory for glacial isostatic adjustment to model Free Air gravity and geoid anomalies over Fennoscandia and Laurentia consistent in general with observations using viscosities of 10^{21} and 4.5×10^{21} Pas, respectively, for the upper and lower mantle.

Lambeck *et al.* (1990) obtained an upper mantle viscosity of $3-5 \times 10^{20}$ Pas and lower mantle viscosity of $2-7 \times 10^{21}$ Pas from inversion of the observations of the postglacial sea level changes in northwestern Europe. Thus there is a general agreement about the order of magnitude of viscosity in the Earth's mantle as a whole, a figure that has changed little since the classical study of Haskell (1935). However, new information and ideas continue to appear and further observations and integrated modelling are still needed. We consider these observations, and their geodynamic significance, in Chapter 7.2.

5.3.3 THE ALPS, MOLASSE BASIN AND PO PLAIN

Data on recent vertical crustal movements from western Germany, Switzerland, and northern Italy compiled for EGT (see EGT Atlas Map 6) allow a comparison to be made of movements from the Black Forest and Molasse basin across the Alps to the Po Plain. Although different reference levels are used for the three levelling systems in this area, comparisons are possible because elements of various networks overlap. The most striking are the subsidence of the Po Plain and the uplift of the Alps. When interpreting detailed variations, of about $1-2$ mma^{-1} which are geodynamically significant, we have to be aware of the reference level problem. Along the German/Swiss border both systems show consistently very small relative movements (<0.5 mma^{-1}, close to the level of resolution). Observations in overlapping networks between the Swiss and Alpine systems to the south indicate an upward movement of the Swiss reference of 1.3 mma^{-1} relative to the Italian reference (reference point Genova) which in turn from tide gauge records shows subsidence of 1.2 mma^{-1} relative to mean sea level. Again assuming $1-2$ mma^{-1} for 'regional eustatic sea level rise' the system used for northern Italy may thus be close to an 'absolute' reference. By adding between 1 and 1.5 mma^{-1} to the relative movements obtained for the Swiss area we can obtain present vertical movements consistent with those obtained for northern Italy. The Swiss Alps generally show an average uplift of about 1 mma^{-1}, locally to about 1.5 mma^{-1}, relative to the comparatively stable northern Switzerland and Alpine foreland. In terms of the Italian reference, the Alps may thus show uplift of about 2 mma^{-1} and the Po Plain a marked subsidence ranging from 2 mma^{-1} west of Milano to 5 mma^{-1} by the Adriatic Sea. Such differences have significant geodynamic implications, as will be discussed again in Chapter 7.

Gravity and isostatic modelling has been carried out by a number of investigators since the early models of Mueller and Talwani (1971). We are faced with the prime question whether significant deviations from isostatic equilibrium occur, that is, whether mass excesses and deficiencies are present, and whether masses are supported by the strength of the lithosphere or by dynamic processes beneath it. Modelling by Kissling *et al*. (1983) is discussed in Chapter 7.2.

Lyon-Caen and Molnar (1989) have analysed Bouguer gravity across the Alps along four profiles (Figure 5-13) in terms of simple isostatic and flexural models. The observed Bouguer gravity anomalies reach their lowest values across the Alps, generally where elevations are highest, and the calculated Bouguer anomalies, assuming local Airy-type compensation of topography, show relatively small (<30 mGal) deviations from those observed over the Alps and the Molasse basin, but significant (>50 mGal) differences NW and SE of the Alps (Figure 5-13a).

Topography along the southern part of the Western Alps and the Po basin shows marked deviations from isostatic equilibrium, and a large positive anomaly is observed over the Ivrea body of high density basic rock material (see EGT Atlas Map 9). Large positive gravity deviations are observed over the flanks of the Rhine graben and the elevated Vosges and Black Forest regions with heights in excess of 1000 m. The simple isostatic model used should not be taken in support of crustal thickness variations following simple Airy-type topographic compensation, but the results do show that the present Alpine topography seems largely to be the isostatic surface expression of the accumulated thick, relatively light materials within the crust. Flexural modelling requires unrealistic values for the effective elastic thickness EET (defined in Chapter 4.2) in order to account for the gravity variations over the Molasse Basin, the sub-Alpine chains, and the elevated areas of the Vosges and Black Forest. Lyon-Caen and Molnar (1989) found that the simple model of an elastic plate loaded by the weight of the overlying topography and by a force and bending moment on its end could not provide a satisfactory explanation of the gravity anomalies across the Alps and their foreland. Their calculations do not mean that a strong European plate does not exist (or lacks flexural strength), but rather that a simple elastic model alone is inadequate to account for the mass excesses and deficits in the Alpine region. The high elevations of the Vosges and Black Forest and their anomalous gravity values were taken to suggest active upwelling and high asthenospheric temperatures beneath these areas, and downwelling is suggested beneath the Po basin. The Po basin downwelling is consistent with observations of low heat flow and recent rapid subsidence. Lyon-Caen and Molnar also suggest that the recent rapid uplift of the Alps might be the result of the removal of a downward force that had maintained the Alpine chain in a state of overcompensation. This removal could be associated with a deep decoupling of the subducted European lithosphere. This point is taken up again in Chapter 7.2. Alternatively, it is possible that the present and recent Alpine uplift is simply the result of ongoing accumulation of relatively light crustal materials (crustal thickening) and the isostatic response to collisional processes.

5.4 RECENT VOLCANISM

Volcanism is one of the most spectacular expressions of the dynamic activity of the lithosphere. Petrological and geochemical studies are now able to characterise volcanism in such a way that mechanisms responsible for volcanism can be deduced. While recent volcanism is absent in the Baltic Shield, a moderate amount is found in central Europe and substantial activity characterises the Mediterranean area. In central Europe, volcanism is mostly alkaline in nature and related to a rift system that includes the Rhenish massif (lower Rhine and Leine grabens) and the upper Rhine graben (Figure 5-14) with its main volcanic

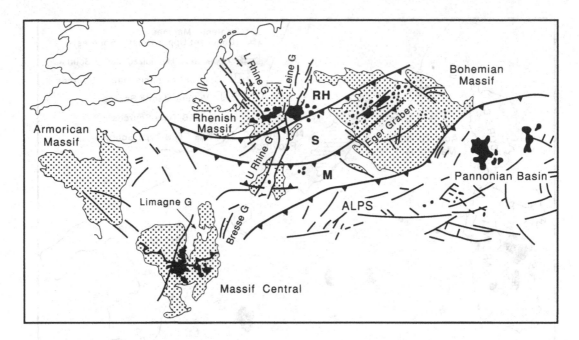

Figure 5-14. Recent volcanic activity in central Europe, after Wilson and Downes (1991).

activity during the Neogene. In the upper Rhine graben, Neogene volcanism has been dated as 19–13 Ma in age and is found along the southern margin of the graben. Quaternary volcanism is restricted to the west and east Eiffel region, with the main activity between 0.7 and 0.01 Ma.

Most of the volcanic expressions are located not within grabens themselves but on the adjoining horst blocks or near the main structural dislocations (see Figure 5-14). This suggests that the volcanism is structurally controlled, at least at upper crustal levels where the dominant mechanism of extension is simple shear (Wilson and Downes 1991). The alternation of periods of subsidence with compressional events (Ziegler 1982) indicate that magmatism may also be related to lithospheric flexuring caused by Alpine activity, resulting in adiabatic decompression and partial melting of the upper mantle (Wilson and Downes 1991). On the basis of detailed geochemical studies, these authors suggest involvement of both lithospheric and asthenospheric mantle source components without extensive crustal contamination, finding no need to invoke the existence of mantle plumes.

Figure 5-15 shows the known volcanism in the Tyrrhenian sea and adjacent regions according to a classification of magma sources by Serri (1990). The evolution of the area, including Corsica and Sardinia, can be constrained by existing studies of Neogene-Quaternary volcanism. The latter, however, embraces such a wide variety of magma types that it is impossible to relate it to a single geodynamic process.

Volcanism related to ocean island basalt-type sources is only found south of 41°N latitude, with large amounts being erupted during the Plio-Quaternary in Sardinia. Geochemical studies show that this volcanism may have been contaminated by subduction and/or crustal effects. Magma related to partial melting of MORB-type sources is limited but present in the Tyrrhenian area (e.g. Vavilov basin).

Volcanism typical of island arcs is common, although it has only been found south of 41°N. This volcanism developed in two clearly separated cycles. The first was Oligocene–Miocene (32–13 Ma) in age and is found in Sardinia. This type of volcanism can be explained by a roughly NNW-dipping subduction of oceanic lithosphere beneath Sardinia. The second cycle was Quaternary in age and is mostly localised around the Marsili basin in the

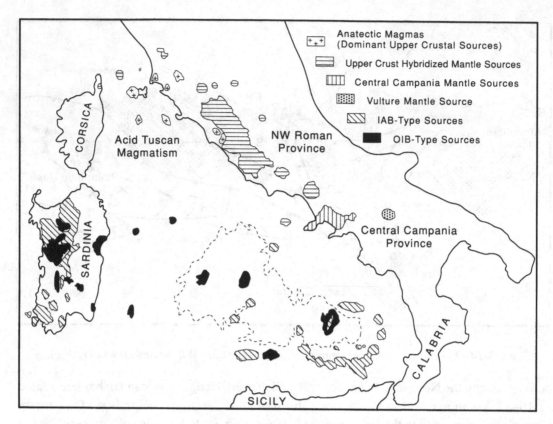

Figure 5-15. Recent volcanism in the Tyrrhenian Sea and adjacent regions, simplified from a classification of magma sources by Serri (1990).

Tyrrhenian Sea. Data obtained from ODP leg 107 indicates that oceanic crust drilled at site 651 in the Vavilov basin and at site 650 in the Marsili basin is consistent with calc-alkaline convergent plate margin basalts. A detailed discussion by Serri (1990) led to the recognition that the subducted plate under the Tyrrhenian region, which is still seismically active, is oceanic in nature.

Widespread volcanism from Tuscany to Campania covers a time span from Tortonian to Recent. Detailed studies of this volcanism have helped to define the roles played by upper crustal and mantle magma sources and by subduction-derived components. As an example, the potassic volcanism of the NW Roman province displays a marked contamination of mantle sources through hybridization with melts derived by subducted upper crustal material. Serri (1990) has argued that intermediate to acid rocks were formed by melting of the upper crustal material subducted within the uppermost mantle during continent–continent collision. In the Tuscan region, granitic and rhyolitic volcanism appears to have been derived by melting of crustal material with typical features of an upper crustal reservoir, although melting may have occurred in the uppermost mantle. This, and the upper crust hybridised mantle sources of potassic volcanism, requires the subduction of a continental lithosphere during collision related to Apennine orogenesis.

Taking all the available data, Serri (1990) found that Neogene–Quaternary volcanism in the Tyrrhenian and adjacent areas could be explained consistently. A lithospheric boundary along 41°N and a transform fault oriented NE–SW which marks a second lithospheric boundary separating the NW Roman province from the central Campania volcanism appears to be related to subduction of the continental Adriatic lithosphere. This subduction carried large amounts of upper crustal rocks to depths of 50–100 km or more below the NW Roman/Tuscany region. In contrast, an old oceanic (Ionian) lithosphere, still seismically active,

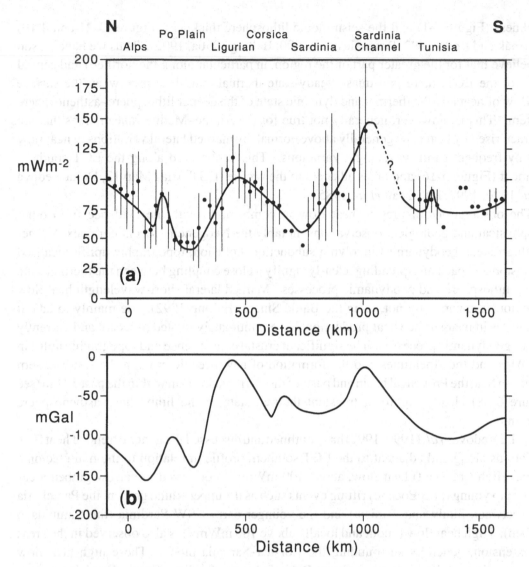

Figure 5-16. *Heat flow along the southern segment of EGT showing significant short wave-length variations, after Della Vedova et al. (1990).*

(a) Average heat flow density (dots) with standard deviations calculated for areas of 100 km width; regional heat flow (continuous line).

(b) Bouguer gravity anomaly. Areas of unusually high heat flow probably represent transient heat flow and correlate with marked positive Bouguer gravity anomalies.

below the Calabrian arc and southern Tyrrhenian Sea is required to explain volcanism south of the lithospheric boundaries just described.

The recent age of volcanism indicates that some of the geodynamic processes responsible for it are still active. This is shown, for instance, by the deep seismicity related with the subduction of the Ionian lithosphere. Therefore it may well be that the currently active processes produce perturbations in the steady state heat flow in the form of transient pulses.

5.5 TRANSIENT HEAT FLOW

Along most of EGT the heat flow variations (see EGT Atlas Map 13 and Chapter 4.1) can be explained by steady-state models. There is overall agreement between thermal lithospheric thickness (Figure 4-1) and the seismologic lithosphere thickness (Figures 3-21 and 4-19) (Cêrmák and Bodri 1992, Pasquale *et al.* 1990, Balling 1990a, 1992). Thus we have reason to believe that for the greater part of the region, in particular along the northern and central parts of the EGT, there is a quasi steady-state thermal equilibrium between the surface outflow of heat and the thermal and dynamic state of the deeper lithosphere–asthenosphere system. This is, however, generally not true for the Alpine–Mediterranean areas that are characterised by heat flow generally above normal, by marked lateral variations in heat flow and by frequent short wavelength variations. This is observed along the EGT southern segment (Figure 5-16) and areas adjacent to the profile (EGT Atlas Map 13, Della Vedova *et al.* 1990, 1992, Lucazeau *et al.* 1985).

The observations of present heat flow in combination with the evidence from other geophysical and geological observations, notably the Neogene to recent complex Alpine-Mediterranean geodynamics involving subduction, collision, topographic uplift, localised extension and back arc spreading, clearly signify a close coupling between the thermal state of the lithosphere and geodynamic processes. Marked lateral short-wavelength heat flow variations are generally not, as in the Baltic Shield (Balling 1992), due mainly to lateral variations in near surface heat production, but are intimately related to recent and currently active geodynamic processes. The significant crustal convergence and topographic uplift in the Alps and the Apennines and the formation of new oceanic crust in the southwestern Ligurian Sea, the Provençal basin and parts of the Tyrrhenian basin within the past 30 Ma (see Figure 6-38) clearly signify a transient thermal state of the lithosphere–asthenosphere system.

Della Vedova *et al.* (1990, 1992) have outlined and discussed the general trend of heat flow variations along and adjacent to the EGT southern profile in relation to the main tectonic units. High (transient) heat flow, above 100 mWm^{-2}, is observed in narrow graben areas subject to young (post Eocene) rifting events such as the upper Rhine graben, the Pantelleria rift (between Sicily and Tunisia) and the younger part of SW Sardinia (the Campidano graben). High heat flow (about and locally above 100 mWm^{-2}) is also observed in the areas of extension, generally surrounding the Corsica–Sardinia block. These high heat flow regions (Ligurian basin, Provençal and Balearic basins, Sardinia Channel and Tyrrhenian basin) are regions of marked extension where sedimentary basins and/or new oceanic crust (see Figures 6-35 and 7-5) are formed. Generally, marked positive heat flow anomalies correlate with positive gravity anomalies (Figure 5-18), shallow Moho (Figure 3-17, compare EGT Atlas Maps 2, 9 and 13) and thin lithosphere (Figure 3-21), thus supporting the view of a generally young transient lithospheric thermal regime closely related to the recent and currently active geodynamic processes.

Very high heat flow, locally above 200 mWm^{-2}, is observed in areas recently affected by magmatic and volcanic activity such as the Tuscany geothermal area and the oceanic Tyrrhenian basin (Della Vedova *et al.* 1992). The moderately high surface heat flow over parts of the Alps and the Molasse basin (about and locally above 80 mWm^{-2}) and the exceptionally low heat flow of about 40 mWm^{-2} in the Po basin (see EGT Atlas Map 13) may also signify transient thermal components related to deep geodynamic processes. Thus kinematic modelling by Werner and Kissling (1985) indicates that the temperatures in the Alpine lithospheric root may be reduced by several hundred degrees relative to their surroundings, the significance of which is discussed further in Chapter 7.

In conclusion, we hope to have demonstrated in this chapter that a good knowledge of current tectonic activity is the key to understanding how forces arise in the lithosphere–asthenosphere system and how they are distributed as stresses to which the Earth responds according to its rheological properties. Stress measurements help to quantify the nature and distribution of forces, as does seismicity which also provides information about the Earth's response to stress, quantified through measurements of energy release and strain rate. Uplift and subsidence rates give an additional factor to quantify the longer term viscous response of the Earth. Gravity and heat flow data add further evidence about mass distributions in the Earth which can give rise to buoyancy forces. Thus in this chapter we have formulated the essential ingredients for the discussion of geodynamic processes in Chapter 7. But first, in Chapter 6, we should address the time element of geology and, from our knowledge of the present state of Europe's lithosphere, reconstruct its past.

6 Tectonic evolution of Europe

In this chapter we discuss the geology and the crustal and lithospheric tectonics as viewed against the background of the information presented in Chapters 3, 4, and 5. Again we proceed from north to south along the EGT, working from the oldest to the youngest units. General reference should be made to the tectonic, gravity and magnetic maps (1, 9 and 10) of the EGT Atlas.

6.1 PRECAMBRIAN EUROPE

B. Windley

Precambrian rocks are unevenly distributed throughout Europe. The Lewisian Complex in NW Scotland was accreted from Laurentia during formation of the Caledonian orogen and is far removed from the European Geotraverse. The main area of Precambrian rocks is in the Baltic Shield. South of this there are few Precambrian outcrops, but a few Precambrian ages are known from isotopic work on zircons. Precambrian rocks of Gondwana crop out in northern Africa, but at the southern end of EGT in Tunisia they are only known in wells. This section therefore concentrates on the Baltic Shield.

6.1.1 THE BALTIC SHIELD

There is an overall decrease in age from NE to SW of rocks and structures within the Baltic Shield, the main tectonic units of which are shown in Figure 6-1 (EUGENO-S Working Group 1988). In the north the Kola–Karelian orogen contains five Archaean terranes, which were amalgamated by collision tectonics in the period 2.0–1.9 Ga. In contrast, in the south the Sveco-Fennian orogen contains no Archaean material, consisting of new crust (<2.2 Ga) that was accreted and underwent collisions in the period 2.0–1.8 Ga and was reworked by crustal melting in the period 1.8–1.54 Ga. Further southwest the Gothian orogen developed at 1.77–1.5 Ga and contains no older material, whilst the Sveco-Norwegian orogen has an age

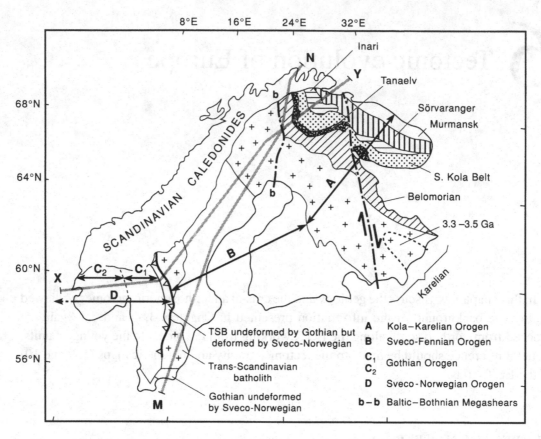

Figure 6-1. Map showing the main tectonic units of the Baltic Shield, in particular the four orogens referred to in the text. XY and MN are the lines of the cross sections of Figure 6-3.

of 1.05–0.9 Ga, and reworked most of the Gothian orogen. Figure 6-2 is a new geotectonic map of the Baltic Shield.

Syntheses of the geology and tectonic evolution of the Baltic Shield include Gaál (1986), Berthelsen (1980, 1987), Gaál and Gorbatschev (1987), Gorbatschev and Gaál (1987), Rundqvist and Mitrofanov (1991) and Park (1991). The FENNOLORA (Guggisberg and Berthelsen 1987, Guggisberg *et al.* 1991), POLAR (Luosto *et al.* 1989), EUGENO-S Working Group (1988) and BABEL seismic profiles (BABEL Working Group 1990) provide two-dimensional sections of the crust–upper mantle across parts of the Baltic Shield as described in Chapter 3. Figure 6-3 is a schematic cross section of the shield, based on these profiles.

The Kola–Karelian Orogen

This orogen has distinctive crustal geophysical properties that are different from those of the Sveco-Fennian orogen (see later). The crust has an average thickness of 45 km, a mainly magnetic dioritic upper layer, and an eclogite facies transition at 38 km (Henkel *et al.* 1990). Magnetic crustal structures are defined by Henkel (1991), as explained in Chapter 4.1.

Archaean terranes

The Murmansk gneiss–granulite terrane consists of predominant tonalitic gneiss, granodiorite, amphibolite and migmatite, and minor granulite, pyroxene gneiss and schist,

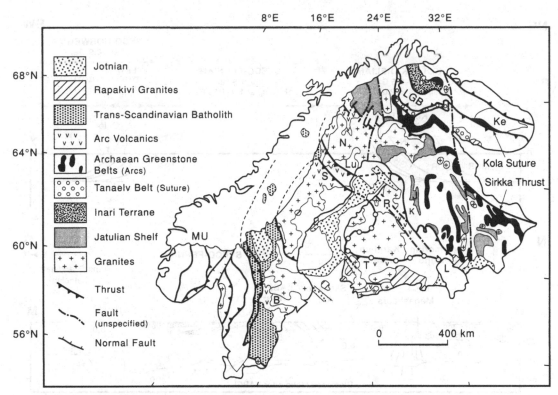

Figure 6-2. A geotectonic map of the Baltic Shield compiled from many sources. The electro-magnetic conductivity anomaly of the Skellefte–Tampere arc-suture is after Hjelt (1991). B: Bergslagen, J: Jormua, K: Kuopio, Ke: Keiv, L: Ladoga, LGB: Lapland Granulite-Gneiss Belt, Lu: Luleå, MU: Mandel–Ustaoset fault, N: Norrbotten, R: Raahe, S: Skellefte, T: Tampere.

and intercalated banded iron formation (Bylinski *et al.* 1977). All these rocks were meta-morphosed in the higher amphibolite or granulite facies. The major structures are large-scale reclined folds that are intruded by plutons of late Archaean granitic rocks. U–Pb zircon ages on gneisses are 2.9–2.7 Ga.

The composite Sörvaranger island arc terrane consists of:

(a) two greenstone belts whose amphibolites, ultramafic rocks and agglomeratic meta-volcanics and meta-psammites, pelites, banded iron formations and quartzites were deformed mostly under amphibolite facies conditions,

(b) amphibolite to granulite facies migmatitic alumino-silicate schists and gneisses that are in thrust contact with the greenstone belts. The discordant Neiden granitic pluton, that intrudes greenstone belt rocks and the gneisses, has a U–Pb and Rb–Sr age of 2.5–2.55 Ga. Berthelsen and Marker (1986a) suggested that the greenstone belts formed in arc and back-arc environments and that the alumino-silicate gneisses are derived from turbidites deposited in an arc-trench accretionary wedge.

The Inari gneissic terrane contains heterogeneous migmatitic trondhjemitic to granitic orthogneisses within which there are conformable layers and lenses up to 10 km wide of amphibolite and mica schist associated locally with calcic gneiss, quartzite and banded iron formation. U–Pb ages of zircons from gneiss give 2.73–2.55 Ga. The composite Belomorian terrane contains

(a) amphibolite-facies meta-pelitic gneisses, orthogneisses, amphibolites and granites. U–Pb zircon ages from tonalitic-trondhjemitic gneisses are 3.11 Ga and Nd isotopes suggest crustal material separated from the mantle by 3.5 Ga (Kröner *et al.* 1981, Jahn *et*

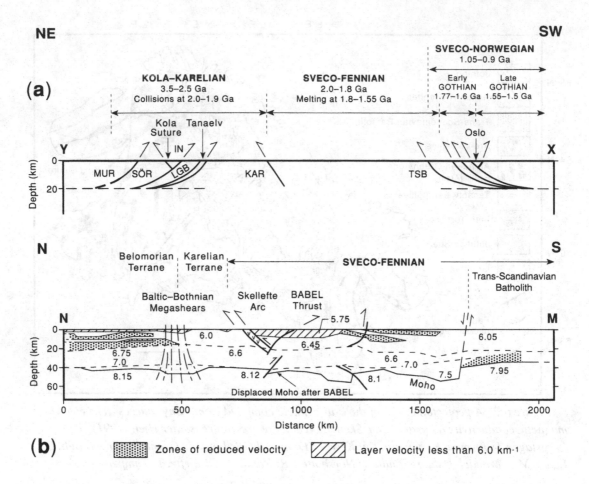

Figure 6-3. (a) Schematic tectonic section across the general strike of the orogens of the Baltic Shield showing main tectonic boundaries and thrust structures in the north and south.

(b) Cross section of the Shield showing the velocity–depth model of the crust along the FENNOLORA profile, modified after Guggisberg et al. (1991). Vertical exaggeration 5:1. Values are velocities in kms⁻¹.

Thrusts that displace the Moho and transect the Skellefte arc, and the position at depth of the Skelefte arc are after BABEL Working Group (1990). For position of these two sections see Figure 6-1.

al. 1984). Other gneisses have ages in the range 2.9–2.4 Ga.

(b) several greenstone belts composed of the Lapponian Supergroup in Finland, which includes a lower unit of komatiitic and tholeiitic basalts, and mafic to felsic tuffs belonging to a lava plateau, a central unit of pillow-bearing amphibolites, arkosic quartzites and aluminous slates, and an upper unit of extensive ultramafic and basaltic komatiites, mafic lavas and tuffs, carbonaceous greenschists and graphitic slates (Saverikko, 1987). The Kittilä greenstone belt (island arc?) is cut by a gabbro dated at 2.44 Ga (Simonen 1980). In the eastern Belomorides a 2.7–2.6 Ga period of collision was recognised by Kratz *et al.* (1978).

The basement of the Karelian composite terrane is made up of Archaean greenstone belts, and gneisses and granites. There are more than 20 major greenstone belts up to 100–150 km long and numerous smaller ones separated by belts of gneisses and granites of different types (Gaál and Gorbatschev 1987). Greenstone belts amount to not more than 15% of the present area. In Russian Karelia the greenstone belts are referred to stratigraphically as Lopian.

Across eastern Finland and Karelia there are four tectonic zones, showing differences in the composition and age of the volcanic rocks in the greenstone belts, in the type, composition and age of the intervening gneisses and granites, and in the general degree of metamorphism and deformation (Lobach-Zhuchenko *et al.* 1986). Tholeiitic basalts make up 40–70 % of most greenstone belts, but some range from komatiite to rhyolite, and others are bimodal of basalt and rhyolite. Overall the greenstone belts resemble modern island arcs (Sokolov and Heiskanen 1985). There is a significant younging of the greenstone belts westwards (U–Pb on zircons) from 3.0–2.9 Ga in eastern and central Karelia, to 2.80–2.75 Ga in western Karelia, to 2.65 Ga in eastern Finland (Lobach-Zhuchenko *et al.* 1986). This strongly suggests progressive westward accretion of successive island arcs, and this in turn implies eastward-dipping subduction zones. The gneisses and granites are less well understood than the greenstone belts. Some consist mostly of paragneisses, and others of orthogneisses and granites, and many show a close spatial and temporal relation with the development of the greenstone belts. Some of the gneisses between the greenstone belts were probably derived from accretionary wedges. The oldest known rocks in the Baltic Shield are gneisses in SE Karelia that have a zircon age of 3.5 Ga. amphibolites and migmatites are dated at 3.2 Ga and tonalites span a time-range from 3.4 to 3.1 Ga (Lobach-Zhuchenko 1989).

Early Proterozoic collision tectonics

In the period 2.0–1.9 Ga the above-mentioned Archaean terranes collided and were amalgamated to form the Kola-Karelian orogen. Early Proterozoic (2.4–1.9 Ga) rocks and structures include island arcs, Andean-type magmatic arcs, sutures and remnant shelf successions (Figure 6-4). The early Proterozoic structure of this orogen is well constrained by geophysical data of the POLAR Profile (Gaál *et al.* 1989, Marker 1990, Chapter 3.2.1).

The southern border of the Murmansk terrane is marked by a thrust zone that is several kilometres wide, dips northwards at 60–80°, deforms the large-scale folds, is marked by a prominent negative linear magnetic anomaly and current topographic depression, and consists of biotite-bearing mylonitic gneisses within which there are remnants of meta-sedimentary and meta-volcanic rocks, tectonic lenses of anorthosite up to 100 km long and 2.5 km wide, and ultrabasic rocks; this is the Keiv–Porosozero suture, in which thrusts truncate the early Proterozoic Keiv Group to the south (Bylinski *et al.* 1977). The Keiv Group in Russia (Figure 6-2) is a 2.0–1.9 Ga rifted passive continental margin succession deposited on gneisses of the Sörvaranger terrane. The successiion passes upwards from basal conglomerates, sandstones and minor carbonates, to andesites, basalts, dacites and rhyolites, overlain by arkosic sandstones, greywackes and aluminous schists that are chemically equivalent to kaolinitic clays. The aluminous pelites are metamorphosed to biotite, garnet, staurolite and kyanite grades and thrust in recumbent folds (Shurkin *et al.* 1980) on this craton margin, and many alkali granites have intruded the Keiv group (Rundqvist and Mitrofanov 1992)

The Kola suture zone (the Polmak–Pasvik–Pechenga belt of Gaál *et al.* 1989) is a south-dippng thrust zone up to 40 km wide that has placed the Inari terrane against and over the Sörvaranger terrane (Berthelsen and Marker 1986a, Marker 1990). The borders of the suture zone are marked by mylonites and the zone contains at least two thrust-bound slices made up of the 2.4–2.0 Ga Pechenga Series (penetrated by the Kola Superdeep Hole, Kozlovsky 1987) containing sediments that range from conglomerates (locally still unconformable on Sörvaranger gneisses), sandstones (rifted continental margin), through stromatolitic dolomites, siltstones and phyllites (shelf-rise transition), to turbidites (trench), and volcanic rocks that

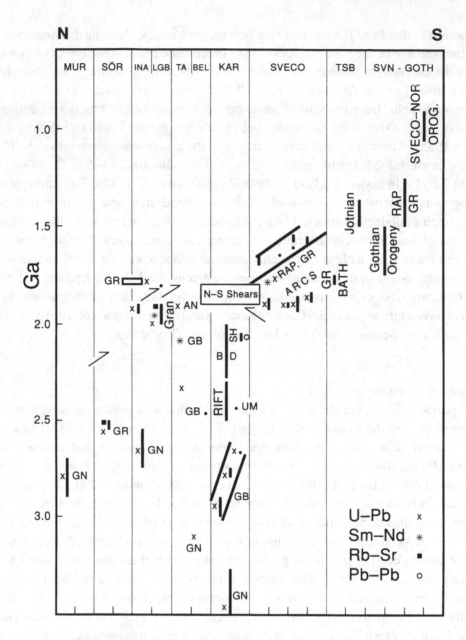

Figure 6-4. An isotope-summary time chart illustrating the Archaean growth and Early
Proterozoic amalgamation of the terranes of the Kola–Karelian orogen, and the progressive
southwestward growth of the Sveco-Fennian orogen, and Gothian and Sveco-Norwegian orogens
during the Proterozoic. GB: greenstone belt, GN: gneiss, GR: granite, RAP GR: rapakivi granite,
BD: basic dykes, UM: ultramafic rocks, AN: anorthosite, SH: shelf carbonates.

pass upwards from andesites through alkalic basalts to tholeiitic basalts (with REE characteristics resembling those of MORB) to ferropicrites that have a Sm–Nd isochron age of 1.99
Ga. These picrites contain gabbro-wehrlite intrusions with Ni–Cu sulphide deposits. On the
south side of the suture zone there is a thrust-bounded, greenschist-grade early Proterozoic
island arc that consists of weakly deformed abundant andesites, basaltic pillow lavas, minor
komatiitic lavas, tuffs, and sulphide-bearing and carbonaceous pelites (Marker 1985,
Berthelsen and Marker 1986a, Gaál *et al.* 1989). A further result of the southward subduction
that gave rise to the island arc was the emplacement into the northern border of the Inari
terrane of an Andean-type magmatic arc represented by 1.95–1.9 Ga (U–Pb, Meriläinen

1976) calc-alkaline plutons (Barbey *et al.* 1984). The 1.79 Ga (U–Pb) crustal melt Vainospää granite stitches the suture (Figure 6-2).

The Lapland Granulite-Gneiss Belt (Figure 6-2) is 80 km wide, consists largely of paragneisses derived from turbidites, and has a distinct magnetic structure (Henkel 1991). Its lower half is highly sheared, and in its upper part meta-sedimentary granulites contain conformable lenses of noritic-enderbitic ortho-gneisses of calc-alkaline parentage (Hörmann *et al.* 1980, Barbey and Cuney 1982). Isotopic ages of meta-igneous granulites suggest emplacement and metamorphism at 1.9–2.0 Ga (Meriläinen 1976, Bernard-Griffiths *et al.* 1984). Post-tectonic crustal melt granites are dated at 1.8 Ga. On Figure 6-2 the granulite Belt of Finland continues eastwards as amphibolite facies gneisses on the western side of the White Sea, where they occur north of rocks characteristic of the Tanaelv belt. The turbidite precursors of the Lapland Granulites may have been deposited in a back-arc basin adjacent to a magmatic calc-alkaline arc (Berthelsen and Marker 1986a), or on a passive continental margin subsequently intruded by arc magmas (Barbey and Raith 1990). A further possibility is that the turbidites accumulated in an accretionary wedge.

The Tanaelv high-grade ductile thrust belt contains intercalated tholeiitic amphibolites and calc-alkaline gneisses, within which there are many lenses of gabbroic, anorthositic, ultramafic and eclogitic rocks (Bylinski *et al.* 1977). The Vaskojoki anorthosite has a U–Pb age of 1.91 Ga (Bernard-Griffiths *et al.* 1984). A U–Pb age of 2.36 Ga suggests the presence of some relict Archaean material. A combination of reflection seismic (Behrens *et al.* 1989), gravity (Elo *et al.* 1989), electromagnetic (Korja *et al.* 1989) and magnetic data (Marker *et al.* 1990) of the POLAR profile show that the Granulite and Tanaelv Belts form a north-dipping thrust wedge, the apex of which continues to a depth of 15-20 km (see Figure 3-3). This wedge is made up of a stack of high-grade thrust sheets on top of a major basal decollement. A result of 1.9 Ga thrusting of a hot slab of Lapland granulites over a cooler slab of Belomorian gneisses is that the Tanaelv belt contains an inverted metamorphic sequence (Krill 1985). Barbey *et al.* (1984) and Berthelsen and Marker (1986a) interpreted the Tanaelv belt as a suture zone containing relict oceanic crustal material located on the site of a closed back-arc basin, that may have been about 125 km wide (Marker 1990).

In the western Belomorian terrane the Karasjok greenstone belt has a Sm–Nd isochron age of 2.09 Ga. This may be an early Proterozoic island arc, consistent with the suggestion of a 1.9–1.8 Ga period of collision in the eastern Belomorides (Kratz *et al.* 1978). The Belomorides were intruded in the early Proterozoic by a variety of granitic bodies that range from the Hetta granodiorites, tonalites and trondhjemites (maximum intrusion age of 1.97 Ga), the Haparanda monzonites (1.88–1.80 Ga), and post-tectonic crustal melt microcline granites (1.8–1.76 Ga) (Gaál *et al.* 1989).

The Sirkka Thrust (Figure 62) is a major tectonic boundary along which the high-grade Belomorian terrane has been thrust southwards under the low-grade Karelian terrane. The boundary is marked by mylonites and refraction seismic data show that it dips ca. 40°S (Luosto *et al.* 1989). There is no positive evidence that this is a suture; more likely it is a post-collisional thrust. The upper crust of the Karelian terrane does not contain a layer of reduced seismic velocity, in contrast to the Belomorian terrane and the Sveco-Fennian orogen to the north and south respectively (Figure 6-3b). The Karelian terrane has considerable evidence of rifting and shelf deposition in the early Proterozoic when it constituted the foreland or rifted passive continental margin of the Sveco-Fennian ocean to the south (Gaál 1986). There are 2.44 Ga mafic-ultramafic layered intrusions, many 2.4–2.3 Ga and 2.2–2.0 Ga NW-trending (coast-suture parallel) basic dykes (Gorbatschev *et al.* 1987), and 2.5–2.3 Ga conglomerates and tholeiitic volcanics (continental rift facies) of the Sumi–Sariola Group.

After erosion and weathering a continental platform developed with deposition of the Jatulian Group of basal quartzites, shelf carbonates (2.05 Ga, Pb–Pb) and banded iron formations (2.08 Ga, Pb–Pb), followed by mafic volcanics, carbonaceous slates and black schists which are the analogue of the deep water facies starved of terrigenous debris typically formed during the break-up of a modern carbonate platform.

The Sveco-Fennian Orogen

This orogen contains no Archaean terranes and litle or no Archaean isotopic material (Wilson *et al*. 1985, Huhma 1987, Patchett *et al*. 1987, Romer 1991). It developed by the growth and collision of 2.0–1.8 Ga juvenile arcs, and by extensive crustal melting in the period 1.8–1.55 Ga (Figure 6-4). This Sveco-Fennian orogen has a mainly paramagnetic dioritic upper crustal layer, an average crustal thickness of 48 km (maximum 54 km), and a thick lower crustal layer with a P-wave velocity of 6.8–7.8 kms^{-1}, the eclogite transition being at 36 km depth (Henkel *et al*. 1990). Also the Sveco-Fennian upper crust contains zones of relatively low seismic velocity (Figures 3-4 and 6-3b).

The Luleå–Kuopio suture zone (named after the eponymous towns) separates the Kola–Karelian orogen from the Sveco-Fennian orogen. Whereas many previous authors (e.g. Gorbatschev and Gaál 1987, Berthelsen 1987) have extended it along the Skellefte island arc, on Figure 6-2 it is placed about 100 km to the north, where the Archaean–Proterozoic palaeoboundary is defined by initial epsilon Nd isolines, a change in aeromagnetic anomaly intensities, and a difference in the chemistry of ca. 1.9 Ga granitoids – those to the north have higher contents of K, Ba, Rb and La at corresponding Si-contents than those to the south (Öhlander *et al*. 1992). The suture zone is displaced by 1.9–1.8 Ga N–S megashears (Berthelsen and Marker 1986b), which have a prominent magnetic structure (Henkel 1991) and which may be the cause of an underlying depression of the Moho of 10 km (Figure 6-3b).

The suture zone contains thrust slices of different types and origin. There are two types of Kalevian turbidites:

(a) Those situated on the craton side of the suture are unconformable on Jatulian shelf sediments, and even on Archaean basement where Jatulian rocks have been eroded. Autochthonous turbidites contain Archaean and Proterozoic detritus, and are locally interbedded with tholeiitic volcanics. Ward (1987, 1988) suggested deposition in transtensional, intra-cratonic rifts near the craton margin.

(b) On the suture zone side ophiolitic rocks are associated with allochthonous turbidites that Ward (1987, 1988) suggested were most likely deposited from debris flows and turbidity currents in submarine canyons on an accretionary margin. These rocks were subsequently thrust northeastwards over the craton.

In the suture zone the Outokumpu nappe contains serpentinites, gabbros, basaltic pillow lavas, non-detrital quartzites, dolomites, Mg-rich meta-volcanics and Cu–sulphide deposits (Park 1984). The Jormua ophiolite (Kontinen 1987) was thrust about 30 km onto the continental margin of the Karelian terrane. It contains an upward sequence of serpentinites; gabbros cut by basic dykes; a sheeted dyke complex with screens of gabbro and serpentinite; basaltic pillow lavas and pillow breccias; tuffites, cherts and carbonate sediments; and turbiditic greywackes and semi-pelites. Zircons from the gabbros give a U–Pb crystallization age of 1.96 Ga. In the Sveco-Fennian suture zone, thrusts have imbricated Karelian basement and its Jatulian cover and have transported Kalevian turbidites and accretionary wedge rocks eastwards over the Archaean craton (Park and Bowes 1983). Bowes *et al*. (1984), Park (1985) and Gaál and Gorbatschev (1987) suggested that the main Sveco-Fennian subduction was directed from the south side of an island arc towards the Karelian foreland, but Ward (1987) proposed that subduction was southwards from the site of the present suture zone towards the orogen. The fact that the thrusts have transported rocks, including the Jormua ophiolite, onto the continental margin, does indeed suggest that the subduction zone dipped to the south, as in modern collisional orogens like the Himalayas.

There are several magmatic arcs within the Sveco-Fennian orogen; the first was recognised by Hietanen (1975). In northern Sweden the Norrbotten arc (1.90–1.87 Ga) consists of porphyritic intermediate and felsic lavas that resemble modern Andinotype high-K, calc-alkaline arc lavas (Pharaoh and Brewer 1990). Along the Sveco-Fennian side of the suture is the NW–SE trending, ore-rich Raahe–Ladoga fault zone (Korsman 1988) within which there are 1.91–1.90 Ga intermediate-acidic lavas associated with base metal mineralisation, 1.9–1.88 Ga gabbros with nickel ores, and 1.89–1.875 Ga granites and gabbros (Vaasjoki and Sakko 1988). The rocks and ores are comparable to those in modern island arcs and intra-arc rifts. Late faulting is often concentrated along hot, young arcs. The Savo schist belt contains highly metamorphosed and deformed arc rocks and post-collisional, crustal melt granites (Nironen 1989a). The 1.89 Ga Skellefte island arc has mature intra-arc volcanics (Vivallo and Claesson 1987) and granodiorite–granite intrusions derived from subduction-related melts (Wilson *et al.* 1987). Magnetotelluric transfer functions along the FENNOLORA seismic profile show that there is an extremely high conductivity and low resistive anomaly across the Skellefte arc (Rasmussen *et al.* 1987), and the BABEL reflection profile (Figure 3-5) shows that S-dipping reflectors project to the surface near Skellefte – Figure 6-3b (BABEL Working Group 1990); these structures are probably related to a S-dipping subduction zone responsible for this arc. The BABEL profile also shows a major N-dipping thrust south of the Skellefte arc and an underlying thrust that has displaced the Moho by about 10 km (Figure 6-3b); these thrusts may be associated with a N-dipping subduction zone and the formation of adjacent arcs (BABEL Working Group 1990). The FENNOLORA profile shows that 400 km south of Skellefte there is a major change in crustal thickness that coincides with a S-dipping thrust zone that displaces the Moho by more than 10 km (Guggisberg *et al.* 1991).

The Skellefte arc–suture zone continues eastwards around a 500 km long arcuate loop outlined by a major geoelectric conductivity anomaly up to 20 km wide to join the Tampere island arc (Figure 6-2, Hjelt 1991). On the surface the regional strike and granite belts parallel this anomaly. The 1.9 Ga Tampere arc contains medium-K basalts, medium- to high-K andesites, and high-K dacites and rhyolites; these calc-alkaline volcanic rocks resemble those in Recent mature island arcs (Kähkönen *et al.* 1989). The structure of the Tampere arc suggests that the derivative subduction zone dipped southwards (Nironen 1989b). In SW Finland the Orijärvi arc comprises submarine alkaline–subalkaline basic–intermediate lavas, felsic pyroclastic rocks associated with massive Cu–Pb–Zn deposits, co-magmatic gabbro–tonalite bodies and a 1.91 Ga (U–Pb on zircon) granodiorite, all formed in a back-arc environment (Colley and Westra 1987). Along strike to the west, arc-type mafic/intermediate to rhyolitic volcanic rocks (Ehlers and Lindros 1990) are associated with a high-grade mélange zone of gneiss (indicated by magnetic anomalies) containing fragments of amphibolite and pyroxenite, formed above a postulated N-dipping subduction zone (Edelman and Jaanus-Järkkälä 1983). At Bergslagen in southern Sweden (Baker and Hellingwerf 1988) there is at least 10 km thickness of 1.9–1.8 Ga felsic pyroclastics, ignimbrites and immature sediments that may have developed in major rifts that were later deformed, inverted and intruded by diapiric granites.

U–Pb zircon data indicate that many of the Sveco-Fennian arc lavas were erupted in the short period of 1.92–1.87 Ga contemporaneously with the intrusion of innumerable 1.91–1.86 Ga, subduction-derived plutons and batholiths of tonalite, granodiorite and granite (Nurmi and Haapala 1986, Welin 1987). Nd data from Sweden indicate that the I-type granitoids and volcanics have ENd in the range -1 to +3 and that they consist of 90% material newly derived from the mantle at 1.9–1.7 Ga (Patchett *et al.* 1987). Many volcanic rocks are associated with massive Cu–Zn–Pb–Ag–Au mineralisation, and the granites with porphyry-type Cu–Mo–Au deposits (Gaál 1990). Seismic refraction data show that the Moho is depressed to 60 km depth below the main batholith area of central Finland (Luosto and Korhonen 1986).

Between many of the Sveco-Fennian arcs there are biotite-bearing granitic gneisses and schists that have been widely regarded as meta-greywackes and meta-pelites within the now-outdated concept of a 'Bothnian Basin'. Within these meta-sediments there are many conformable kilometre size (maximum) lenses of amphibolite, meta-gabbro and meta-ultramafic rocks. Important Ni–Cu deposits occur in some peridotite–dunite–pyroxenite–gabbro lenses, that have been interpreted as intrusions in precursor sediments made conformable to the derived gneisses by deformation (see Papunen and Gorbunov 1985). However, I suggest that the greywackes and pelites were deposited in accretionary wedges between island arcs and that the mafic–ultramafic lenses are slices of ocean floor that were thrust into the sediments of the accretionary wedges. Such a model accounts satisfactorily for the sediments that would be expected to occur in the zones between accreted arcs, like those between Japanese or Archaean arcs (Hoffman 1989), and thus the mineralisation is not surprisingly comparable to that in modern obducted ophiolites.

Following amalgamation of the arcs, syn- and post-collisional deformation took place. Thrusting and folding was associated with high amphibolite facies metamorphism that locally reached granulite grade. Rb–Sr data suggest late kinematic activity ceased by about 1.77 Ga (Welin and Stålhös 1986). Crustal thickening led to the formation of abundant crustal melt granites (maximum age of crustal sources was 1.9 Ga, Wilson *et al.* 1985), of which there are three main types (Nurmi and Haapala 1986):

(a) The peak of regional metamorphism gave rise to partial melting of paragneisses and formation of 1.85–1.808 Ga, water-saturated, near-minimum melt granites as *in situ* plutons and dyke networks. These are comparable to the Miocene leucogranites of the High Himalayas (Crawford and Windley 1990).

(b) Deep-crustal dehydration melting promoted by mantle input led to non-minimum melt, 1.80 Ga granites in far-travelled plutons associated with ultrapotassic, mantle-derived lamprophyre dykes. These are similar to the Miocene Karakoram batholith in the upthrust western end of the Tibetan plateau (Crawford and Windley 1990).

(c) By 50–200 Ma after the last tectonic activity internal slow heating of the thickened crust had led to its final extension and collapse, and thus to decompression melting of the mantle and melting of depleted granulitic lower crust left over by removal of type 2 magmas (however, note Emslie 1991), and the result was the formation of 1.65–1.54 Ga metaluminous subsolvus rapakivi granites and coeval mantle-derived gabbros, anorthosites and basic dykes (Haapala and Rämö 1990, Rämö 1991).

The last event in the history of the Sveco-Fennian orogen was the deposition (maximum age, 1.5 Ga) of Jotnian sandstones in a major elongate basin below the Gulf of Bothnia (Figure 6-2), which locally extends into exposed basement as Jotnian rifts associated with mafic dykes. This basin could have been the final result of the extension and thinning of the Sveco-Fennian crust.

Along the southwestern margin of the Sveco-Fennian orogen is the 1600 km long,150 km wide, 1.84–1.75 Ga Trans-Scandinavian Batholith (Figure 6-4) that ranges from early arc-type monzo-diorite to quartzo-monzodiorite and granite and late within-plate, peraluminous, crustal-melt leucogranites (Zuber and Öhlander 1990). The batholith may have developed above an E-dipping subduction zone on the western margin of the Sveco-Fennian orogen (Larson *et al.* 1990, Andersson 1991). The FENNOLORA seismic profile shows that there is a major depression in the Moho from 40 km to 52 km below the batholith (Figure 3-4, Henkel *et al.* 1990), a fundamental lithospheric transition zone between the batholith and the Sveco-Fennian orogen, and that the mantle low velocity zone dips eastwards below the batholith (Clowes *et al.* 1987). The FENNOLORA profile also shows that below the batholith there is a thick low-velocity zone in the lower crust that may be connected with the development of the granites of the batholith (Figure 6-3b, Guggisberg *et al.* 1991). The batholith is overlain by 1.7–1.65 Ga (U–Pb, zircon) sub-Jotnian conglomerate and sandstone and ash-flow ignimbrites and tuffisites which are co-magmatic with some post-tectonic granites and that formed in calderas and graben in the volcanic roof of the batholith. These rocks are overlain unconformably by Jotnian continental conglomerates, sandstones and basalts. In a zone along its western side the Trans-Scandinavian Batholith has been reworked by Sveco-Norwegian deformation (Figure 6-1).

The Gothian Orogen

A small part of the Gothian orogen is well preserved in Blekinge in SE Sweden and on Bornholm island in Denmark because it has escaped Sveco-Norwegian reworking (Johansson and Larsen 1989). In this area there are acidic metavolcanics (U–Pb age of 1.705 Ga), metabasites, quartzites, mica schists and gneisses (U–Pb age of 1.69 Ga), and intrusive granites.The remainder of the former Gothian orogen is difficult to analyse and describe because it has been heavily overprinted by the Sveco-Norwegian orogeny. The oldest kown Gothian rocks here are 1.77 Ga diorites, and 1.758 Ga amphibolites associated with paragneisses (Åhäll and Daly 1989). These supracrustal rocks were deformed and meta-morphosed under amphibolite facies conditions and intruded by many calc-alkaline tonalite–granodiorite bodies. The Kongsbergian orogen of Starmer (1991) can be subdivided (Berthelsen 1980) into eastern early (1.77–1.6 Ga; this includes the Blekinge–Bornholm area) and western late (1.55–1.5 Ga) Gothian parts (C1 and C2 respectively in Figure 6-1, and Figure 6-3a) that might have undergone collision at about 1.5 Ga. During a tectonically quiescent period (1.5–1.25 Ga) rapakivi granites, gabbros and basic dykes were intruded, bimodal volcanics were extruded, and clastic sediments were deposited in rifts in the eroded basement. These rocks formed in the terminal extensional event of the Gothian orogeny, that may be comparable in principle to the terminal Sveco-Fennian rapakivi granite event.

The Sveco-Norwegian Orogen

This 1.05–0.9 Ga orogeny gave rise to a N–S Cordilleran and collisional belt (that reworked the Gothian crust) and later intrusive rocks. There are meta-andesites, tuffs, agglomerates and volcanic breccias, and syn-orogenic calc-alkaline plutons and post-tectonic granites. The best isotopic age of the supracrustal rocks is 1.25–1.2 Ga. The geochemical features of the earliest basalts, acid volcanics and gabbro intrusions suggest they formed in an extensional, continental margin setting related to subduction processes that were ultimately responsible for the Sveco-Norwegian orogeny (Atkin and Brewer 1990).

Metamorphism and associated deformation reached the upper amphibolite facies. The subduction zone may have dipped either to the east from off the present western coast (Torske 1977, Falkum and Petersen 1980), or to the west from near Oslo Fjord (Berthelsen 1977); according to the latter model the Permian Oslo graben may have been structurally controlled by the late Proterozoic suture zone. After deep crustal metamorphism and deformation, nappes were thrust eastwards, giving rise to the Protogine zone, which is a complicated, north-trending, 20–30 km wide zone of mylonitic and ductile faults and thrusts along the western border of the Trans-Scandinavian Batholith.

The EUGENO-S seismic profile 6 (see Figure 6-7) and a Swedish reflection profile both show low angle W-dipping reflective structures of this thrust zone (Figure 6-3a, Green et al. 1988, Dahl-Jensen et al. 1991). The zone has been interpreted variously as a collisional suture, a long-lived zone of thrusting and crustal weakness (Larson et al. 1990), and the easternmost thrust (the Sveco-Norwegian frontal thrust) as the extension of the Grenville Front. Wahlgren and Stephens (1990) demonstrated that the zone has a fan-like structure and suggested that it was caused by rotation of shear zones after their formation. This would be consistent with formation of a thrust system with a hinterland to the west and a foreland to the east, confirming a suggestion of Berthelsen (1980). According to the Berthelsen (1977) model the Dalsland thrust, which carried a high-grade crystalline slab eastwards over the low-grade Dal Group, would be equivalent to the Main Central Thrust in the Himalayas. Late extensional structures in the Protogine zone (Andréasson and Rodhe 1990) developed as the expected consequence of the collapse of a thrust-thickened crust, as in modern collisional orogens.

The Precambrian rocks of southern Norway are divided into two blocks by the north-trending Mandal–Ustaoset fault (Figure 6-2) along which there are ultracataclasites and mylonites. The fault separates the mid-Proterozoic Telemark supracrustals in the east from a belt of post-tectonic granites on the west. First movements on this little-known structure took place before the deposition of 1.2 Ga supracrustals, but the latest normal fault movements occurred after the intrusion of 0.9 Ga post-tectonic granites.

6.1.2 CENTRAL AND SOUTHERN EUROPE

Useful information on Precambrian rocks south of the Tornquist Line comes from four studies of zircons: Detrital red zircons from Cambrian and Ordovician rift sediments from the Saxo-Thuringian belt have apparent $^{206}Pb/^{238}U$ ages of 1.8–2.2 Ga, and colourless zircons show 0.6–1.5 Ga. According to their model ages the former may be derived either from Baltica or from Gondwana, but red zircons are unknown from Baltica. The model ages of the colourless zircons point to some Gondwana-related source (Dorr et al. 1991).

Ion microprobe analyses by Gebauer et al. (1989) show that a detrital zircon in Moldanubian paragneiss from Bavaria has U–Pb ages of 3.84, 2.59, 1.94 and 0.46 Ga. Zircons from the Massif Central provide ages of 3.15, 2.9, 2.76, 2.65, 2.3, 2.10,1.75, 1.0 and 0.6 Ga. The common presence of Pan-African-age detrital zircons in these Massif Central meta-sediments suggests that detritus was derived from Gondwana.

Von Hoegen et al. (1990) studied 57 detrital zircons from Cambrian sandstones of the Ardennes–Brabant Massif. Colourless zircons have ages of 0.53–1.65 Ga and their U–Pb characteristics suggest a strong Cadomian/Pan-African influence. Reddish zircons with ages of 1.75 and 2.4 Ga are dominantly derived from Archaean to Early Proterozoic crystalline

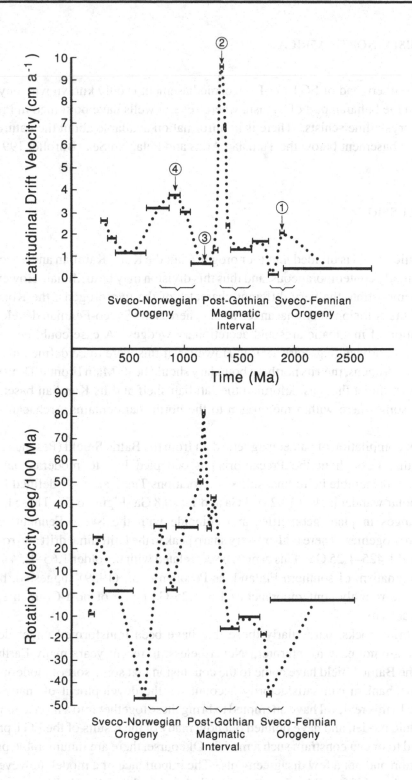

Figure 6-5. Variations in latitudinal and drift velocity and rotation velocity of the Baltic Shield with suggested correlations with the Sveco-Fennian and Sveco-Norwegian orogenies and the post-Gothian magmatism (modified after Pesonen et al. 1989).

rocks. These results suggest that the Ardennes–Brabant massif was a coherent part (Avalonia) of the Gondwana continental plate before the Caledonian orogeny.

U–Pb age patterns of zircons from meta-igneous and meta-sedimentary rocks from the Lepontine area of the Swiss Central Alps suggest that some meta-sediments were deposited probably during the Late Precambrian to Cambrian (Köppel *et al.* 1980).

6.1.3 TUNISIA: NORTH AFRICA

At the southern end of EGT pre-Palaeozoic basement is only known with any degree of certainty in the Saharan part of Tunisia where several wells have bottomed in Precambrian granite or crystalline schists. There is no information available about the nature, age, and depth of the basement below the Tunisian Atlas and Pelagian Sea (Burollet 1991).

6.1.4 DISCUSSION

The Baltic Shield is divided into four orogens, but the Kola–Karelian and Sveco-Fennian were essentially contemporaneous, and thus this division may be artificial. However, it does serve to demonstrate the very different character of the two orogens: the Kola–Karelian formed by the collision of Archaean terranes, whereas the Sveco-Fennian developed by the amalgamation of magmatic arcs and accretionary wedges. A case could be argued for a Svecokarelian orogen (e.g. Park 1991). However, if this were to be defined in conformity with modern orogens, then its northern boundary should be its Main Frontal Thrust, being the northernmost thrust that has deformed the Jatulian shelf and its Karelian basement. This should lie somewhere within the orogen to the north that contains Archaean terranes or basement.

A recent compilation of palaeomagnetic data from the Baltic Shield (Pesonen *et al.* 1989) indicates that throughout the Precambrian it occupied low to moderate latitudes and underwent considerable latitudinal shifts and rotations There are well-defined loops in the apparent polar wander path at 1.92–1.7 Ga and 1.1–0.8 Ga (Figure 6-5). These loops reflect major changes in plate geometries and coincide with the Sveco-Fennian and Sveco-Norwegian orogenies. There is also a very sharp peak in the latitudinal drift and rotation rates in the period 1.425–1.25 Ga. This cannot be correlated with the older 1.65–1.54 Ga rapakivi granite magmatism of southern Finland, as Pesonen *et al.* (1989) suggested, but may be related to the roughly contemporaneous 1.5–1.25 Ga phase of post-Gothian epeirogenic magmatic activity.

Precambrian rocks, particularly those that have been transformed under deep crustal conditions, are not easy to interpret. Nevertheless, in recent years many Earth scientists studying the Baltic Shield have come to the conclusion that some sort of modern-style plate tectonic mechanism can satisfactorily account for the development of many rocks and structures. In this review I have attempted to bring these together into a more comprehensive plate tectonic model, and have pointed out how many of the results of the EGT project have contributed to or can constrain such a model. Of course, there are innumerable problems of interpretation and not a few disagreements. The importance of a model, however, is that it can be tested, and if found inadequate, it can be modified or discarded.

6.2 FROM PRECAMBRIAN TO VARISCAN EUROPE

A. Berthelsen

6.2.1 AROUND THE TORNQUIST ZONE

During the Early Palaeozoic, the Precambrian crust of the western part of the Baltic Shield became deeply involved in the Caledonian orogeny due to Baltica's collision with Laurentia. Since then the shield has reacted as one large, resistant craton which has suffered little subsequent tectonic activity, although from time to time reactivation of old structures caused long-range epeirogenic movements, and rifting and magmatic activity were felt locally. In present southern Norway, the Oslo–Skagerrak rift system and the associated alkaline igneous rocks of the Oslo region were formed during the Late Carboniferous and Permian (Wessel and Husebye 1987, Ro *et al.* 1990, Lie and Husebye 1991). The Oslo rift trends slightly obliquely to, and partly masks, important basement structures: the Old to Young Gothian terrane boundary, a 1.4 Ga tholeiitic dyke swarm, and a high angle shear zone of late Sveco-Norwegian age (Hageskov 1980, 1985).

South of the present Baltic Shield, in the Danish foreland area (Figure 6-6), Baltica's Precambrian basement is hidden beneath a thick Phanerozoic sedimentary cover. As demonstrated by the EUGENO-S deep seismic surveys, the concealed basement underwent significant crustal thinning during the Palaeozoic and Mesozoic. Locally it is divided into fault blocks in the neighbourhood of the so-called Tornquist zone (Figure 6-6). Seismic mapping and deep wells both on land and offshore indicate that buried Precambrian basement belonging to Baltica underlies Jutland, Fynen, and Sealand. It probably wedges out at mid-crustal levels against the E–W trending Trans-European Fault in northern Germany, south of the Caledonian deformation front. North of the Caledonian front, a basement high with a relatively shallow cover and a positive Bouguer gravity anomaly, the Ringkøbing–Fyn high, separates the post-Caledonian North German basin from the Permian and younger Norwegian–Danish basin. The Norwegian–Danish basin extends from the Norwegian and Danish sectors of the North Sea over northern Jutland to Sealand, shallowing ESE. Another and stronger positive Bouguer anomaly, the Silkeborg gravity high (SGH in Figure 6-6), is situated over the southern part of the basin in central Jutland, extending towards NW Sealand.

The 20–50 km wide Sorgenfrei Tornquist zone which, like the Teisseyre Tornquist zone of Poland, was affected by Late Cretaceous to Early Tertiary inversion tectonics, trends through the northern part of the Norwegian–Danish basin, parallel to the depocentre axes of the Permian and Mesozoic fill of the basin. In the Bouguer gravity map, the Mesozoic to Early Tertiary Sorgenfrei Tornquist zone elements follow a belt of strong gradients along the southwest border of a broad high from Scania through Kattegat.

Preliminary 1-D interpretational models for M-T measurements (Balling 1990b), show lateral variations in crustal conductivity structure related first of all to the thick sedimentary fill of the Norwegian–Danish basin but also in the underlying basement and in the Ringkøbing–Fyn high where resistivity values are less than a tenth of those of the Baltic Shield. The presence of downfaulted Cambro-Silurian strata within the basement and structural and textural changes related to thinning of the crystalline crust might provide an explanation for this reduction in resistivity. None the less, some gross scale features of Precambrian origin appear to have been preserved.

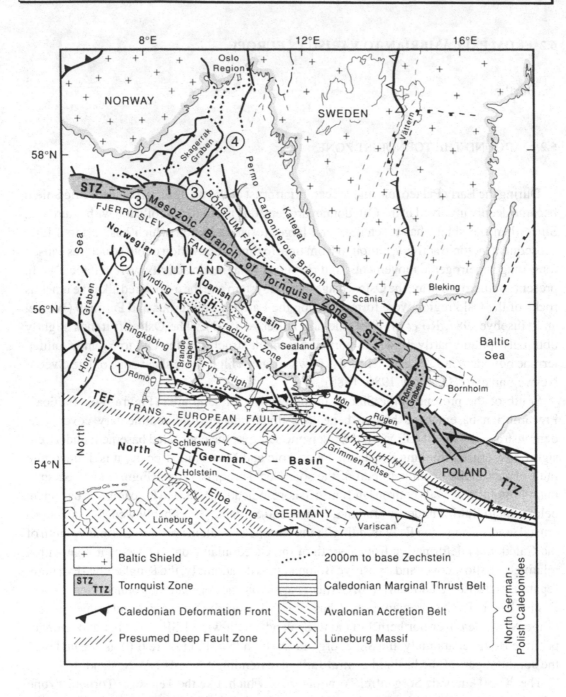

Figure 6-6. Tectonic sketch of the Danish area and neighbouring parts of Norway, Sweden, Germany and Poland. Abreviations used are STZ Sorgenfrei Tornquist zone; TTZ Teisseyre Tornquist zone. The 'Tornquist fan' comprises TEF the Trans-European fault. (1) The Rømø fracture zone, (2) The Vinding fracture zone, (3) the Fjerritslev and Børglum faults, and (4) the Permo-Carboniferous branch.

Contrasting primary Precambrian crustal structures

As mentioned in Chapter 2, most of the Sveco-Norwegian orogen in the southwestern part of the Baltic Shield displays a two-layered crust, which is 30–36 km thick and does not include a lowermost, high velocity layer. The Gothian and Sveco-Fennian provinces further to the east, however, exhibit a three-layered crustal structure with thicknesses of about 40–45 km. The transition from one to the other takes place along a N–S trending zone, located

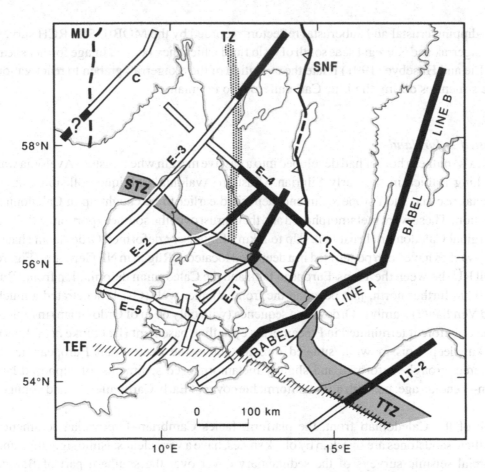

Figure 6-7. Locations of EUGENO-S sesimic profiles E1 to E5, line C (Cassel et al. 1983), line LT-2 (Guterch et al. 1986) and BABEL lines A and B, showing the transition zone (TZ) from two-layered crust (white) to three-layered crust (black) in the southern part of Baltica. SNF: Sveco-Norwegian front, STZ: Sorgenfrei Tornquist zone, TEF: Trans-European fault, TTZ: Teisseyre Tornquist zone.

about 100 km west of the Sveco-Norwegian deformation front. This has been taken to indicate that a two-layered Sveo-Norwegian reworked crust, resulting from overthrusting along the W-dipping Sveco-Norwegian deformation front, came to overlie a three-layered Sveco-Fennian and Gothian crust (EUGENO-S Working Group 1988, Green *et al.* 1988).

Although reduced in crustal thickness, both two- and three-layered types of crust have also been recognised in the concealed Precambrian basement of the Danish region, the two-layered type being found in EUGENO-S Profiles 3, 2, and 1, and the three-layered in BABEL Line A to the southeast, as shown in Figure 6-7. Similar two-layered and three-layered crustal provinces have been described previously by Beloussov and Pavlenkova (1984) from, respectively, Phanerozoic Europe and the East European platform, that is to either side of the Teisseyre Tornquist zone This characteristic has generally been taken to indicate that Precambrian and Phanerozoic crustal structures are somehow fundamentally different in origin.

In the light of findings of the EUGENO-S and BABEL projects, however, it now seems reasonable to assume that the point when a two-layered crust began to develop, instead of a three-layered one, occurred within the Late Precambrian after the Gothian orogeny and before or during the Sveco-Norwegian orogeny between 1.5 and 0.95 Ga ago. This difference in crustal layering is discussed further in Chapter 7.

A Sveco-Norwegian origin has also been suggested by Pedersen *et al.* (1990) for gently

SE to S-dipping crustal and subcrustal reflectors recorded by the MOBIL SEARCH survey in the Skagerrak and Kattegat seas south of Oslo fjord which they believe image former shear zones. Lie and Husebye (1991) relate the evolution of the Skagerrak graben to reactivation of these structures during the Late Carboniferous to Permian.

From shelf to foreland

Baltica's southern border had developed into a passive margin when Eastern Avalonia was approaching Baltica in the early Silurian. Eastern Avalonia's oblique collision caused continental rise and shelf slope sediments deposited earlier to be caught up in Caledonian deformation. These were metamorphosed and then thrust over the southern portion of Baltica as a marginal Caledonian thrust belt. Up to 3 km of intensely deformed Ordovician shales and greywackes have been penetrated in a deep well located at Rügen in NE Germany (Figure 6-8, well RÜ) between the Trans-European fault and the Caledonian deformation front. But about 50 km further north, just beyound the front, another well (G-14) penetrated a much reduced Vendian(?) Cambro-Ordovician sequence (with only 60 m of Ordovician) in typical shelf facies before it terminated in Precambrian cystalline basement (D. Franke *et al.* 1989). The 7 km deep Loissin well, situated directly above the deep Trans-European fault, encountered strongly deformed and slightly metamorphosed sediments, of supposed late Riphean–Vendian age, beneath an unconformable cover of Early Carboniferous and younger strata.

North of the Caledonian front, the platform facies Cambrian–Ordovician sediments, where silicic sandstones are overlain by black shales, have a very clear seismic signature, and commercial seismic surveys of the sedimentary cover over the southern part of Baltica demonstrate their wide occurrence in the Danish Scanian area. They form the lowermost part of a 3–6 km thick Cambrian–Silurian sequence preserved in downfaulted positions. A regional sub-Permian unconformity is developed over the downfaulted blocks.

Most of the Cambrian–Silurian sequence is made up of Late Silurian turbidites, which were deposited over the Danish Scanian area in a foreland basin, north of the north German–Polish Caledonides. The drilled upper 130–330 m of the Upper Silurian sequence contains basaltic 'intercalations' which amount to 30–50% of the total. It is still debated, however, whether some or all of these basaltic rocks may be of Late Carboniferous–Early Permian age (Abrahamsen and Madirazza 1986, EUGENO-S Working Group 1988).

Conclusive evidence for volcanism accompanying the formation of the Caledonian foreland basin is given by Bjerreskov and Jørgensen (1983) who reported on early Late Silurian volcanic ash turbidites off the south coast of Bornholm. The ash is assumed to stem from a K-dominated volcanic centre located on a palaeoslope about 200 km west of Bornholm, close to the Caledonian front.

During the Devonian, most of the Danish Scanian area belonged to the Old Red Continent and supplied detritus to the sea which, with the close of the Early Devonian, spread northwards over the collapsed north German–Polish Caledonides. In adjacent regions of Poland, there is evidence of Devonian faulting, and possibly this activity also affected the Danish area. The general lack of stratigraphic well control from the Devonian and Carboniferous to the north of the Ringkøbing–Fyn high, makes it difficult to distinguish any late Caledonian unconformity from the ever present sub-Permian unconformity.

We know that by the end of the Carboniferous, the original structure of the Caledonian foreland basin over Denmark had been destroyed by faulting, uplift, and erosion. It is therefore no longer possible to identify specific crustal structures associated with Late

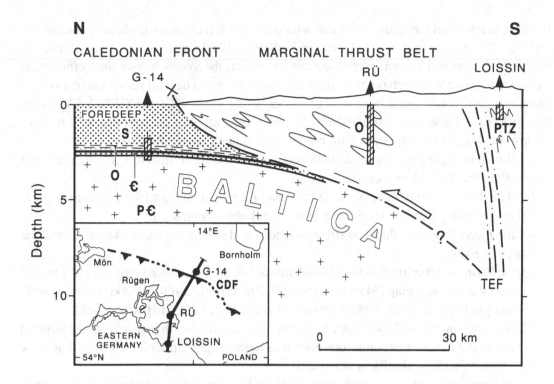

Figure 6-8. Tentative tectonic reconstruction of the Caledonian marginal thrust belt along a profile through Rügen, eastern Germany. Devonian and younger cover strata, and fault displacements displayed by these, have been removed, based on D. Franke (1991) and E. Hurtig (pers. comm.). Wells G-14, Rü and Loissin; TEF: Trans-European fault.

Silurian flexuring of the crystalline crust beneath the foreland basin. Assuming that the basin was deepest close to the Caledonian deformation front in the south, the depth to basement adjacent to the front could well have been 7–8 km, considering that some 100 km further to the north it was between 4 and 6 km.

The Tornquist fan: a Carboniferous–Permian splay

In Late Carboniferous to Early Permian time, a splay of major faults and fault zones, the so-called 'Tornquist Fan' developed over the Danish Scanian and western Baltic area. The fan-shaped splay started to open in NW Poland, branching off the Teisseyre Tornquist zone. It widened towards WNW over the southern, now concealed, part of Baltica, dividing its cover and basement into a number of fault blocks.

The north German part of the Trans-European fault formed the southernmost branch of this splay. During Early Permian time it defined the northern border of the rapidly sinking north German basin, shown in Figure 6-15. Since then, this branch lost influence, even though reactivation occurred at intervals, and flower structures were developed in the Mesozoic cover.

The formation of the Permo-Carboniferous fault splay and pull-apart basins over the Danish Scanian area can be regarded as a local expression of Late Variscan wrenching in Europe that was typical of much of the Variscan foreland (Vejbœk 1985, Liboriussen *et al*. 1987, Thybo and Schönharting 1991). Well developed flower structures at higher stratigraphic levels bear witness to repeated post-Permian reactivation (Cartwright 1987, 1990). None the less, only modest strike-slip displacements can be estimated. Including the effects of younger faulting, the net horizontal displacement within the entire fault splay probably amounted to less than 50 km of dextral offset. However, both dextral and sinistral movements did occur. Where the Sorgenfrei Tornquist zone in Scania intersects the Sveco-Norwegian deformation front, the latter appears to have been displaced 10–15 km in a dextral sense where it occurs again in a narrow basement horst next to the SE border of the STZ (EUGENO-S Working Group 1988). However, a sinistral displacement of 4 km can be demonstrated to have occurred along the northwestern fault border of the horst (Sivhed 1991).

As shown in Figure 6-6, the main fault zones of the Carboniferous to Permian Tornquist fan north of the Trans-European fault are:

(a) The Rømø fracture zone between the Trans-European fault and the Caledonian front,

(b) the Vinding fracture zone north of the Ringkøbing–Fyn high,

(c) the Fjerritslev fault that was strongly reactivated during the early Mesozoic, and the Børglum Fault,

(d) the Permo-Carboniferous fault branch through Kattegat, marked at the shield's border zone on the tectonic map (Map 1) of the EGT Atlas (EUGENO-S Working Group 1988, Cartwright 1990, Ro et al. 1990, Liboriussen et al. 1987, Lie and Husebye 1991).

The Rømø, Vinding and Fjerritslev faults continue into the North Sea, whilst the Børglum fault (?) and the Permo-Carboniferous branch in Kattegat join up with the Oslo–Skagerrak graben system, ending blindly in the shield.

It has been suggested previously that the Oslo–Skagerrak (alias Oslo–Bamble) graben system, which now appears to abut the Sorgenfrei Tornquist zone, was originally continuous with the Horn graben (Ziegler 1982). Alternatively, the Oslo–Skagerrak graben, together with the Horn graben, and N–S trending grabens in Jutland, may have formed a triple junction (Thybo and Schönharting 1991). In contrast, the grabens south of the Sorgenfrei Tornquist zone underwent major subsidence during the Triassic (Madirazza et al. 1990). So, too, did the Glückstadt graben south of the Trans-European fault in Schleswig Holstein.

Permo-Carboniferous magmatic activity resulted in locally marked changes in the crustal structure. Figure 6-9 shows the integrated interpretation of a well resolved seismic velocity model for EUGENO-S line 2 across the Silkeborg gravity high, suggesting that this +50 mgal anomaly is caused by a major intrusion of mafic rocks into the lower crust plus a subvolcanic horizon at 6–8 km depth. The mafic body in the lower crust beneath the anomaly apparently caused a 4 km rise of both the Conrad and the Moho discontinuties (Thybo et al. 1990, Thybo and Schönharting 1991). The fan shape of the Late Carboniferous to Permian fault splay, the common occurrence of listric normal faults, the intrusion of a dyke swarm in Scania, and the opening of pull-apart structures and aborted rifts all indicate net crustal extension in a NNE–SSW direction, increasing in amount to WNW. The resultant crustal thinning will be discussed in the next section.

Late Carboniferous to Permian tectonic and magmatic activity and associated crustal thinning fundamentally changed the tectonic character of the southern part of Baltica. It no longer acted as an integral part of the rigid Baltic Shield. As in the more southerly Variscan foreland, it progressively subsided and formed the basement to Permian, Mesozoic and Cenozoic successor basins. Thus, considering only the Permian and younger development of the Danish area and SW Scania, it would seem as if they formed part of Phanerozoic Europe. This explains the long held view that the northwestern part of the Tornquist zone (Sorgenfrei Tornquist zone) should define the crustal boundary between Precambrian and Phanerozoic Europe. However, this view became obsolete when Precambrian basement was drilled beneath the Ringkøbing-Fyn high (Noe-Nygaard 1963, Sorgenfrei and Buch 1964) and it can now be understood to be part of Baltica which has subsequently been modified in

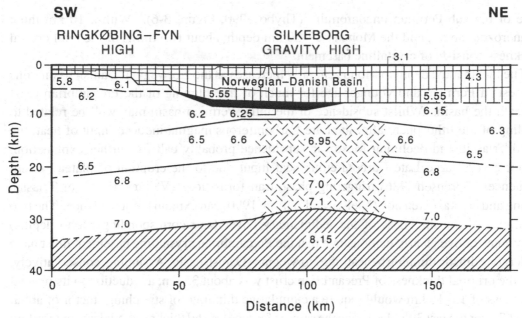

Figure 6-9. Integrated model cross section for the southern part of EUGENO-S line 2 (see Figure 6-7) across the Silkeborg gravity high, based on Thybo (1990). Lower lithosphere is shaded, the pre-Zechstein subvolcanics are black and the presumed deep intrusive body is marked with an open ornamentation. Numbers refer to P-wave velocities in kms[-1]*.*

character. As Brochwicz-Lewinski *et al*. (1981, 1984) pointed out, and which was later confirmed by others, the southwest border of Baltica's Precambrian basement must lie further to the south, presumably at the Trans-European fault (Figure 6-6).

Crustal thinning beneath the Norwegian–Danish basin

After the formation of the Permo-Carboniferous fault splay and the development of the North German basin, Early Mesozoic rifting, following the trend of the Fjerritselv fault (Figure 6-6), produced increased subsidence in the Norwegian–Danish basin resulting in the deposition of thick Triassic sequences (Knudsen *et al*. 1991). Sedimentation continued almost uninterrupted for the rest of Mesozoic time. In northern Jutland, a magnetic anomaly associated with the Fjerritslev fault suggests associated magmatic activity (Madirazza *et al*. 1990), perhaps coeval with the intrusion of Jurassic (and Early Cretaceous?) alkali-olivine dolerite plugs in Scania. Sedimentation in the Danish basin persisted into the Earliest Tertiary until a change in the regional tectonic regime occurred, caused by Atlantic opening and Cenozoic rifting in the North Sea region (Ziegler 1988, 1990).

In the central part of the Norwegian–Danish basin over Jutland, the post-Permian sequence now shows a maximum thickness of 9.5 km (Britze and Japsen 1991). Underlying Zechstein evaporites were originally about 1 km thick, but this was changed radically by the growth of salt pillows and ascent of diapirs during Triassic to Tertiary time, so that in between the piercement structures, the Zechstein was often completely removed by migration. Thus, disregarding the Zechstein thickness, but including the Rotliegend and Cambro-Silurian strata in underlying fault blocks, the total thickness of the sedimentary column amounts to 10–15 km. Since the Moho lies at a depth of 26–28 km, the Precambrian crystalline basement makes up only about half of the total thickness of the crust.

The Norwegian–Danish basin thins to about 4 km under northern Sealand, but up to 6 km of Cambro-Silurian sediments (with V_p between 5.2 and 5.7 kms[-1]) appears to occur at the base of the sub-Permian unconformity (Thybo 1990, Figure 3-6). With a 10 km thick Phanerozoic cover, and the Moho at 29–30 km depth, about two thirds of the total crustal thickness consists of crystalline basement.

The fact that the depth to Moho generally increases as the Norwegian–Danish basin thins suggests a Permian and younger age for much of the thinning of the Precambrian crust beneath the basin. Whilst subsidence of the intial Permian basin may well be related to cooling of the lithosphere after the Late Carboniferous magma-induced input of heat, the rapid Triassic and declining Mesozoic subsidence probably call for further explanation. Either a very large Late Carboniferous heat input due to the eruption of plateau basalt sequences (Sørensen 1986) triggering phase transformations (Vejbœk 1989), or Triassic rifting and crustal extension (Nielsen and Balling 1990) can explain the subsidence. The two examples of the thickness of the sedimentary cover (and corresponding Moho depths) mentioned above clearly demonstrate that the Precambrian crystalline basement must have undergone drastic thinning beneath the Norwegian–Danish basin. Assuming, conservatively, that the original thickness of Precambrian crust was about 35 km, a reduction to its present thickness of 13–16 km would require a cumulative thinning (or stretching) factor of about 2.2–2.7 over the last 395 Ma. Using the same reference crustal thickness, Nielsen and Balling (1990) calculated a thinning factor of 1.3–1.7 for the last 250 Ma, a significant part of which was attributed to Triassic rifting. The difference in these estimates, which refer to time spans of different duration, might suggest that thinning had already started over the Danish area during the formation of the Caledonian foredeep.

It must be borne in mind, however, that these thinning factors are approximate and may have been underestimated. For reasons explained below, we have neglected the effects of Late Cretaceous to Early Tertiary inversion tectonics.

Alpine-induced basin inversion: rise of Tornquist

During Late Cretaceous and Early Tertiary phases of inversion, the Sorgenfrei Tornquist zone acquired its principal near-surface inversion structures with a more or less uplifted central zone, limited by reverse faults. The intensity of the inversion inceases from Jutland towards Scania and Bornholm. In Scania and along BABEL Line A (Figures 3-7 and 6-10), vertical uplift amounted to at least 2–3 km. Inversion of the nearby Rönne graben caused reversal of its eastern border fault, and open to overturned folds of kilometre scale were developed in the Mesozoic sequence on and offshore western Bornholm (Hamann 1989).

Inversion of the Polish trough was even stronger. The Pomeranian–Kujawic Wall rose from the deepest part of the trough and Late Cretaceous–Early Tertiary basins developed on both flanks. In NW Poland these inversion structures were formed directly above their presumed deep-seated counterpart, the anomalous lower crustal structure and Moho trough of the Teisseyre Tornquist zone. In southern Poland, however, the inversion axis was offset from that of the anomalous deep crustal structure and uplift occurred to the west, creating the Holy Cross Mountain uplift. Late Cretaceous to Early Tertiary inversion also affected nearby parts of the Trans-European fault in eastern Germany. There the basement was uplifted 1–2 km along the so-called Grimmen Achse/Axis (Figure 6-6 and Figure 4 in D. Franke et al. 1989).

In northern Jutland, where the EUGENO-S Profile 3 traverses the Sorgenfrei Tornquist zone, the Late Cretaceous to Early Tertiary inversion amounted to only about 1 km uplift (Japsen, 1992). The fact that a gentle Moho uplift was modelled along the EUGENO-S profile here could be taken as support for the hypothesis that a rise rather than trough in the

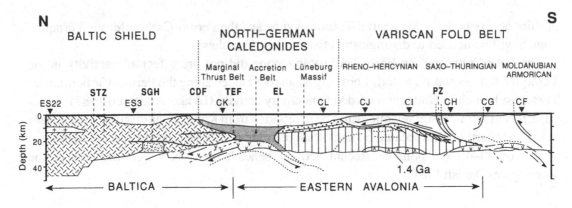

Figure 6-10. From Precambrian to Variscan Europe: Tectonically interpreted crustal struc-
ture along the EGT. Shotpoints are included to facilitate comparison with Figures 3-6 and 3-9.
STZ: Sorgenfrei Tornquist zone, SGH: Silkeborg gravity high, RHF: Ringkøbing–Fyn high, CDF:
Caledonian front, TEF: Trans-European fault, EL: Elbe line, PZ: northern phyllite zone. Epi-
Cadomian cover is stippled. Underlying Cadomian and older crystallines of the Lüneburg massif
have open vertical ruling.

Moho and a thick crust–mantle transition layer has been developed because the Late
Cretaceous to Early Tertiary transpression did not fully counteract earlier crustal thinning
caused by the Permian and Mesozoic rifting and basin development (Ziegler 1982, 1988,
1990, BABEL Working Group 1991).

This model is attractive as it concurs neatly with stratigraphic evidence, and if it is assumed
that the depressed lower crust of the anomalous deep structure became subjected to eclogite
facies metamorphism, it also explains the unusually high P-wave velocities of the crust–
mantle transition layer within the Moho trough, through the mechanisms proposed by
Mengel in Chapter 4.3. But there could be complications. A 10 km reflection seismic profile
coincident with a part of EUGENO-S line 3 across the Sorgenfrei Tornquist zone (Balling
1990b) has revealed clear sub-Moho reflections beneath it, raising the question whether they
might represent the base of a thickened crust-mantle transition layer. If so, could the
EUGENO-S refraction profile have failed to record a deep transition layer and a Moho trough
beneath the North Jutlandic Sorgenfrei Tornquist zone? Hardly. Or could an anomalous
lower crustal structure have been developed beneath one of the other fault zone branches of
the Tornquist fan? This last question is more difficult to ignore. The Bornholm–Scania part
of the Sorgenfrei Tornquist zone is characterised by a positive Bouguer anomaly, which is
most probably caused by high density material in the lower crust (Balling 1990b). This
positive anomaly does not continue into the WNW-trending part of the Sorgenfrei Tornquist
zone across Jutland; but instead follows the NNW-trending Permo-Carboniferous Tornquist
branch through Kattegat. Recalling that Permo-Carboniferous dykes are found on the
Swedish Kattegat coast, the Bouguer anomaly trend could suggest either that the supposed
deep-seated high density body was created by Permo-Carboniferous magmatic activity, or
that Alpine-induced inversion structures were developed with a considerable lateral offset.
Provided that during the Alpine-induced transpression, low-angle shear zones were operative
at mid-crustal level as proposed in Figure 3-7b, laterally offset near-surface inversion
structures and lower crustal subversion structures could well have been formed more or less
simultaneously (Blundell and the BABEL Working Group 1992). This seems to be the case
in southern Poland where the uplifted Holy Cross Mountains are offset from the deep Moho
trough of the Teisseyre Tornquist zone. Further acquisition of deep reflection seismic

profiles across both the Sorgenfrei Tornquist zone and the Permo-Carboniferous Tornquist branch will be needed to distinguish between these hypotheses.

The Late Cretaceous to Early Tertiary inversion did not bring tectonic activity in the Tornquist fan to a complete end. Local sag basins developed along the Permo-Carboniferous Tornquist branch in Kattegat recorded Quaternary activity (Lykke-Andersen 1987). Likewise, recent intra-crustal seismicity has been reported from Kattegat, as discussed in Chapter 5.3. Other earthquakes located on and offshore around the west coast of Jutland with focal depths of 30–40 km point to Recent activity beneath the WNW-trending axis of the Norwegian–Danish Basin.

6.2.2 THE CALEDONIAN OROGEN CROSSED BY THE EGT

The marginal thrust belt of the north German Caledonides

South of the Danish area of concealed Precambrian basement, the European Geotraverse intersects the British–North German–Polish Caledonides of East Avalonian origin, crossing from north to south; the marginal thrust belt, a buried accretionary belt, and the Cadomian (and older) Lüneburg massif, shown in Figure 6-10. The Caledonian deformation front at the northern border of the marginal thrust belt is nowhere exposed at surface but can be traced in reflection seismic profiles both west and east of EGT. However, a seismic profile along strike over the Horn graben shows clear SSW-dipping reflectorsthat are interpreted to outline thrust structures at the front. BABEL Line A, offshore Møn, reveals similar S-dipping structures. Together with the constraints supplied by deep wells in southern Denmark and northern Germany, the location of the deformation front can be defined with a fair degree of accuracy, as indicated in Figures 6-6 and 6-8.

Where drilled to the west of EGT, the Caledonian marginal thrust belt consists of low grade schists, phyllites, and slates, whereas fossiliferous Ordovician shales and greywackes have been encountered in deep wells in NE Germany and NW Poland (Frost *et al.* 1981, Ziegler 1982, D. Franke *et al.* 1989, Dadlez 1990). The depth to the top of the Caledonian-deformed basement varies from 1.5 to 7 km. Where EUGENO-S line 1 crosses the marginal thrust belt in Schleswig Holstein, the position of the low grade metamorphics of the marginal Caledonian belt is fairly well expressed in the velocity model, outlined by P-wave velocities between 5.5 and 6.1 kms^{-1} shown between shotpoints CK and CL in Figure 3-6 (Thybo 1990).

A buried Caledonian accretionary belt?

We suggest that the deep crystalline basement beneath the north German lowland south of the Trans-European fault and north of the Elbe line is made up of a Caledonian accretionary belt of truly East Avalonian origin as shown in Figure 6-10.

It has been suggested repeatedly in earlier literature that Precambrian massifs occur within this belt such as the so-called Elbe and Pritzwalk massifs or that the entire belt has a Precambrian basement (the so-called north German massif). We prefer to believe, however, that the available geophysical data and regional geological features are better explained by assuming that the belt is composed of Caledonian deformed and metamorphosed post-Cadomian sediments, volcanics, and ophiolitic melange with remnants of oceanic crust possibly being preserved in the lowermost crustal layer having $V_p = 6.9$ kms^{-1} (Figure 3-6).

This primary crustal structure may have been modified by post-Caledonian events, not least by locally intense Late Carboniferous to Early Permian magmatic activity (Trappe 1989).

On the Bouguer anomaly map (EGT Atlas Map 9), the belt exhibits two substantial gravity highs (about 75 km in diameter) caused by a local rise in the Moho or by high density bodies in the lower crust. At the southern boundary of the belt along the Elbe line, the average crustal density and the magnetization of the crust both decrease along with a marked change in the P-wave velocity above the Moho (from 6.9 to 6.4 kms^{-1}).

Coincident with these changes, a deep-lying conductivity anomaly (Figure 4-6) shows up to the south of the Elbe line, rising gently southwards, from 15 km to 5 km depth at the Variscan front. This anomaly is stratigraphically controlled and has been explained by the occurrence of a fairly thin layer of altered, supposedly Upper Cambrian, black shales within the slightly deformed epi-Cadomian cover of the Lüneburg massif. According to Jödicke (1990), a primary bitumen content in the black shales, and subsequent deep burial and strong coalification with formation of graphite, explain the observed very high conductivity of this layer. We believe that the absence of a similar conductive layer to the north of the Elbe line is due to different lithologies and stronger Caledonian deformation and metamorphism of the Caledonian sequences there. Caledonian deformed and metamorphosed rocks have been drilled in the North Sea and in southern Poland but, in between, there are no deep well data other than the Loissin well directly above the Trans-European fault. The 10 km depth to top basement has been prohibitive. But Nature has itself sampled the concealed basement: Rheno-Hercynian Middle Devonian clastic deposits of northern provenance contain spinel grains of ophiolitic origin (Press 1986), conceivably derived from the accretionary belt of the North German Caledonides that had just recently collapsed.

The Lüneburg massif, a Cadomian relic in Avalonia

Like the Midlands massif in southern England, the Lüneburg massif apears to be of Cadomian and older origin. It is probably bordered to the west in the North Sea by the Loke Shear, and the Elbe line forms its northern border. It becomes deeply buried to ESE under the Variscan thrust belt that is arcuate around the Bohemian indenter, but it may reappear in the basement of the East Silesian massif of SW Poland. Thus the southern part of the Lüneburg massif is hidden beneath overthrust Variscan units. Northward tectonic transport of the Rheno-Hercynian sequence over the Lüneburg massif was probably facilitated by the presence of the graphite-bearing (conductive) shale layer in the epi-Cadomian cover of the massif. The DEKORP 2-N deep seismic reflection profile, close to the EGT (Figures 3-10 and 6-20), suggests that the massif underlies all of the Rheno-Hercynian zone, and that at lower crustal levels its wedge shaped, but imbricated, southernmost part just comes into contact with the Saxo-Thuringian thrust unit as shown in Figure 6-10. This means that the crustal xenoliths sampled by the Tertiary basalts of the North Hessian Depression (Mengel, Chapter 4.3) could actually stem from the basal part of the Lüneburg massif. If so, the lower crust of the Gondwana-derived Lüneburg massif could be of pre-Cadomian origin with an island arc stage in which granulite facies was reached about 1.4 Ga ago, long before the main Cadomian events and the later Variscan evolution. The only Cadomian age obtained so far for the crust of the Lüneburg massif is a zircon U–Pb lower intercept age of around 600 Ma for the Egger gneiss, which was brought up from lower crustal levels as a raft during the intrusion of the 295 Ma old Harzburger gabbro in the Hartz Rheno-Hercynian (H in Figure 2-4, Grauert et al. 1987).

Caledonian lithospheric structures

In Chapter 2 we raised the question as to whether Avalonia docked as an 'orogenic float' or as a wholesale lithospheric terrane. Relying on the observations presented and interpretations made in the intervening chapters, we now feel convinced that priority should be given to the latter possibility. We see that the uniform seismic Moho depth is more likely to be a result of a scavenging eclogite facies front than detachment at the base of a 'float'. This view (which we also consider valid for the Variscan part of EGT) is in agreement with the marked lateral hetrogeneity revealed by the P-wave velocities at sub-Moho levels (Figures 3-9 and 3-19).

As outlined in Figure 6-10, docking of Avalonia alongside Laurussia's southern margin probably caused oblique subduction both north and south of the last remains of the Tornquist ocean. The lower crust under the Caledonian accretion belt with Vp of 6.9 kms^{-1} is assumed be made up of oceanic crust from the Tornquist Sea and other remains may be found at the base of the flake structure where Baltica protudes southwards at midcrustal levels. In the upper part of the flake structure, the marginal Caledonian belt was thrust over Baltica and, due to tectonic loading and flexure, a foreland basin was developed over nearby parts of Baltica. South of the Caledonian accretion belt, at its boundary with the Lüneburg massif, another, oppositely directed, flake structure was probably developed (cf. Figure 1 in D. Franke *et al.* 1989). The epi-Cadomian cover of the Lüneburg massif was hardly influenced by Caledonian deformation. With the Early Devonian collapse of the north German Caledonides, a stable passive margin began to build up over Avalonia's southern margin, but in different ways on either side of the Loke Shear. Early Devonian strike-slip displacement along the Loke Shear (now situated beneath the Lower Rhine graben), probably explain why the English–Belgian branch of the Caledonian fold belt (west of Loke) now ends blindly against the southwestern part of the Lüneburg massif (east of Loke). Strike-slip displacements might also explain why the Eifel and Lower Hesse volcanoes, although placed in line along the Variscan strike, have sampled different xenolith associations from deep crustal levels. During the ensuing Variscan evolution, further terranes derived from Gondwana became welded on to Avalonia. In Chapter 2 a fanciful tectonic model for the pre-collisional history of these Variscan terranes was sketched. In the following section, a less putative and more authoritative model will be presented.

6.3 PHANEROZOIC STRUCTURES AND EVENTS IN CENTRAL EUROPE

W. Franke

In order to make the complex development of Variscan events more understandable, this section is presented in a narrative style. It begins with an account of the geodynamic evolution of central Europe, as seen by the author. Then the main geological structures in the pre-Mesozoic basement are outlined to emphasise features correlatable with geophysical observations. More detailed discussions of the geotectonic development during the Palaeozoic are available in Franke (1989a), Neugebauer (1989) and Matte (1986), and in thematic volumes edited by Dallmeyer *et al.* (1992), Martin and Eder (1983), Matte (1990), Matte and Zwart (1989), and Meissner and Gebauer (1989). The works edited by Ziegler (1987c, 1988, 1990) and Fuchs *et al.* (1983) provide access to the Mesozoic and Cenozoic development.

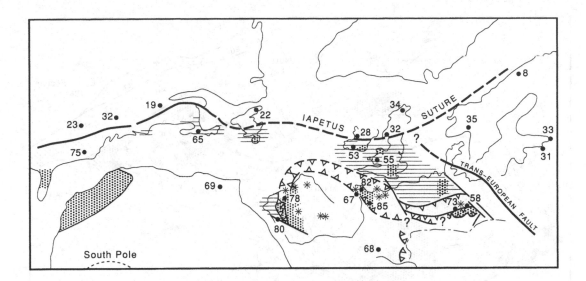

Figure 6-11. Plate-tectonic affinities of Variscan Europe and neighbouring regions. Stipple: Cadomian basement at outcrop or in the subsurface. Horizontal ruling: parts of the Avalonia microplate. Numbers refer to Ordovician palaeomagnetic inclinations in degrees, after literature cited in Figure 2-2 and Kent and Van der Voo (1990), Khattach (1989) and Soffel et al. (1992). Asterisks: diamictites with dropstones relating to the Saharan glaciation (Ordovician/Silurian boundary).

6.3.1 GEODYNAMIC EVOLUTION

Proterozoic and Early Palaeozoic: Setting the Stage

Palaeozoic development of Europe was controlled by the mutual approach and eventual assemblage of three major plates: Laurentia, Baltica and Gondwana. As discussed in Chapter 2, the collision of Laurentia and Baltica led to the formation of a new continental block (Laurussia) and brought about the Scottish–Norwegian Caledonide orogeny, which was terminated in the Silurian. Likewise in the Silurian, the northernmost part of what is now continental Europe (a microcontinent mostly referred to as 'Avalonia') was accreted to the newly assembled Laurussian block, thereby closing 'Tornquist's Sea' and producing the 'north German–Polish Caledonides', which are now largely covered by late Palaeozoic and Mesozoic sediments (Figure 6-11, see Chapter 2.4.7 and Section 6.2.2). This docking event failed to produce a major orogenic belt, and had no detectable influence on the European crust further south.

The crustal structure of central and southern Europe (including the Mediterranean) is largely controlled by the Devonian and Carboniferous closure of basins which had formed, during Cambrian through to Devonian time, within the northern part of Gondwana. Figure 6-11 illustrates the major plates involved, and the arguments pertaining to the kinship of tectonic units in Europe. Everywhere in Europe there are outcrops of crust with a late Proterozoic (ie. Cadomian or Pan-African) imprint, which is completely lacking to the west of the Iapetus suture and to the north of the Trans-European and Tornquist fault zones. 'African' affinities are also reflected in the U–Pb ages of detrital zircons (Gebauer *et al.* 1989, Hoegen *et al.* 1990).

Close relations with Gondwana are also documented, especially for the Armorican and Bohemian massifs and areas further south, by high palaeo-inclinations and by traces of the

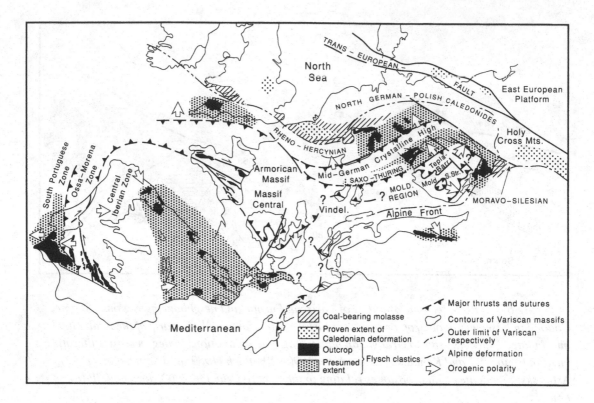

Figure 6-12. Structural subdivision of the European Variscides (after Franke 1989a).

'Saharan' glaciation, which is responsible for the faunal break defining the Ordovician/ Silurian boundary (Figures 2-2 and 6-11, Robardet and Doré 1988). It is apparent that Europe, as far north as Bavaria, was situated so close to the south pole in West Africa that it received striated stones – today, glaciogenic sediments are not found beyond 40° latitudinal distance from the pole (Hambrey and Harland 1982). Lastly, the Ordovician of Armorica– Bohemia and areas further south is characterised by the Armorican quartzite and trilobites of the *Neseuretus* province (Sadler 1974). As explained below, most of the basins within Variscan Europe attained their widest extent during the Ordovician. The relationships just described may therefore be taken to rule out the existence of wide oceans, at least in the central and southern parts of Variscan Europe.

Cambrian–Ordovician rifting: break-up of Gondwana

During the Cambrian and Ordovician, the northern part of Gondwana underwent a major phase of crustal extension, which is documented in widespread sedimentary and volcanic sequences. Rifting produced, in some areas, pronounced crustal thinning that partly evolved to the stage of narrow oceans. This applies, for example, to the 'Massif Central ocean' of Matte (1986), the Variscan basement of the Alps (see Chapter 6.4), the Saxo-Thuringian belt of central Germany, and probably also to parts of the Moldanubian region (see Section 6.3.2, for possible correlations with France see Figure 6-12 and Matte *et al.* 1990). These basins were formed within the northern part of Gondwana, which was moving northwards during that time, thereby encroaching upon the northern continents (Laurentia plus Baltica). This scenario may be taken to suggest that the Cambro-Ordovician basins were formed by back-arc spreading (Cogné and Wright 1980).

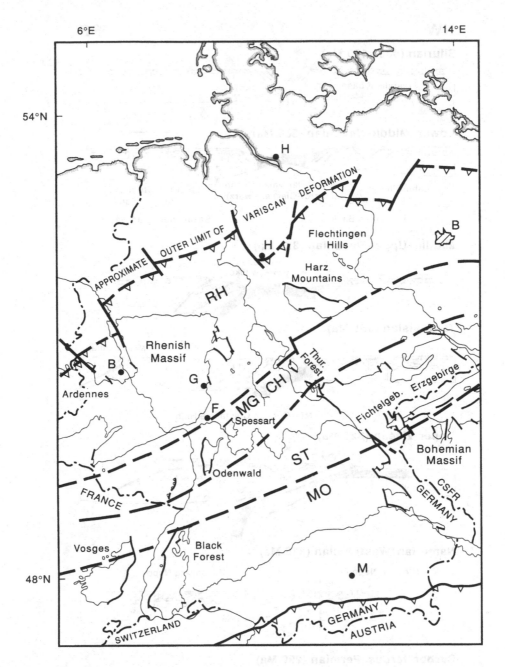

Figure 6-13. Structural boundaries in the Variscan basement and basement outcrops (shaded) in the EGT Central Segment. Outer limit of Variscan deformation is marked by the saw-toothed line; subsurface positioning after D.Franke (1991), D. Franke et al. (1989). RH: Rheno-Hercynian belt, MGCH: Mid-German crystalline high, ST: Saxo-Thuringian belt, MO: Moldanubian region. B: Bonn, Berlin; F: Frankfurt, G: Giessen; H: Hannover, Hamburg; M: Munich.

Devonian Rifting

Thick Cambro-Ordovician sequences are detectable in the Rheno-Hercynian Belt west of the River Rhine, but appear to be missing from the eastern part of the Rhenish massif. Anyhow, the most important extension in both areas occurred during the Devonian, after SE England and the northern parts of continental Europe (Avalonia) had collided with Laurussia. Detritus was supplied to the Rheno-Hercynian basin from this northern block (the 'Old Red Continent

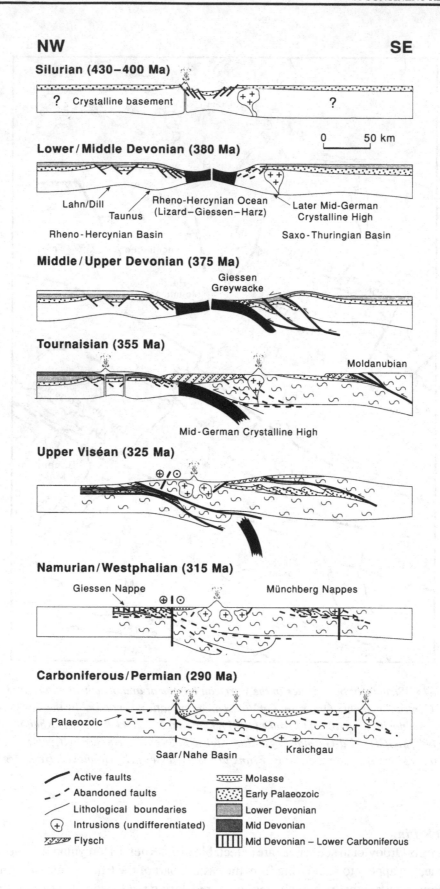

Figure 6-14. Cartoon displaying the opening and closure of the Rheno-Hercynian basins and the closure of the Saxo-Thuringian basin between 430 and 290 Ma ago, from Franke and Oncken (1990).

of the older literature). This is also reflected in the record of detrital zircons (Haverkamp *et al.* 1991). Devonian extension is documented in the onlap of Devonian shelf sediments of locally more than 5 km thickness, giving way up-section to hemi-pelagic deposits, and bimodal volcanism. Rifting probably commenced during the Silurian (Figure 6-14) and attained at least the narrow ocean stage. This is documented in remnants of Devonian oceanic crust contained in nappes in the Rhenish massif and along-strike in SW England (Lizard area, Floyd 1984). In both areas, volcanic activity persisted into the Early Carboniferous. A short-lived extensional episode is also documented in the early Upper Devonian of the Saxo-Thuringian Belt (Franke 1984b, 1989a,b, c).

Devonian to Early Carboniferous extension requires some special explanation. As explained below, closure of the Rheno-Hercynian sea had already been accomplished before the Tournaisian (Early Lower Carboniferous). Plate convergence in the active, southern margin of the basin and extension in the basin itself must have occurred, therefore, at the same time in closely adjacent regions. To reconcile these contrasting processes, one might envisage that the Rheno-Hercynian basement, during its northward drift, overrode the spreading axis of Tornquist's Sea, and underwent extension in a setting comparable with the Gulf of California at the present time (Matthews 1978).

The sedimentary record of the Rheno-Hercynian Belt bears no indication of a pre-Devonian passive margin situation. It is possible, therefore, that the opening of the Rheno-Hercynian basin effected the separation between the Avalonian and Armorican microplates. It should be noted, however, that this is not in accord with the palaeomagnetic data presented by Torsvik and Trench (1991), which suggest an earlier separation. Devonian extension in areas further south (along the northern margin of Gondwana) may be deduced from palaeomagnetic data but still requires further substantiation.

Silurian to Early Carboniferous subduction

The earliest indications of Variscan plate convergence have been detected in eclogites in the northwestern part of the Bohemian Massif, dated at approximately 430 Ma (Ordovician/Silurian boundary; Quadt and Gebauer 1988). Metamorphic ages around 380 Ma (Early to Middle Devonian) occur locally in the mid-German crystalline high at the southern margin of the Rheno-Hercynian basin and are very widespread in the Bohemian massif (Hansen *et al.* 1989, Kreuzer *et al.* 1989, Nasir *et al.* 1991). Synorogenic clastic sediments (flysch) are detectable from the late Middle Devonian onwards. Subduction of oceanic crust in all Variscan basins was probably accomplished during the Devonian, so that the tectonic development, at least from the basal Carboniferous onwards, can be taken to represent the collisional stage.

Tectonic polarity during the collisional development is indicated by the directions of tectonic and sedimentary transport (Figures.6-12 and 6-14), as well as by the migration of foreland basins, tectonic and metamorphic activities. These features indicate that the Rheno-Hercynian and the Saxo-Thuringian crust was subducted grossly to the SE, whereas the southern part of the Moldanubian zone shows the opposite transport direction. Such a bilateral array of largely coeval tectonic activities has also been described from France, Iberia, and southeastern USA/Mauretania (Matte 1986), and is taken to indicate bilateral underthrusting under a median microcontinent (Armorican massif and central part of Bohemian Massif, see Section 6.3.2).

Tectono-metamorphic activities at the Rheno-Hercynian, Saxo-Thuringian and southern Moldanubian fronts occurred simultaneously throughout much of the Devonian and early

Carboniferous, but can be shown to have terminated earlier in the more internal parts of the Variscan Belt. In the externides on both sides of the belt, in the Rheno-Hercynian and Mediterranean forelands, deformation did not cease before the Westphalian (Engel and Franke 1983, Engel 1984). The pre-collisional shortening is poorly constrained. For the Carboniferous collisional stage, conservative estimates based upon balanced cross sections amount to a minimum of several 100 km in a N-S direction in central Europe (Behrmann *et al.* 1991).

Important strike-slip displacements prior to and in the early stages of collision probably took place but are difficult to assess because of the intense later overprint. The main sources of evidence are palaeogeographic 'misfits', such as within the Rheno-Hercynian of SW England where the Devonian to the north of the Culm synclinorium is developed in Old Red Sandstone facies, while the time-equivalent pelagic sequences further south are entirely devoid of sandstones. Similar problems arise at the southern margin of the Rhenish massif west of the River Rhine, where Late Devonian flysch greywackes (Meyer 1970) contrast with time-equivalent limestones in the notional source area further south (Krebs 1976). Indications of early strike-slip movements are also present at the SE margin of the mid-German crystalline high and at the SE margin of the Saxo-Thuringian belt (see Section 6.3.2).

The early stages of deformation coincide with high- to medium-pressure metamorphism. In late Lower Carboniferous time, low-pressure metamorphism took over. This is associated with and followed by the intrusion of large amounts of granite, most of which is post-tectonic with respect to the ductile deformation.

There are but few plutons whose chemical composition can be related with subduction (see Section 6.3.2). This might be due to the narrow extent inferred for the Palaeozoic basins; alternatively, it is possible that those segments of the plate-boundaries observed now in central Europe originated as transcurrent faults rather than subduction zones.

Permo-Carboniferous: late- and post-orogenic magmatism, strike-slip and minor extension

The high heat flow observed during the late stage of the Variscan orogeny continued throughout the Late Carboniferous and Permian. It brought about the extrusion of large masses of bimodal volcanic rocks, derived from hot mantle and heated crust beneath the Variscan edifice (Lorenz and Nicholls 1984). With respect to the slightly older granitoids, there appears to be a trend with time toward deeper crustal sources, relatively dry melts (e.g. Emmermann 1977, Flick 1987), and an increased mantle component in the Permian. There is, however, no marked break in the magmatic development. Within the area of Variscan deformation, the Permian volcanics occur in largely the same internal portions of the orogen as the earlier granites. Hence, both the late- and post-collisional magmatism may relate to one and the same heat-source. Since it is difficult to supply the heat required in a thickening crust, and Permian crustal extension was only moderate, we might envisage delamination of lithospheric mantle and heat advection from the asthenosphere (e.g. Sandiford and Powell 1991), as will be discussed in Chapter 7.

In any case the Permo-Carboniferous heating probably had a paramount influence on the thickness and composition of the crust, and especially so of the lower crust. Magmatic acitivity most probably brought about underplating and intrusion of basaltic melts at the crust–mantle boundary and in the lower crust, as exemplified by a Permian mafic body in the lower crust of the Ivrea zone in the Southern Alps (Voshage *et al.* 1990). This process may also be held responsible for the 'layered lower crust' (see Chapter 3.2) encountered in many

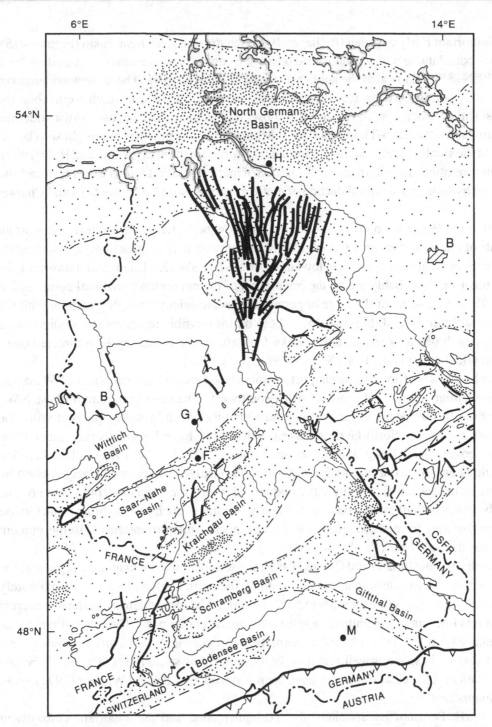

Figure 6-15. Sedimentary basins in Early Permian (Rotliegend) time. Volcanic deposits outside the basins are not represented. The synsedimentary graben structures in northern Germany are only known from oil industry wells. Dense stipple: depocentres. Compiled after Bachmann et al. (1987), SW Germany; Gast (1988), graben structures; Brink (pers.comm.), northern Germany; D. Franke (pers. comm.), NE Germany; Emmert (1981), SE Germany. Completed after Brinkmann (1986). City initials as in Figure 6-13.

seismic sections of Europe, where it clearly post-dates the Variscan convergence and probably pre-dates the Tertiary extension and volcanism (Fuchs *et al.* 1987, Wenzel *et al.* 1991). The DEKORP-1 reflection seismic profile has detected re-equilibration of the Moho under the Permo-Carboniferous Saar/Nahe basin (DEKORP Research Group 1991).

An important Early Permian (Rotliegend) depocentre, the Kraichgau basin (Figure 6-15), is clearly correlatable with a pronounced positive Bouguer gravity anomaly (extending from about 49°N, 8°30'E towards the northeast, see EGT Atlas Map 9). The large wavelength of the anomaly points to a deep-seated mafic intrusion in the lower crust, which is corroborated by the seismic velocity profile (Figure 3-9 shows high velocities in the lower crust under the southern part of the Saxo-Thuringian belt). Minor intrusions into the roof might well be the source of the smaller-scale magnetic anomalies observed in the area (Wonik and Hahn 1989). This interpretation is an alternative to the notion of mafic granulites which were produced and emplaced by southeastward subduction of mafic materials (see also Figure 6-18 and Chapter 4.3).

Most of the Permian volcanism occurred in narrow belts, which also received important amounts of clastic sediment (Figure 6-15). Many of these depocentres trend approximately N–S (grabens in the north German lowlands) or NNE–SSW (Zechstein and Triassic of the Hesse Basin) and probably record the orientation of the maximum horizontal compressive stress. However, some of the more important basins are oriented roughly parallel with the Variscan structural trend (Figure 6-15). These basins possibly represent strike-slip related features developed at Variscan along-strike fault zones reactivated by N–S directed compression (Bachmann *et al.* 1987, Schäfer 1989, Stets 1990).

Other basins were formed along important transverse fault zones, such as at the SW margin of the Bohemian massif and in areas further southwest. The most important of these NW–SE trending basins occur in the southwestern vicinity of, and roughly parallel with, the Tornquist zone (*e.g.* North German basin), where they have been detected, under thick younger cover, by drilling and geophysical surveys (D. Franke 1991). The Permian depocentres are, for example, well correlated with positive gravity anomalies revealed by stripping techniques, which imply mafic bodies underlying the basins at depth. These basins possibly represent pull-apart structures formed by dextral wrenching (Bachmann and Grosse 1989), or else were formed above a low-angle detachment fault in a transtensional environment with some dextral shear (Brink *et al.* 1990).

According to Bachmann and Grosse (1989), the sedimentary and magmatic regime of the North German basin is terminated in the south by the north German line, an approximately E–W trending gravity gradient at about 52°N latitude. The higher density of the lower crust north of this line can be attributed to a substantial mafic component. The orientations of the sedimentary basins and especially the magmatic bodies of the Permo-Carboniferous record a high diversity, in space as well as in time, of the controlling stress fields. This might be due to important en-bloc rotations which have been deduced, by Edel and Wickert (1991), from palaeomagnetic data.

The Early Permian basins along the Tornquist zone and an important embayment branching off toward the SSW (Hesse Basin) also guided the influx of the hypersaline Late Permian (Zechstein) Sea, which left behind important evaporite deposits. Their later displacement in the form of numerous diapirs has created the majority of the hydrocarbon traps in the Mesozoic beds of northern Germany. Transgression of the Zechstein sea marked the end of the Variscan and post-Variscan era and the onset of the Mesozoic development of the central segment of EGT.

Triassic to Early Cretaceous: steps towards modern Europe

Some of the important Permian depocentres were maintained during the Triassic; the North German basin to the SW of the Ringkøbing–Fyn High, and the SSW–NNE trending embayment into Hesse (*Hessische Senke*).

A new tectonic and palaeogeographic regime began in Jurassic time. In the Middle Jurassic (Dogger), an E–W trending land area emerged in central Germany, separating, from this time onward, the embyonic North Sea from a south German basin, a shallow, epicratonic sea, which was at times connected with the Alpine Tethys. To the north of this land bridge, an E–W trending depocentre (Lower Saxony basin) was formed during the Late Jurassic through to the Early Cretaceous, probably by dextral wrenching, connected with the opening of the North Atlantic (Betz *et al.* 1987, Ziegler 1988). Some deep-seated (?)Early Cretaceous intrusions at the southern margin of the basin, which are only known from their thermal effects, probably relate to the same tectonic regime.

Late Cretaceous to Recent: Alpine repercussions

The Late Cretaceous was a period of basin inversion everywhere in Germany, with the exception of the North German basin. Certain segments were uplifted along NW–SE trending fault zones (Figure 6-16) which probably date back to the Variscan deformation. Upthrusting occurred alternately toward the SW or to the NE, partly with a component of dextral strike-slip. The main areas of uplift are the Harz Mountains, the *Teutoburger Wald*, the SW margin of the Bohemian massif, and that part of the Bohemian massif (*Lausitz*) NE of the Elbe lineament (Zulauf 1990a,b, Drozdzewski 1988, Ziegler 1987b). The uplifted areas provided detritus for co-related local basins. These activities are attributed to the impact of the Alpine collision between Europe and the Adriatic plate (Chapter 6.4).

The Alpine collision is likewise held responsible for the Cenozoic development of the EGT central segment. NNE–SSW trending, Early Tertiary graben structures (Upper Rhine graben and minor structures within the Hesse basin, Figure 6-17) probably record the axis of main horizontal compressive stress in the Alpine foreland (Sengör *et al.* 1978, Dewey and Windley 1988). A later, anticlockwise rotation of the stress field brought about NW–SE trending grabens in the Middle and Lower Rhine areas, while the Upper Rhine graben was transformed into a sinistral wrench-zone (e.g. Ahorner *et al.* 1983). Alternatively, the Cenozoic graben formation might signal a large-scale reorganisation of plate boundaries (Ziegler 1990). The stress field of the Alpine collision is also documented in roughly N–S trending horizontal stylolites in the Mesozoic carbonates (e.g. Beiersdorf 1969). The Late Cenozoic structural pattern also accounts for most of the seismic activity observed in the central Segment (Chapter 5.3). It is important to point out that all outcrops of Variscan basement in the central European 'Mittelgebirge' are not erosional remnants of a Variscan mountain range but owe their present elevation to Alpine processes such as compression-related inversion, uplift of graben shoulders, or plateau uplift over an asthenospheric high (Fuchs *et al.* 1983).

The Alpine orogeny is somehow responsible for the Late Cretaceous to Recent volcanism in central Europe. Geochronological investigations have revealed that the earliest events date back to approximately 110 Ma (Lippolt 1983). Volcanic activity attained its widest extent during the Oligocene to Miocene and the youngest eruptions are only 11000 years old. Frequent emanations of CO_2 and helium isotopy in source waters (Oxburgh and O'Nions 1987) are proof that volcanism must not be regarded as extinct. The geochemistry of the Cenozoic volcanic rocks is typical of intra-plate settings (e.g. Wedepohl 1987, Fuchs *et al.*

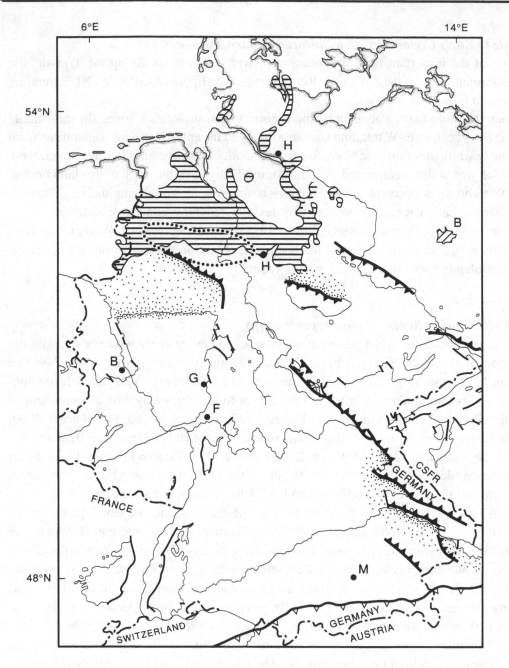

Figure 6-16. Cretaceous reverse faults (saw-toothed lines) and associated clastic sediments (stipple); Lower Saxony basin (horizontal ruling); contact aureole of the Bramsche pluton (dotted contour). City initials as in Figure 6-13.

1983). Xenoliths from the basalts have provided useful information for the interpretation of geophysical profiles (Chapter 4.3, Figure 6-18).

There is no systematic shift with time in the position of the volcanic centres. Cenozoic volcanism in central Europe forms a belt approximately 100 km wide whose axis parallels the Alpine front at a distance of about 300 km (Figure 6-17). Similar relationships exist in France, where Cenozoic grabens and volcanic chains occur in a N–S trending belt roughly parallel with the Western Alps. In Germany, there is no systematic relationship between volcanism and graben formation: outside the E–W trending volcanic belt the grabens are almost devoid of volcanic rocks. For their ascent, the volcanic materials made use of all kinds of pre-existing fractures so that the orientation of dykes and volcanic chains is highly

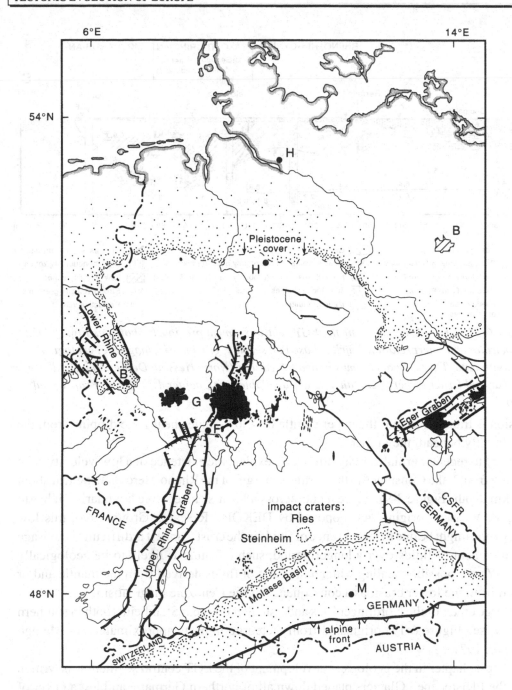

Figure 6-17. Tertiary volcanic rocks (black) and graben structures. City initials as in Figure 6-13.

variable. Coeval fractures appear to trend N–S and E–W. The geotectonic relationship between the Alpine orogeny and the volcanic phase in the foreland remains enigmatic. However, a possible mechanism for this is proposed in Chapter 7.2.2.

The Tertiary rifting and magmatism also brought about, in some areas, a profound re-equilibration of the crust. The Upper Rhine graben is marked by a pronounced Moho high and correlated effects such as a gravity anomaly, a fairly shallow brittle/ductile transition at approximately 15 km deduced from seismicity, and high heat-flow (Chapter 4.1, Zeis *et al.* 1990). It is again puzzling to note that the volcanic belt does not coincide with, but cuts across this Moho high well to the north of its apex. A crustal model for the Lower Rhine graben produced by Meier and Eisbacher (1991) suggests that the upper and lower crust are thinned by separate systems of similarly oriented shear zones, with a mid-crustal detachment.

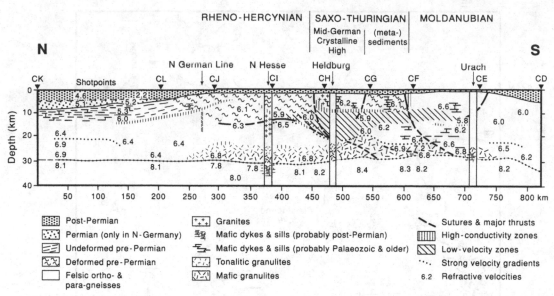

Figure 6-18. Crustal section along the EGT central segment previously shown as Figure 4-22 with a petrological interpretation which is based on seismic, gravimetric, magnetic, geological and xenolith data. The vertical columns indicate locations (North Hessian Depression, Heldburg Gangschar and Urach) where xenolith data are available (see Chapter 4.3). From Franke et al. (1990b).

Extension is transferred into the upper mantle by a low-angle shear zone dipping from the Moho directly beneath the graben.

Tertiary to recent thermal activity might also explain the existence of a low-velocity zone extending from Lake Constance to the northern margin of the Rheno-Hercynian belt, at about 10–15 km depth (Figure 3-9). Since it cuts across all Variscan tectonic boundaries, it has to be of post-Variscan origin. As proposed by DEKORP Research Group 1990, this low velocity zone might be due to the presence of fluids in the crust. Since it is difficult to envisage the containment of fluids over a longer period, such a feature is likely to be geologically young. We are possibly dealing with an invasion of fluids derived from the mantle and/or expelled from the lower crust, as a consequence of the Cenozoic magmatism.

The large crater of the Nördlinger Ries and a smaller one near Steinheim (both in southern Germany, see Figure 6-17) originated from meteorite impacts approximately 15 Ma ago (Gall *et al.* 1977).

The latest chapter in the geological development of the EGT central segment was written during the Pleistocene. Glaciers planed down all of northern Germany and left a cover of clastic sediments, which has done much to stimulate the development of drilling and geophysical exploration techniques directed at hydrocarbon traps in the underlying Mesozoic. Recent subsidence of the North Sea coast is related to the reduction of the elastic forebulge after the removal of the Scandinavian ice cap.

6.3.2 MAIN STRUCTURAL UNITS IN THE PRE-MESOZOIC BASEMENT

The subdivision of the Variscan Belt goes back to Kossmat (1927) and is based mainly upon the metamorphic and tectonic differentiation. Modern geological and geophysical investigations have confirmed the validity of this subdivision, which can easily be translated into plate tectonic terms. As explained in Section 6.3.1, Kossmat's Rheno-Hercynian and

Saxo-Thuringian zones represent Palaeozoic rift basins which were closed by southward subduction during the Devonian and Carboniferous (Figures 6-14 and 6-18). Major amendments have to be made only in the crystalline areas of the Moldanubian region, which not only contains Proterozoic basement, but also has a strong Variscan overprint and even metamorphosed Palaeozoic rocks. Deciphering these crystalline complexes is still at an early stage.

Rheno-Hercynian belt and north German foreland

Beneath the north German lowlands, and as far as the North Sea, unfolded Devonian and Early Carboniferous sediments have been encountered in wells under Permo-Carboniferous and younger cover (D. Franke 1991). To the south, the area of Variscan deformation is approached. East of the River Rhine, at least, there is no well defined Variscan front. Deformation appears to die out gradually northwards. The Rhenish massif and its along-strike continuation in the Harz Mountains expose mainly Devonian to Late Carboniferous rocks. The Devonian consists of thick neritic clastic sequences, which represent the southern, passive margin of the Old Red Continent. Older beds (down to the Cambrian) are restricted to a few anticlines, especially west of the River Rhine. Southern parts of the Rheno-Hercynian belt are overlain by a nappe system, which contains Devonian MORB-type basalts, pelagic sediments and Late Devonian flysch (Giessen and south Harz nappes, see Engel et al. 1983, Floyd et al. 1990).

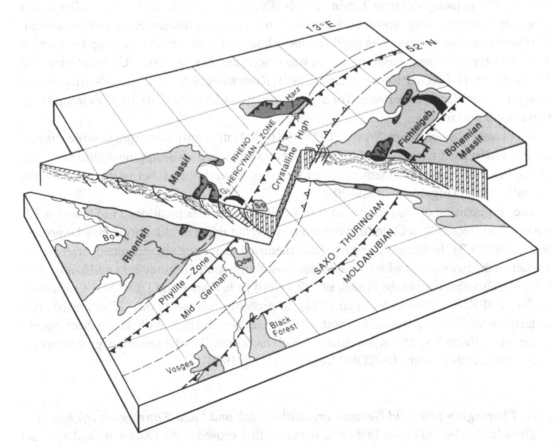

Figure 6-19. Interpretative 3-D diagram of the main Variscan structures in the EGT central segment, after Martin and Franke (1985). Bo: Bonn, Gi: Giesse, Odw: Odenwald, Sp: Spessart.

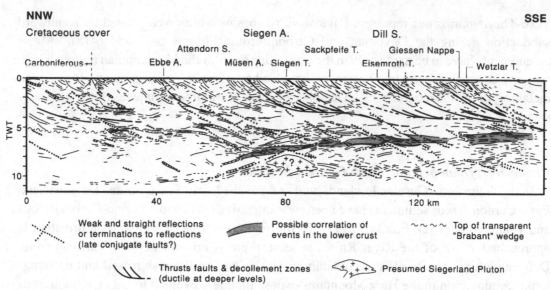

Figure 6-20. Line-drawing and structural interpretation of the main part of the DEKORP-2N seismic profile shown in Figure 3-10 across the Rhenish Massif, from DEKORP Research Group (1990).

A greenschist-grade belt of rocks along the southern margins of the Rhenish massif and Harz Mountains contains a mix of Ordovician to Devonian rocks including Late Ordovician calc-alkaline volcanics (Meisl 1990, Sommermann *et al.* 1990), MORB-type and intra-plate basalts. The palaeogeographic kinship of this 'Phyllite-zone' (Figure 6-19) is still a matter of debate, but it probably represents a more metamorphosed, southeastern part of the Rheno-Hercynian passive margin. The phyllite zone is bounded to the south by a steep fault which probably originated as a major thrust, but was reactivated during Permo-Carboniferous and Tertiary times (DEKORP Research Group 1991, Korsch and Schäfer 1991). This major fault is clearly identified by a narrow belt of high electrical conductivity (ERCEUGT Group 1992, Figures 4-7 and 6-18).

As already pointed out by Weber (1978, 1981) and later confirmed from seismic studies, the Rhenish massif represents an external fold and thrust belt. SE-dipping, listric faults are rooted in a mid-crustal décollement zone which dips towards the SE below the mid-German crystalline high (Figures.6-19 and 20).

The composition of the lower crust is difficult to assess. As mentioned in Section 6.3.1, it has been affected by a Cadomian orogenic event, but relics of older crust may be present (see Chapter 4.3). In the northern part of the Rhenish massif there is a seismically transparent domain in the lower crust which wedges out towards the south (Figures 3-11 and 6-20). This wedge is also observed in the French ECORS Nord de la France and DEKORP-1 reflection profiles (DEKORP Research Group 1991). It probably represents old (?Cadomian) crust underlying the Brabant Massif. Its southern edge may be correlated with the northern margin of the area affected by Devonian rifting, which has controlled the seismic signature in the lower crust further south (DEKORP Research Group 1990).

Saxo-Thuringian belt (mid-German crystalline high and Saxo-Thuringian rift basin)

The adjacent belt of crystalline rocks to the south, exposed in the Odenwald and Spessart Mountains and in the Thuringian Forest, is known as the 'mid-German crystalline high', and was included by Kossmat into his Saxo-Thuringian zone. The crystalline high acted as an active margin with arc magmatism during the Early Carboniferous collisional stage, and is the source region of the Rheno-Hercynian flysch. The protolith ages of the metamorphic rocks in the crystalline high are largely enigmatic, but metamorphosed Silurian is present in the Spessart Mountains. Metamorphism occurred in the Devonian and Early Carboniferous. Early Carboniferous syn- and post-tectonic plutons include subduction-related rocks (Henes-Klaiber *et al.* 1989, Hirschmann and Okrusch 1988, Nasir *et al.* 1991). In the southeast, the crystalline high has been backthrust over the Palaeozoic rift sequences of the Saxo-Thuringian basin, exposed in NE Bavaria, Thuringia and Saxony, as well as in the northernmost parts of the Black Forest and Vosges. Since the Devonian to Early Carbon-iferous uplift of the crystalline high has left no unequivocal traces in the Saxo-Thuringian sedimentary record, this backthrust might well mask an important strike-slip displacement.

The Saxo-Thuringian rocks have been overthrust, in their turn, by a pile of nappes which contains unmetamorphosed Palaeozoic deeper water facies and crystalline rocks derived from the NW margin of the Moldanubian region, to the southeast. The metamorphic nappes contain eclogites derived from Cambro-Ordovician MORB-type mafic rocks, which may be taken to represent a Saxo-Thuringian ocean floor. They were emplaced in a zone of dextral transpression along the Saxo-Thuringian/Moldanubian boundary.

Seismic profiling across this boundary has provided evidence of the nappe thrusting over the foreland, and revealed a slice of strongly reflective high-velocity rocks in a SE-dipping zone, which probably represents the suture between these units and can be traced down to the Moho. The German Continental Deep Drilling Programme (KTB) borehole has been positioned with the intention of penetrating this suture zone. Reflection seismic surveys have also revealed listric, NE-dipping thrust faults which relate to Alpine transpression (see Section 6.3.1). Details of the geological development and seismic coverage have been published by Emmermann and Wohlenberg (1989), DEKORP Research Group (1988), Dallmeyer *et al.* (1992), and Franke (1989a). The southern margin of the Saxo-Thuringian zone is indicated by an electrical high-conductivity zone (Figure 6-18).

During the Carboniferous collisional deformation, the NW margin of the Moldanubian region was an active margin which shed flysch into the Saxo-Thuringian basin. Magmatic rocks in the northern Vosges have been interpreted as relating to southward subduction of the Saxo-Thuringian basin (Volker and Altherr 1987, Holl and Altherr 1987).

Moldanubian region

The internal structure of the Moldanubian region is complex and as yet poorly understood. A central part of the Bohemian massif (Tepla–Barrandian = Bohemian) is unconformably overlain by Cambrian through Middle Devonian sediments and apparently represents Cadomian basement. Nevertheless, deeper and marginal parts of the Bohemian terrane have undergone Early Variscan tectonothermal reactivation. Further to the southeast, in the Moldanubian *sensu stricto*, SE directed nappe thrusts emplace high-grade over medium grade metamorphic sequences (Tollmann 1982, O'Brien *et al.* 1990, Carswell 1991). Similar features have been described from the Black Forest (Flöttmann 1988, Flöttmann and Kleinschmidt 1989) and may be suspected, therefore, also to be present in the intervening parts of southern Germany where the Variscan basement lies hidden under a thick Mesozoic cover. At least the southern part of the Moldanubian Region belongs to the SE flank of the Variscan orogen. From the SE part of the Moldanubian region, Finger and Steyrer (1990) have reported I-type granitoids. The Moravo-Silesian unit to the southeast of the Moldanubian was probably linked with the Rheno-Hercynian belt by an arc structure around the central part of the Bohemian Massif. It is now separated from the Moldanubian core region by an important zone of dextral transpression (e.g. Schulmann *et al.* 1991). To the south, the Moldanubian *sensu stricto*, probably continues into the Variscan basement of the Alps, which likewise contains a Variscan tectonic front with south or SE directed tectonic polarity (see Raumer and Neubauer 1992). Seismic, gravimetric and magnetic surveys of the Moldanubian crust have not revealed a clear internal subdivision and structure. This is probably due to polyphase deformation, multistage metamorphism and granitoid intrusions, all of which have acted to 'homogenise' the primary structures.

6.4 ALPINE OROGENY

A. Pfiffner

The EGT crosses the Alps in eastern Switzerland in what will be referred to as the Central Alps. The Central Alps extend all the way across Switzerland and are characterised by a regional strike oriented E–W to ENE–WSW. In this transect the entire nappe pile making up the Alpine orogen is exposed at surface due to important axial plunges of the various structures. In the adjoining Eastern Alps of Austria the regional strike remains more or less E–W and the lower tectonic units of the nappe pile outcrop in isolated windows only. These lower tectonic units represent the southern margin of the European plate, stretched during the Mesozoic and subsequently compressed in the course of the Tertiary Alpine collision. Towards the west the Central Alps grade into the Western Alps of France with the regional strike oriented roughly N–S.

The tectonic evolution of the Alpine orogeny is discussed in a sequence of snapshots, proceeding forwards in time to the present. This was achieved in practice by progressively retrodeforming the present-day structure of the lithosphere backwards through time.

6.4.1 PRE-TRIASSIC EVOLUTION

Pre-Alpine basement rocks outcrop in scattered areas throughout the Alps. They are most abundant in the central part where high grade crystalline rocks are exposed due to extensive uplift and erosion in the course of the Alpine orogeny. This basement contains a mix of pre-Variscan basement, Late-Variscan granitoids associated with clastic and volcaniclastic rocks, and post-Variscan volcaniclastic rocks.

The paleogeographic map shown in Figure 6-21 shows the position of the major basement blocks containing pre-Triassic rocks in the Alps. It was obtained by a palinspastic reconstruction of the movements suffered during the Alpine orogeny and Mesozoic rifting (for detailed discussion see Pfiffner 1992).

Apart from the Moldanubian zone (Black Forest–Vosges in Figure 6-21) Late-Variscan granitoids are approximately lined up in three E–W trending belts, the northern, central and southern Granite belts. Permo-Carboniferous volcaniclastic sequences were deposited in narrow furrows some which are parallel to these granite belts. The volcaniclastics were in part intruded by these granites and intensively folded as indicated by the angular unconformity with the overlying Triassic strata. The igneous volcanic and plutonic rocks of the calcalkaline series are interpreted as being subduction related (Oberhänsli *et al.* 1988). The magmatic

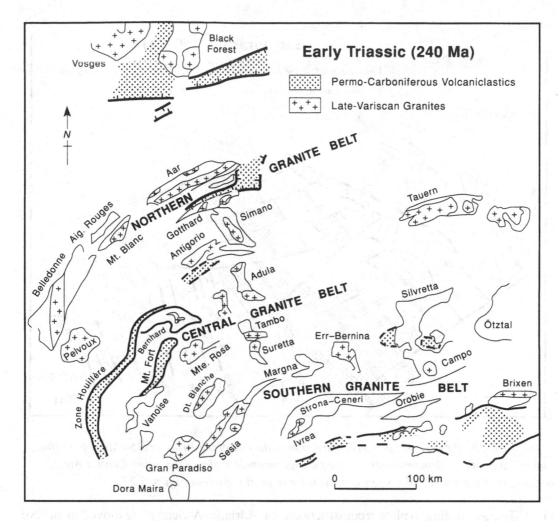

Figure 6-21. Paleogeographic reconstruction of the Alpine segment of EGT in Early Triassic time (240 Ma). The names refer to future Alpine nappes or basement blocks.

activity is related to a phase of thermal updoming and subsequent stretching of the thickened crust (Lorenz and Nicholls 1984). The angular unconformities observed within the Permo-Carboniferous troughs suggest that these furrows formed in a transpressive regime (Laubscher 1987) which is related to a dextral transform zone between the Urals and the Appalachians (Matte 1986, Ziegler 1990, Franke Section 6.3). In Permian times, post-Variscan stretching and subsidence broadened the troughs. Typically red beds accumulated in these basins and are conformably (or with a slight unconformity) overlain by Triassic sediments.

6.4.2 MESOZOIC RIFTING PHASE

During the Mesozoic the lithosphere underwent a phase of stretching and thinning associated with sinistral strike slip between the European and the Adriatic–African plate as explained in Chapter 2.4.13. This rifting phase was particularly active in Jurassic times during the separation of Gondwana and Laurasia and the associated opening of the Tethys and Atlantic ocean.

Figure 6-22 is a paleogeographic map illustrating the situation at the end of the Jurassic

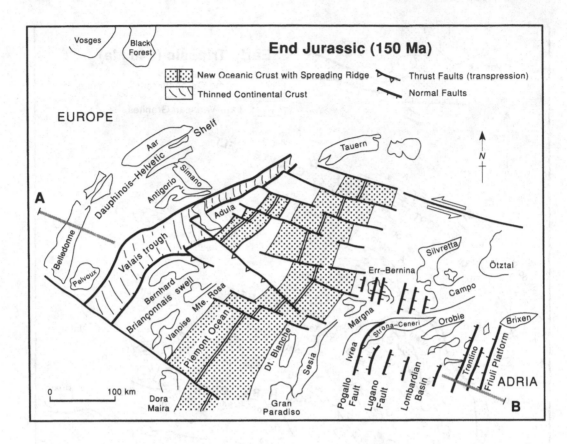

Figure 6-22. Paleogeographic reconstruction at the end of the Jurassic (150 Ma) after rifting and opening of the Piemont ocean. In the paleogeographic realm of the future Central Alps, opening is oblique with offset ridges. A–B: trace of profile given in Figure 6-23.

(150 Ma). According to plate reconstructions, the Adriatic–African plate moved in an ESE (Dewey *et al.* 1989) to SE (Savostin *et al.* 1986) direction generating a basin dominated by transform faults in the future Central Alps (Weissert and Bernoulli 1985, Lemoine *et al.* 1989, Stampfli and Marthaler 1990).

This Tethyan basin is of a true oceanic nature in its central part (Piemont ocean) and is bordered by the thinned passive margins of the Adriatic and European continents (see Figure 6-23).

Stretched European margin

Stretching of the European passive margin resulted in a complex maze of basins and swells. The largest basin, the Valais trough, is characterised by intercalations of basaltic material in a thick clastic sequence. This basin might have formed in a transtensional regime as a pull-apart basin. The continental shelf (Helvetic Dauphinois domains) to the NW of the Valais trough is marked by slow subsidence increasing basinwards. Stepwise subsidence with characteristics of extension are typical in the Lower Cretaceous sediments (Funk 1985). The Briançonnais swell situated to the SE of the Valais trough is bounded by steep scarps with scarp breccias. South of the Adula block the swell is dissected by a transcurrent fault. The sediments associated with this zone are exposed in the Schams nappes of eastern Switzerland. Angular unconformities and sediment transport directions observed within them point to a transpressional scenario (Schmid *et al.* 1990). The crust underlying the Valais trough is likely to be thinned continental crust; locally it was possibly cut by extensional fractures which

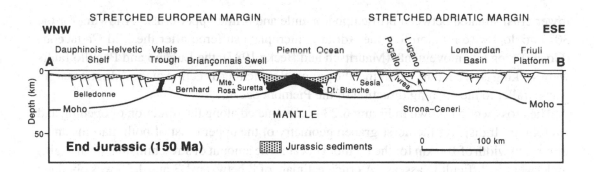

Figure 6-23. Cross section illustrating rifting and thinned continental margins of the Piemont ocean at the end of the Jurassic. Trace of cross section is given in Figure 6-22. Pogallo and Lugano are major syn-sedimentary normal faults extending into the basement.

served as pathways for the ascent of basaltic melts from the mantle. The small volume of ophiolitic material found in the corresponding nappes (e.g. Martegnas mélange) speaks for incipient spreading rather than a wide ocean (see also Schmid *et al.* 1990). Based on a comparison with modern oblique rift systems, Kelts (1981) proposed a two-stage model with a phase of subsiding continental crust segmented by deep vertical faults followed by oblique rifting associated with the formation of a narrow belt of oceanic crust.

Opening of the Piemont ocean

The opening of the Piemont ocean is coeval with the opening of the central Atlantic. The original width of the Piemont ocean is difficult to assess. Estimates range from 100 to 500 km. In any case the opening was oblique in the transect of the Central Alps as opposed to the situation in the Western Alps. The importance of the transform faults is indicated by the occurrence of (a) pelagic sediments in stratigraphic contact with serpentinites (indicating serpentinite protrusions along fracture zones), (b) ophiolite breccias formed by fracturing of oceanic crust coeval with sedimentation in the fractures and (c) pebbles of oceanic and continental provenance (crystalline basement and oolitic limestones) contained within massflow conglomerates in the deep sea sediments (radiolarian cherts) of the Piemont ocean (Lagabrielle *et al.* 1984, Lemoine 1980, Weissert and Bernoulli 1985). In Figure 6-22 these transform faults are drawn parallel to the movement direction between the European and Adriatic–African continents.

Stretched Adriatic margin

Along the Adriatic margin, stretching and thinning of the continental crust led to a pronounced, rugged morphology with NNE–SSW-trending swells and basins. In the Lombardian basin, for example, the distal continental margin submerged and became increasingly starved. Differential subsidence associated with synsedimentary faulting resulted in a sequence of silicic limestones up to 4 km thick in the basin in contrast to a thin shallow water sequence on the swell to the west. Two normal faults can be followed into the basement; the Pogallo fault, which separates the Ivrea from the Strona–Ceneri basement block (Handy 1987), and the Lugano line that marks the western end of the Lombardian basin (Bertotti 1990). Both indicate crustal attenuation by listric master faults dipping towards the continent, similar to the modern example of the Bay of Biscay margin (Le Pichon and Barbier 1987) where attenuation of the lithosphere is achieved by a ductile zone of decoupling in the lower crust situated between brittle upper mantle and brittle upper crust. Allowing for the 30° anticlockwise rotation that the Adriatic microplate suffered after the Mid-Cretaceous (Gosau) orogenic movements (Mauritsch and Becke 1987), the Lugano and Pogallo faults were striking NNE. It thus seems that the principal stretching direction of the Adriatic margin was parallel to the opening direction of the Piemont ocean.

The cross section shown in Figure 6-23 is constructed along the direction of opening and stretching. It displays the horst-graben geometry of the upper crust of both margins and a minimum width of 100 km for the Piemont ocean. The amount of stretching that the margins underwent is difficult to assess. A crude estimate of β between 1.3 and 1.65 was obtained in the Austroalpine nappes (Campo block in Figure 6-22) based on fault geometry (Froitzheim 1988).

6.4.3 CRETACEOUS CONVERGENCE

The building of the Alps implied at least two distinct episodes, one in the Cretaceous, affecting primarily the Eastern and Western Alps, and one in the Tertiary (split into two or more events by some authors) and of prime importance for the Central Alps. The Cretaceous (or eo-Alpine) orogenic episode corresponds to NE–SW directed convergence between the European and African plates (Dewey *et al.* 1989, Chapter 2.4.14). The Adriatic microplate was separated from the African plate between 130 and 80 Ma and suffered a 30° anticlockwise rotation during that period (Dercourt *et al.* 1986, Platt *et al.* 1989, Heller *et al.* 1989). For the plate boundary on the northwestern margin of the Adriatic microplate, this rotation corresponded to a dextral transpressive movement relative to the European plate. Its expression is found in west to northwest oriented thrusting in the western and Eastern Alps which led to the formation of the eo-Alpine chain (see below).

One of the primary constraints on palaeogeographic reconstructions is the fact that sedimentation persisted and outlasted this orogeny in the realm of the future Central Alps. In fact, fragments of this eo-Alpine orogen are at present in an even higher structural position than these Cretaceous–Cenozoic sediments as a result of later N–S oriented compression. Figure 6-24 is an attempt to sketch the paleogeography at Mid-Cretaceous times (90 Ma). The SE part of the Adriatic microplate is displaced over a distance of 270 km in a westerly direction in this reconstruction. An E to SE-dipping subduction zone and associated accretionary wedge developed in the Eastern and Western Alps. A roughly E–W trending dextral transcurrent fault zone links the leading edges of the advancing nappe piles between Western and Eastern Alps and one branch delimits the Eastern Alps to the north. In the area to the northwest, sedimentation on the carbonate platform of the European margin (the Dauphinois-Helvetic shelf) and on the Briançonnais swell persisted. In the immediate vicinity of the transcurrent faults and the accretionary wedges a number of flysch basins formed, accumulating detritus from the rising cordillera. Provenance from the south and east is particularly conspicuous for Late Cretaceous times (Wildi 1985). Some of these flyschs contain conglomerates with pebbles of granites and volcanic rocks pointing to the existence of elongate basement highs. These highs may have formed in a transpressional regime along certain segments (constraining bends) of the transcurrent faults.

A metamorphic event dated at the Mid-Cretaceous overprinted Penninic and Austroalpine basement units in the Western and Eastern Alps (see Frank *et al.* 1987 and Hunziker *et al.* 1989 for a review). In the Western Alps it is a high-pressure event related to a subduction zone

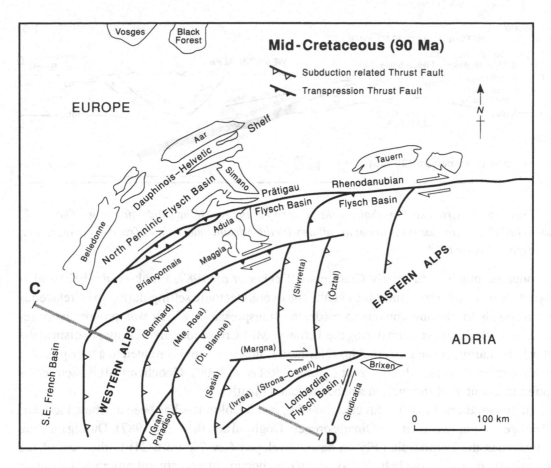

Figure 6-24. Paleogeographic reconstruction at Mid-Cretaceous times (90 Ma) illustrating eo-Alpine plate tectonic setting. A mountain chain extends from the western to the Eastern Alps. To the north, sedimentation continues on the accretionary wedge and flysch basins develop on either side of the evolving orogen. C–D: trace of profile given in Figure 6-25.

(Polino *et al.* 1990). In Figure 6-25 this subduction zone is taken to dip in an E to SE direction, that is in the direction of plate convergence and in a direction favourable to a rapid consumption of the Piemont ocean of the Western Alps. The basement blocks of the European margin (Dora Maira, Mte Rosa and Bernhard) and the Piemont oceanic crust are shown subducted beneath the Adriatic plate and undergoing high-pressure metamorphism. The Grand Paradiso and Sesia–Lanzo blocks of the Adriatic are also brought to greater depths and undergoing high-pressure metamorphism through underplating. One of the major difficulties associated with this eo-Alpine orogeny is the observation that the pre-Triassic basement rocks of the Bernhard nappe were seemingly undergoing high-pressure metamorphism (Hunziker *et al.* 1989) while its sedimentary cover, which constitutes the Briançonnais swell, was still at a site of sedimentation which continued up into the Cenozoic. We are therefore forced to conclude that the cover was already detached from the basement and incorporated into the accretionary wedge (which contradicts observations by Sartori 1987) or that in the case of the Bernhard nappe the high-pressure metamorphism is of Tertiary age. The length of the subducted slab of Piemont oceanic crust in Figure 6-25 corresponds to the minimum estimate used in the paleogeographic map of Figure 6-24. There is no evidence of any volcanism associated with this Cretaceous subduction zone. It might be that the subducted slab of oceanic crust was too short to generate significant amounts of melt.

In the Eastern Alps WNW-directed thrusting (Ratschbacher and Neubauer 1989) occurred

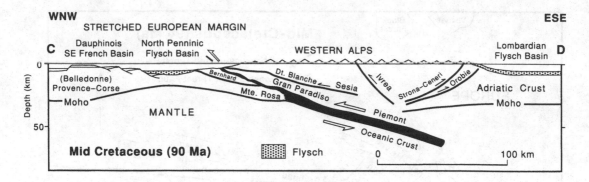

Figure 6-25. Cross section showing the geometry of the eo-Alpine subduction zone. The stretched European margin is about to collide with the Adriatic microplate. Trace of cross section is given in Figure 6-24.

throughout much of the Early Cretaceous (Decker *et al.* 1987, Frank *et al.* 1987) and is synchronous with a metamorphic event. The overall tectonic setting during the Cretaceous corresponds to oblique convergence (dextral transpression) which was progressively replaced by N–S compression during the Tertiary. Mid-Cretaceous alkaline magmatism at the northern margin of the Eastern Alps (Northern Calcarous Alps) is related to a transpressive regime rather than a subduction zone (Trommsdorf *et al.* 1990). Geochemical characteristics point to a source of the melt in a subcontinental mantle.

In the Southern Alps, W-directed thrusting in the Dolomites has been attributed to later, Paleogene compression in the Dinaric orogen (Doglioni and Bosellini 1987). During the Late Cretaceous the Lombardian Flysch basin developed (see Figure 6-24) to the west of the Giudicarie transpressive belt. Paleocurrent data document a westward progradation of this submarine fan system and provenance of the terrigenous material from the northeast. Within the Austroalpine nappes of the Eastern Alps, angular unconformities in small Gosau basins document orogenic movements in the Late Cretaceous (see Faupl *et al.* 1987). Paleocurrents indicate a northern to western source area for most of these basins.

6.4.4 TERTIARY COLLISION

The second phase of convergence during the Tertiary implied N–S to NNE–SSW directed plate motions between Europe and Africa in the Eocene, and NW–SE directed motion since the Miocene (Dewey *et al.* 1989). For the Adriatic promontory of Africa a somewhat independent motion path must be assumed (see also Section 6.5.4). Kinematic indicators in the Western Alps point to more westerly directed thrusting (Platt *et al.* 1989) which can be explained by the Adriatic microplate acting as a rotating indenter (Vialon *et al.* 1989). This rotation is thought to have affected the southwestern margin of the European margin as well as can also be deduced from the orientation of the paleomagnetic vectors (Heller *et al.* 1989).

As a consequence of this more N–S directed plate convergence a bivergent collision orogen evolved along the transect of the EGT. The thinned continental crust of the European plate margin was delaminated. The upper crust was peeled off and thrust northward into the nappe pile of the Penninic nappes (Suretta, Tambo, Adula and Simano in Figure 6-26). Similarly the thinned margin of the Adriatic plate was shortened by southward imbrication of upper crustal flakes (Orobie and its underlying units in Figure 6-26). The timing of these imbrications is not well constrained, but a progression of thrusting from structurally higher

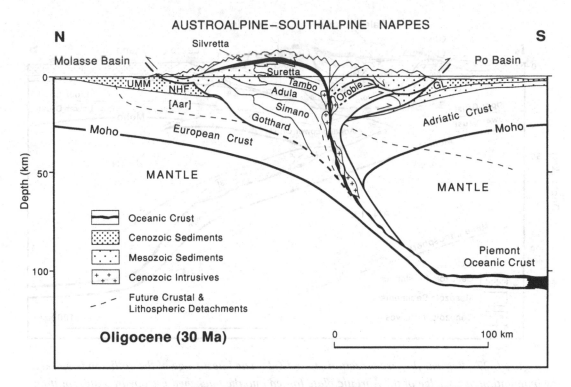

Figure 6-26. Cross section along EGT reconstructed for Oligocene times (30 Ma). The European and Adriatic plates have collided. Lower crust of the thinned margins is subducted, whilst the upper crust is piled up into a bivergent thrust belt. Melts are shown to rise along the suture zone (Bergell and Novate intrusives). Foredeeps to the north and south receive clastics of the rising Alpine chain.

to lower units (i.e. in-sequence thrusting) seems at least plausible. In any case the youngest imbrications involve the northernmost (Helvetic zone) and southernmost (buried Southalpine thrustbelt beneath the Po Plain) units of the Alps (cf. Figures 6-26 and 6-27).

The lower crust of the thinned continental margins was subducted during this Tertiary collision. The cross section of Figure 6-26 showing an asymmetric S-dipping subduction zone was obtained by retrodeforming the present day structure (displayed in Figure 6-27) which is constrained by geophysical data (see Figures 3-11 and 4-23 and Pfiffner *et al.* 1990). The lengths of the slabs of European and Adriatic lower crust are constructed by a combination of volume and line-length balancing. The subducted lower crust of the thinned margins is estimated to have been reduced to 50% of its original thickness by Mesozoic extension. The slab of oceanic crust attached to the European lower crust represents the Piemont ocean. It was subducted in the Cretaceous convergence, oblique to the profile plane of Figure 6-26. This fact is responsible for the break in slope of the subducted slab. The lower crust of the Adriatic margin is shown to be steepened and overturned due to the effect of the asymetric subduction geometry. Sandwiched between the European and Adriatic margins, small pieces of obducted oceanic material are shown to straddle the plate boundary. The sedimentary rocks, deep sea sediments of the former Piemont ocean, associated with these pieces could be held responsible for the production of melts at depth. These melts rose possibly along the suture zone and intruded near the southern end of the Adula and Tambo nappes (Figure 6-26) at around 30 Ma (Bregaglia and Novate intrusions with associated dykes, Trommsdorff and Nievergelt 1983).

To the north of the Alps a foreland basin had developed in Late Eocene times. Its depocentre migrated northward in time in response to loading and flexure of the lithosphere

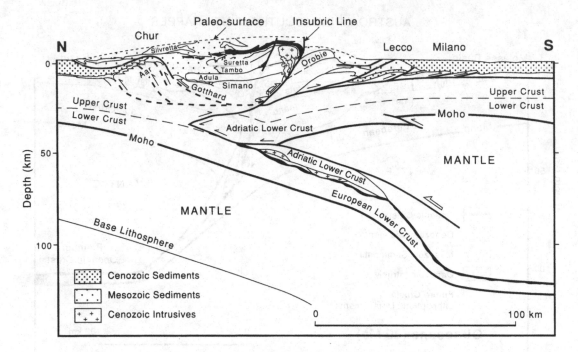

Figure 6-27. Present-day cross section along EGT. The latest stage of the collision is marked by indentation of a wedge of the Adriatic plate forced into the thickened European plate, splitting it apart. The central part (Penninic basement nappes Adula, Tambo and Suretta) is uplifted and eroded in the process with backfolding in the vicinity of the Insubric line. The detached cover of the Adriatic indentor now forms the buried Milan fold-and-thrust belt.

due to the collision event (Pfiffner 1986, see also Section 6.2.2). The southern margin of this foreland basin was controlled by the toe of the advancing Alpine nappe pile and the accumulating clastic sediments reflect the important erosion in the rising Alpine chain. This unroofing stage seems to have had a climax at around 30 and 20 Ma (Pfiffner 1986, Hurford *et al.* 1989). The early stage of the foreland basin was a flysch basin (North-Helvetic Flysch in Figure 6-26) with thick marine clastic sequences that contain volcanic detritus of andesitic and basaltic nature. The source of this volcanism, which is more or less coeval with the Bregaglia and Novate intrusions mentioned above, is thought to have been in the immediate vicinity of the basin (Siegenthaler 1974) but its exact location remains unclear. In any case, this volcanism must be seen in the context of the volcanic activity of central Europe as discussed in Section 6.3.1 and Chapter 7.2.2.

To the south of the Alps, in the Po basin, a foredeep evolved in which syntectonic sediments accumulated, linked to south vergent thrusting in the Southern Alps (Roure *et al.* 1990). The thin Oligocene Chiasso formation is unconformably overlain by the thick, clastic Gonfolite Lombarda group which is of Oligocene–Miocene age and already containing boulders of the Bregaglia intrusives (Gunzenhauser 1985). Geochronological studies suggest that the rapid subsidence of the Po basin during the deposition of the Gonfolite Lombarda group was coeval with rapid uplift in the adjoining Alps to the north. A deceleration of this uplift occurred during the Miocene (Giger and Hurford 1989). The climax of the unroofing stage around 20 Ma is coeval with dextral transpressive movements along the Insubric line (Figure 6-23; see also Schmid *et al.* 1989). Vertical uplift due to steep south vergent thrust faulting affected the northern block (Penninic nappes) and led to the juxtaposition of amphibolite facies grade and anchimetamorphic rocks on either side of the Insubric line. The largest amount of uplift occurred in the Toce-Lepontine area, more or less on the transect of EGT. In response to the updoming of the Toce-Lepontine area, the upper part of the orogen seemed to undergo extension. To the west of this dome the Simplon line, a west dipping crustal-scale normal fault, indicates E–W stretching at a late stage (19–6 Ma) of the collision (Mancktelow 1985, Merle et al. 1989). Similarly the Brenner line east of the Tauern window in the Eastern Alps indicates E-W extension in relation to unroofing of the Tauern thermal dome (Selverstone 1988). There, eo-Oligocene ductile thinning was followed by Miocene low-angle faulting. On the eastern side of the Lepontine dome no similar normal fault exists and data on kinematic indicators are as yet circumstantial.

In the Western Alps the rotating WNW motion of the Adriatic microplate was associated with transcurrent faulting (Vialon et al. 1989). Large scale sinistral strike-slip movements were postulated by Ricou and Siddans (1986) to accommodate the N–S convergence in the Central Alps. Such N–S striking transcurrent faults could also explain the rapid uplift of the high-pressure assemblages: steep reverse faulting on constraining bends of the transcurrent faults would represent an efficient mechanism for vertical uplift. In addition, the relief produced over the developing positive flower structures would form a narrow belt. The associated volume of rocks needed to be removed by erosion to unroof high-pressure assemblages would be much smaller than that required by uplift due to underplating as envisaged by Platt (1986). The relatively small and few flysch basins in which the eroded material accumulated also points to a narrow orogenic belt as source area.

The late stages of the collision produced some spectacular large-scale backfolds in the Penninic nappes of the Central Alps. In Figure 6-27 they include the folds in the Suretta nappe and the antiformal structure of the Adula nappe intruded by the Cenozoic granites. The Adula backfold is associated with steep S-directed reverse faulting on the Insubric line. Foliation and nappe contacts were steepened and overturned in the process (southern steep belt of 'root zone' of the older literature, Milnes 1974). These movements are related to a wedge of Adriatic crust which was forced into the European crust. As Figure 6-27 shows, the European crust was split apart. The upper crust, namely the Penninic nappe pile, was uplifted in the fashion of a pop-up structure. The uplift in the south was along the Insubric line, and the one in the north was along thrusts within and beneath the Aar and Gotthard massifs. All these crustal scale thrusts are shown as listric, decoupling the upper crust in accord with the straight profiles shown in Chapter 4.2. The lower European crust was overridden by the Adriatic microplate. Within the latter, deformation led to the formation of the Milan fold belt (e.g. Roeder 1989a) which is buried beneath the Cenozoic infill of the Po basin. Shortening of the entire south Alpine fold-and-thrust belt is on the order of 70–115 km, but only some 30 km are related to the buried Milano fold belt, which formed last in the kinematic sequence. Its balancing determines the length of the indenting wedge of lower crust. A material balance suggests that the indenting wedge also overrode the northernmost stretched part of the Adriatic crust, which had been stacked up in a southerly direction in the earlier history of the collision.

The latest movements of the collision were coeval with a NW motion of the African plate relative to Europe since the Miocene (Dewey et al. 1989). Within the collision zone the convergence pattern is quite complex. In the EGT transect, the late dextral movements along the E–W striking Insubric line have been interpreted as lateral escape (Schmid et al. 1989). In the Western Alps NW-directed late thrusting accounts for the development of crustal scale basement uplifts (Aiguilles Rouges and Mont-Blanc massifs, Mugnier et al. 1990, Butler et al. 1986). This late thrusting seems to post-date NNW-directed thrusting in the Central Alps. Hinge lines of large scale folds of the Morcles nappe can be traced from the Central to the Western Alps (Ramsay 1989). They are related to the early NNW-directed shortening and subsequently warped by the updoming of the Aiguilles Rouges and Mont Blanc massifs during the northwest oriented late convergence.

In summary it follows that the Tertiary collision in itself represents a polyphase and complex interplay of crustal scale normal, thrust and strike-slip faulting which continues to the present. Recent crustal movements expressed by vertical uplift along the EGT transect reach a maximum value of 2 mma^{-1} near Chur (in Figure 6-27, see also Chapter 5.3.3). In plan view this maximum extends along the basement uplifts of the external massifs (e.g. Aar massif, Figure 6-27). Does this mean that this recent uplift is at least partially triggered by presently continuing collision and associated decoupling of upper crust? The data on recent seismicity is not in conflict with this idea. We explore the possibilities further in Chapter 7.2.2.

6.5 THE FRAGMENTED ADRIATIC MICROPLATE: EVOLUTION OF THE SOUTHERN ALPS, THE PO PLAIN, AND THE NORTHERN APENNINES

P. Giese, D. Roeder, and P. Scandone

This section of the EGT extends from the topographic backbone of the Alps to the north tip of Corsica. It is only 430 km long, but it displays a spectacular architecture of plate interaction. The suture between Europe and the Adriatic microplate is crossed twice by this segment; once in the Alps with a deep subduction slab, and again in the Apennines with a Moho overlap at shallower depth. This EGT section contains the Adriatic microplate warped under subduction load and the thrust loads of the Alps and the Apennines. At its southern end, it contains the tip of the Ligurian rift which functioned during orogeny-related lithospheric extension, but had already evolved during the separation of Laurasia and Gondwana.

6.5.1 TECTONIC SETTING

The crustal silhouettes of the two opposed thrust fronts of the Alps and Apennines are strikingly different and demonstrate two orogenic regimes. Both orogens are in a post-collisional state, and they face each other head on.

The Alps are still largely in a compressional regime with thickening crust, but in the Central Alps, rising topography is already counteracted by the beginnings of extensional tectonics. Fold belts at the Alpine north front record the latest supracrustal trace of the Europe-vergent collisional suture which intersects the Moho a short distance north of the Alpine south front. The latter is a south-vergent backthrust antithetic (conjugate) to the collision suture. It breaks through the Alpine crust too far to the north to show a Moho overlap.

The Apennines are clearly in the extensional stage, with mid-crustal and deeper detachments, metamorphic core complexes, and with new oceanic crust being formed within the the main body of the orogen. Its north front shows a magnificent Adriatic-vergent Moho overlap in the footwall of the original collision suture. It is not older than Pliocene and is oriented synthetic (parallel) to the older suture. Seismicity and Neogene stratigraphy show that the north front of the Apennines is in active compression, as shown in Chapter 5.3.

The common foreland of the orogens is the Adriatic microplate. This cratonic terrane has been part of the field of transtension between Europe and Africa since the Jurassic, when Atlantic opening reactivated the east-facing Tethyan passive margin. Its cover of Mesozoic sediments is now deeply buried beneath Neogene clastic foredeep fill, shed from the rising Alps and Apennines, and its lithosphere is deflected into an antiformal ridge by the thrust loads and, mostly on the Apennine side, by subduction hinge retreat. Near the EGT section line, the northwest spur of the Adriatic microplate has been indenting the European crust since the late Eocene collision.

Southern Alps

Three structural elements comprise the topographic south slope of the Alps. Furthest to the north, a belt of steeply dipping layers is composed of a stack of north-vergent basement nappes above the ophiolite-lined Penninic (Tethyan) collision suture. During Neogene backthrusting, this stack was tilted backward to form the steep zone. Near its southern edge, the steep zone is truncated along the Insubric line, which is a major terrane boundary with a complex history of dip slip and strike slip. Its kinematics and its considerable depth extent are recorded in mylonite fabrics and contrasting cooling histories. South of the Insubric line, a south-Alpine crustal block and its sediment cover contain south-vergent polyphase compression. Seismic data show that a detachment at a mid-crustal level merges with the Insubric line at 15–20 km depth. It transports the south-Alpine crustal body southwards over the load-warped Adriatic lithosphere, displacing the Po Valley basin fill, and forming a classical fold-thrust belt which involves upper crust, Tethyan-margin sediments, and foredeep fill. The south-Alpine fold-thrust belt is being tilted southwards by the Apennine subduction. Therefore, its northern, internal zones are being eroded while its southern front is buried beneath the braided, oxbowed, and swampy gravels of the Po River and its Alpine tributaries.

South Alpine thrusting includes Paleogene, Miocene, and Plio-Pleistocene spasms. Recent thrust tectonics is demonstrated in devastating earthquakes (1976 near Gemona in the Tagliamento valley and possibly 1348 near Villach in the Gail valley) for the extreme eastern part, but so far, the area of the EGT section appears quiet. Figure 6-27 shows the structural style in a representative cross section. A compressionally emplaced thrust sheet involves 10 km of pre-tectonised crust intruded by arc magma, 5–7 km of Mesozoic passive-margin sediments, and 3–7 km of Oligocene to Pleistocene foredeep fill. Its frontal part is a blind thrust within the Miocene section. Its foreland imbrications contain several of Italy's major oil and gas fields.

The south Alpine bulk-strain estimate, of about 100 km, is made from a mix of published data and speculation. It consists of 70 km of main-thrust overlap and 30 km of combined foreland and main-thrust imbrication. Unknown subsurface details may widen the range to between 80 and 150 km, but is still consistent with the EGT Moho data shown in Figure 3-13.

Po basin

The crust of this double foredeep is known from numerous refraction seismic experiments, as indicated in Figure 6-28. Its supracrustal details are known by petroleum exploration. Elastic load flexural models of the basement top have been used to emulate the Moho at its shallowest depth of 28 km and at its antiformal flanks, deepening beneath both thrust fronts.

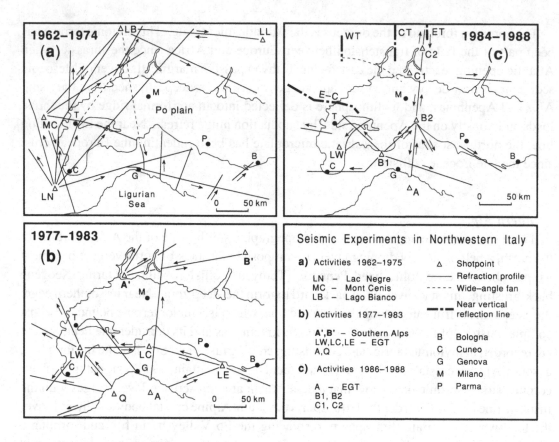

Figure 6-28. Map of seismic experiments across the Southern Alps and northern Apennines.

The sediment cover of the Adriatic crust is composed of the same elements as known from both fold-thrust belts. There, as well as in the Po basin, E–W extension of Jurassic age has generated rifts and has reduced the net crustal thickness to about 20 km.

The south slope of the Po basin contains an Apennine foothills fold belt forming three subsurface lobes or promontories (Pieri and Groppi 1981). These syndepositional folds are of Pliocene age, blind-thrusted and buried beneath undeformed cover sediments, and are displayed in superb detail on reflection seismic profiles.

Apennines

Along the EGT section, the Apennine orogen is composed of four geodynamic units. Polyphase thrust imbrication during the Tertiary has assembled three pre-orogenic paleogeographic units, from top to bottom and from south to north:

(a) the Liguride ophiolites and associated marine sediments of Mesosoic age,

(b) the Tuscanides, a pre-Mesosoic basement with thin Mesozoic cover rocks derived from the stable edge of the Tethyan passive margin and fragmented off Pangea during the Jurassic,

(c) the central Adriatic microplate and its Mesozoic sediment cover.

The latter sequence is a northern correlative of the outcropping and strongly detached Umbrian series, but is known only from wells drilled into mildly detached structures (Pieri and Groppi 1981). The thrust imbrication of the northern Apennines also includes an Upper Cretaceous and Tertiary polyphase accretionary wedge. It covers the three lower units unconformably, it contains their erosional products, shows evidence of repeatedly cannibal-izing perched basins, and is progressively involved in the thrusting and folding. In this polyphase flysch wedge, stratigraphy and sedimentology document the tectonic assemblage of the Apennine stack of thrust sheets (Giannini and Lazzarotto 1975, Bally *et al*. 1988, Patacca and Scandone 1989).

Flysche deposits date a NE-vergent collision or obduction of the oceanic Liguride units with the Tuscan part of the Adriatic plate as Oligocene. NE-vergent thrust stacking continued by successively piggybacking the stack. The present front of Apennine deformation involves Pliocene strata and meets the buried Alpine foothills front head-on near the Po river (Cassano *et al*. 1986). In at least one cross section (Pieri and Groppi 1981, section 6), the Alpine detached units are overridden by an Apennine thrust front.

Ligurian Sea

In the Ligurian hinterland of the Apennines, polyphase extension has accompanied and superposed Apennine thrusting since the Oligocene. Offshore, the European continental crust is progressively thinned southward toward the oceanic crust of the Provençal basin. Its top shows articulation into several basins filled with Neogene sediments. The crustal profile of the EGT shows a complex rift zone with thinned crust and a field of magnetic bodies which can be traced southeastwards into the Tyrrhenian Sea. Onshore, the tectonics of this margin is beginning to be understood as synkinematic backarc spreading. Tortonian and younger extension is affecting at least the middle crust, and probably the entire lithosphere, of the Adriatic microplate. Deep detachments form antiformal viscous pillows or metamorphic core complexes in the Cordilleran sense. Carmignani and Kligfield (1990) describe the Alpi–Apuane as an example. The extended supracrustal material is the thrust belt itself. Compression and thrust overlap near the front and extension in the main orogenic body are contemporaneous. The domain boundary migrates outward and generates complex super-positions of compressional and extensional strain.

Corsica

This relatively thick continental crust (Figure 3-15) with Variscan granites carries a cap of Alpine rocks emplaced westward during the Eocene collision. The cap documents it as a fragment of the European foreland, rotated anticlockwise away from a site adjacent to the coastal Provence (SE France) during the Oligocene and Lower Miocene.

6.5.2 GEOPHYSICAL CONSIDERATIONS

Bouguer gravity

The Bouguer anomaly profile shown in Figure 6-29 (see EGT Atlas Map 9) portrays the lateral variations in crustal thickness interpreted in terms of a density model which has been constrained by the seismic data of Figure 3-12. A gravity low along the crest of the Alps is consistent with 50 km of composite crustal thickness. A gravity high follows the Southern Alps and fringes the curved west edge of the Po basin. It is the potential field expression of the celebrated Ivrea body whose sparse outcrops support the idea of a crustal slab exposed to near the base of the crust. The Apennine foredeep, its low-density fill, and its doubled crust are reflected in another of the deeper gravity minima of the EGT profile. The Ligurian coastal

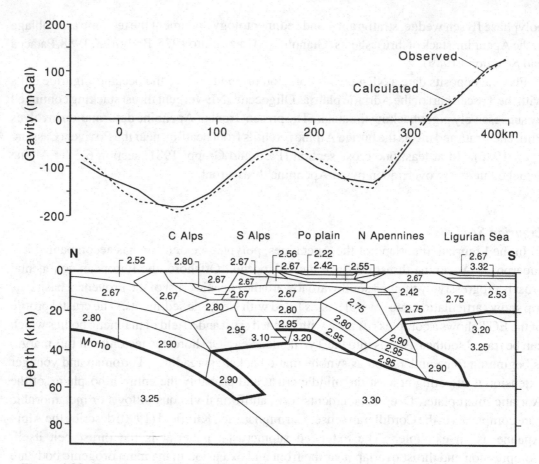

Figure 6-29. Density model for the Alps to Ligurian Sea based on the seismic profile given in Figure 3-12, showing fit of calculated gravity to observed. Numbers show densities in gcm^{-1}.

zones and their extensional structures are associated with intermediate gravity values grading into the oceanic gravity range of the Ligurian and Provençal seas. The continental block of Corsica and Sardina shows a zero level of Bouguer anomaly.

Seismic data

About 6000 km of deep refraction seismic lines recorded since 1956 (Figure 6-28) provide a solid data base for modelling the lithosphere in the Po basin and its Alpine–Apennine frame. Following a number of experiments of both local and international scope, the EGT projects described in Chapter 3.2.3 included a composite N–S line from the Alpine north front to the coastal waters of Corsica, E–W lines in the Apennine region, and fans across the Po basin and the Southern Alps. Deep events recorded in many of these lines can be interpreted as coming from the Moho, and our accumulated wealth of Moho data shows the base of the crust in an unsuspected degree of complication, as illustrated in Figure 3-13.

Rheological stratification of the lithosphere

The seismic data base of dipping and stacked Moho segments can be interpreted tectonically by applying to the lithosphere the model of a rheological stratification with a strong upper crust, a weak middle and lower crust, and a strong upper mantle, as suggested in Chapter 4.2 by the lithospheric composition and the geotherm (Meissner 1986). By analogy with supracrustal compressional tectonics (Suppe 1985 and many others), thrusts

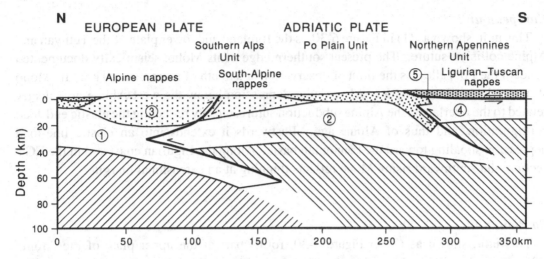

Figure 6-30. Cross section along EGT from the Alps to the northern Apennines showing five separate plate units; 1: European unit, 2: Po Plain unit, 3: Alpine unit, 4: northern Apennines unit, 5: sub-Monferrato unit.

cutting the stratified lithosphere can be expected to show ramped segments in the strong layers and flat detachments in the weak layers.

Intracrustal thrusts cannot generally be seen on seismic data. However, for ramped thrusts within a stratified lithosphere, the finite cutoff angle should be much lower than the expected Coulomb shear angle. Figure 6-30 shows that in 25 km thick stratified crust, we can expect thrust fronts at the basement top to be located up to 150–200 km ahead of the Moho break, instead of 60–80 km as predicted for homogeneous crust.

6.5.3 STACKED SLABS OF LITHOSPHERE

The discussion of Alpine lithospheric imbrication started when seismic studies suggested that the Ivrea zone contains mantle thrust over the European crust (Giese 1968). During the following decade, the discussion became organised in plate-tectonic terms of Benioff (B-type) subduction and quickly dominated the field of geodynamics (Dewey *et al.* 1973). However, the complexity of slabs twisted and stacked in three dimensions is not predicted by the classical plate tectonic model. Its full understanding required a new theory (Pavoni 1961, Tapponnier 1977) and another decade of seismic surveys. The Moho data from scores of seismic surveys were patiently mapped, correlated, and interpreted in several incremental syntheses (e.g. Giese *et al.* 1982). At present, and based on much unpublished work, we can recognise five crustal and lithospheric units and their tectonic interaction. For several of them, we can map the configuration of the Moho (Figure 3-13). For all of them, we can also describe the configuration at the top of the crust. From the bottom up, the stack includes (1) the European unit, (2) the Po Plain unit, (3) the Alpine unit, (4) the northern Apennines unit and (5) the sub-Monferrato unit. Figure 6-30 schematically locates these units relative to each other and relative to the geography. Some degree of paleogeographic and plate tectonic coherence is suggested for the Adriatic microplate since the Jurassic opening of the Atlantic. However, the Alpine collision and indentation since the late Eocene fragmented the Adriatic microplate into separate units (2), (3), (4), and (5).

European unit

This unit, shown as (1) in Figure 6-30, is the foreland and lower plate of the Tethyan and Alpine collision suture. The present southern edge of its Moho, seismically documented beneath the Po Valley, is the limit of observation at a depth of more than 70 km. Its Moho structure is dominated by a nearly circular depression below the west-Alpine arc, clearly related to the overlap on the Alpine subduction suture, closely reflecting the elastic end-load configuration, and thus of Alpine age. Eastwards it extends into an elastic line-load depression shoaling toward the Pannonian basin. Along the Ligurian coast and in the EGT section, the southward rising Moho reflects thinning at an extensional margin.

Po Plain unit

This unit, shown as (3) in Figure 6-30, forms part of the upper plate of the Alpine subduction, but it is fringed by crustal complexities and additional subductions. The southwest and northeast flanks of the Po Plain unit are downwarped and form lower plates of Neogene subduction zones beneath the Apennines and the Dinarides–Hellenides (Moretti and Royden 1988). The crestal line of this double downwarp, not everywhere documented by Moho data, extends from Milano along the Adriatic median high and ends with the south cape of Apulia at the heel of the boot of Italy, where present deep water and thick sediments suggest oceanic crust of Mesozoic age.

Moho stacking in the Po Valley–Ligurian area suggests a subduction-related overlap of at least 100 km beneath the northern Apennines. This overlap may continue northwestward into the Piemont area, but a decrease to zero overlap near the trace of the EGT has also been suggested (Moretti and Royden 1988). Subduction beneath the Dinarides is not yet supported by data on Moho stacking. However, both the Apennine subduction and the Dinaride subduction can be traced southeastwards into seismically active Benioff zones below the Tyrrhenian sea and the Aegean sea.

Mesozoic sedimentary facies belts in the Dinarides and the Apennines suggest that both of these overlaps are located within continental crust. However, the overlaps may also have developed at the sites of Mesozoic failed rifts with thinned or even oceanic crust within the Adriatic microplate.

Alpine unit

Shown as (2) in Figure 6-30, this unit includes the crustal core of the Alps assembled from the original north front of the Po Plain unit or the Adriatic microplate, but it also incorporates substantial parts of the European crust. Its base is defined by shears rather than outlined by Moho segments. Its conceptual definition is largely based on the geological interpretation of Alpine basement complexes, as expressed in the previous section, 6.4, but seismic data clearly outline its shape.

Along the EGT profile, the present base of the Alpine unit is outlined by the south-Alpine backthrust and the Insubric line, and by a segment of the collision suture in its late-Alpine position. Under the Alpine body proper, the base is not particularly visible, but ambiguous seismic data still allow interpretation, including detachments at the base of the Aar massif and below internal foreland units. If geologically real, these elements collectively form an active detachment which is absorbing the present trans-Alpine convergence. It is broadly synclinal in shape and 23–30 km deep. Surface geology would suggest that this detachment cuts discordantly through an assemblage of crustal units, suture segments, and sediments. At its southern tip beneath the Piemont area, the Alpine unit may or may not continue into the Apennines.

Northern Apennines unit

This unit constitutes the crustal underpinning of the Apennine orogen as a mappable area with coherent Moho events. It is distributed between the Apennine crest and the Ligurian coast, with a nearly flat Moho at about 24 km depth. From its leading edge below the Apennine crest, the sub-Moho wedge of material with mantle velocities thickens to about 18 km at the coastline. The upper crust and sediment cover of this segment contain stacked thrust sheets and are affected by late Miocene and younger crustal extension. The supracrustal northward extension of the Moho overlap displays the north-vergent architecture of subsurface thrust lobes and folds explored for oil and gas (Pieri and Groppi 1981). The structural assemblage produced during the supracrustal thrust succession and the Moho stacking suggests that all thrust imbrications on top of the northern Apennines unit are older than the documented Moho overlap and must have been powered by lithospheric thrusts further south or west. It also suggests that the documented Moho overlap is dated by the subsurface foothills folds as Pliocene.

Sub-Monferrato unit

This lithospheric slab is similar to the northern Apennine unit, and both may either be coherent or separated by a fault contact of unknown structure. The Moho of the sub-Monferrato unit is documented along a line of dip of at least 50 km extent, at a depth between 20 and 32 km. Its surface expression is the Monferrato complex of deformed Neogene accretionary wedge. Its thrust emplacement is recent enough to deviate the Po River northward by about 15 km. The sub-Monferrato unit describes an Apenninic crustal slab, presumably emplaced upon European crust during the Pliocene and Pleistocene. Its relationship to the Alps of Piemont immediately to the west is not understood. Apparently the sub-Monferrato unit is separated from the Piemont Alps by a N–S striking fault mapped at the surface as the Sestri–Voltaggio line.

6.5.4 PLATE PATHS, OROGENIES AND LITHOSPHERIC MASS BALANCE

To gain insight into the dynamics of the lithosphere, we are restoring structural cross sections to the depositional state of their sediment cover. In the central Mediterranean area, three problematic topics can be approached by using restorations. First, why are the overthrust vectors of the Neogene orogenic belts not all parallel to the independently obtained direction of convergence between Europe and Africa? Smith (1971) has shown that the documented pattern of intra-Mediterranean small-plate motions cannot be resolved by Eulerian vector addition, a fundamental assumption of the plate tectonic theory. Instead they form a self-contained system. This topic is part of a newly emerging geodynamic model to be discussed in Chapter 7.2.

Secondly, why is the amount of Neogene convergence between Europe and Africa, measured across the Alps as the predominant site of indentation, two or three times greater than the documented and dated Alpine bulk strain? A partly satisfactory answer is that the

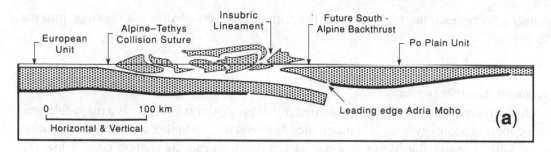

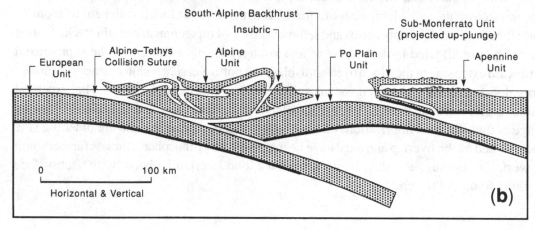

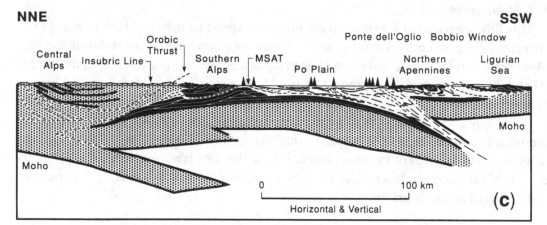

Figure 6-31. Schematic cross sections to illustrate the evolution of the Adratic microplate in relation to the Alps. (a) pre-collision stage, (b) collisional stage, (c) present day.

Adriatic microplate, serving as one jaw of the vice of Alpine convergence, is performing non-Eulerian movements independently of Africa. As another partial answer, the Alpine convergence mismatch suggests that there have been unrecognised oceanic basins contained in the convergence path. These basins have been subducted without geophysical or petrological trace.

Thirdly, where is the disposal site of the lithospheric surplus generated when trans-Alpine convergence is applied to normally thick, thermally defined lithosphere? This problem can be partially solved since seismic Moho mapping has confirmed an asymmetric model of subduction. Thermally defined lithosphere in the lower (European) plate of the Alpine subduction can be eliminated by thermal assimilation dependent on the thermal constants, its thickness, and the convergence rate (Oxburgh and Turcotte 1970) although this may not be sufficient. More likely, lower lithosphere becomes detached and sinks, as explained in Chapter 7. As is shown schematically in Figure 6-31, the base of the upper (Adriatic) plate is a northward and westward rising tectonic surface cutting successively through mantle, crust, and sediment cover. The upper mantle in the upper plate is affected neither by the Alpine subduction nor by the associated compression and backthrusting, because the south-Alpine backthrust merges with, and intersects, the base of the upper plate north of, and above, the leading edge of the (Adriatic) Moho. Therefore, no upper mantle is participating in the south-Alpine convergence. Figure 6-31a shows a restoration, in cross section, of the Adriatic microplate to pre-collisional state after deposition of its cover sediments. To make this restoration, we have assumed that the convergence acted in the plane of section, and we have used the abundant data on supracrustal tectonics (Castellarin *et al.* 1985, Cassano *et al.* 1986, Bally *et al.* 1988, Roeder 1989a). We have matched this restoration against the available seismic and depth converted Moho data. The match between both types of palinspastic data is imperfect, but it does support the tectonic interpretation of stacked Moho segments. In the Apennines, crustal data are strongly affected by insufficiently charted extension, but the base of the crust does show significant tectonic overlap. In the Alps, crustal compression is better understood, but the original base of the crust is incomplete. The regular wedge shape of the north Alpine crustal front suggests that it originated in Mesozoic time as an extensional detachment, although it became modified during a long history of convergence between Europe and the Adriatic microplate, subduction, wedge compression, and mechanisms of blueschist re-emergence.

6.6 SARDINIA CHANNEL AND ATLAS IN TUNISIA: EXTENSION AND COMPRESSION

D. Roeder

At its southern end, the EGT intersects for a third time the Neogene plate boundary between Europe and Africa. A composite fold-thrust belt with detached foothill folds and with docked exotic fragments of the Penninic accretionary wedge overlies a crustal wedge which resembles a passive extensional margin. Although there is neither trench nor Benioff zone nor magmatic arc, this convergent margin is consistent with the Mediterranean orogen model if we appreciate that major parts are not intersected by the EGT profile, and if we make allowance for some shortcomings in our documentation.

Figures 3-19 and 4-19 reveal little evidence of a N-dipping slab of seimic high-velocity lithosphere beneath Tunisia and the Sardinia Channel which could represent a subduction zone. Yet parallel sections of seismic tomography both east and west of the EGT transect (Spakman 1986, 1990a) show such a feature clearly. There is the suggestion in Figure 3-19 of a N-dipping high-velocity slab at greater depth that has become detached from the lithosphere above. Figure 6-32 shows in more detail the P-wave velocity variations in this section of the EGT taken from an earlier model by Spakman (1986). It indicates that asthenosphere or very thin lithosphere may be in contact with the Moho of the Atlasian foreland or Sahara platform. It shows that the possible break between surface lithosphere and the detached, sinking slab is very shallow, perhaps less than 150 km depth. This leads to the intriguing conclusion that the downgoing slab may not be Alpine in age but possibly Variscan or Caledonian. The Atlasian system must have the Mediterranean type of architecture with spreading just behind the thrust front.

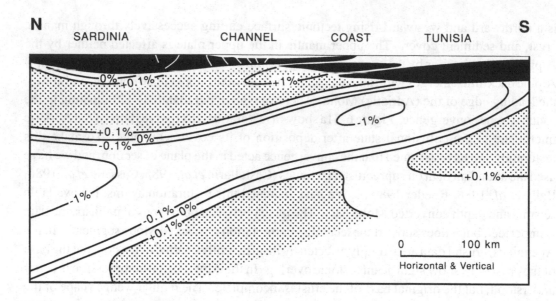

Figure 6-32. Cross section along the EGT trace between the southern tip of Sardinia and the Saharan end of the EGT near Gabes, obtained by combining the crustal section (Figure 6-33) with the seismic tomography data of Spakman (1986), redrawn. The contours indicate the percentage variation in P-wave velocity from a standard Earth model.

Lithospheric data

As interpreted by identification of phases and by raytracing (Figure 3-15), the EGT profile shows the normally thick continental crust and 70 km thick lithosphere of Sardinia thinning southward, a crustal rift under the Sardinia Channel, and a southward thickening crustal wedge under the mountainous part of Tunisia. In the Sardinia Channel, the crust is thinned by the Moho rising from 31 km to 20 km depth, and by up to 5 km of layers with seismic velocities typical of unconsolidated sediments. An 8 km thick body wih high P-wave velocity is most probably a pillow of basic intrusive rock, as have been commonly found beneath rift systems, and the high seismicity and heat flow (Chapters 5.1 and 5.5) are testimony to current activity. In Tunisia, the EGT profile (Figure 3-16) suggests a continental crust thickening southwards at a steady rate of about 90 m per km. Recomputed for an azimuth across the strike direction of the Atlas belt, the crust of the African craton thickens at a rate of 150 m per km, similar to the extensional margin of the European continent against the Ligurian sea.

In Algeria, no published Moho data are available, but the attitude of the surface topography suggests a southward thickening, as well as a segmentation into three slabs: a belt of exotic terranes along the coast, a northward tilted imbricate crustal slab under the Oran Meseta, and the northern edge of the Saharan block beneath the Sahara Atlas.

Supracrustal geology

In its present state, the island of Sardinia is a pre-Alpine crustal and orogenic complex with a discontinuous cover of Tertiary sediments. In the Sardinia channel, the gently synformal base of sediments is interrupted by several crustal fault blocks. One of them outcrops in the island of La Galite in Tunisian waters. The Tunisian shelf and continental slope appear undisturbed and tectonically quiet (Figure 6-33).

Onshore Tunisia, as the site of the Numidian and Atlas fold-thrust belt, shows a major

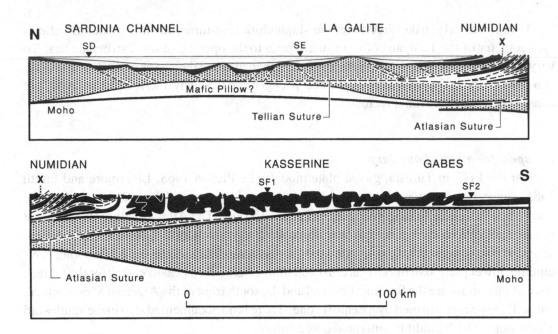

Figure 6-33. Crustal cross section along EGT trace from the south tip of Sardinia to Gabes in Central Tunisia, showing a geological interpretation of the seismic section of Figures 3-15 and 3-16, based on supracrustal data supplied by Zaghouani and mapping by Burrollet (1967). Point X at the left of the lower section is coincident with X at the right end of the upper section.

southeastward thinning wedge of Triassic to Neogene sediments deposited on a north-facing passive extensional margin of Africa. In its present state thickened by folding, the wedge may be as thick as 15 km at the north coast and less than 5 km at the edge of the Sahara platform. In the Tunisian coast ranges, the African series is overlain by the allochthonous Numidian series, which is an accretionary wedge of Saharan provenance, assembled north of its present site, and emplaced during the Miocene.

In the African series, compressional detachment is evident in thrust sheets, box folds, and open folds with a southeastward decrease in strain. The folds have wavelengths of 2–10 km and have not yet reached locking position. Diapirs, core exposures and, perhaps, thrust lamellae of Triassic evaporites show that the main detachment is located near the base of the sediment wedge.

The Numidian series is a stack of thrust sheets with imbricate or chaotic internal structure. Interspersed slices of Triassic evaporite and the basal contacts show that the Numidian terrane was emplaced on top of an undisturbed African sediment complex, and that both parts were detached and deformed after emplacement. The structural style of the Tunisian Atlas suggests between 25 and 50% bulk strain (most likely 30%), and between 60 and 120 km of detachment (most likely 80 km) after emplacement of the Numidian series. Somewhere to the north of the Numidian belt, this detachment should merge with a southeast-vergent intracrustal thrust. A more southerly intracrustal thrust or suture, that is, the trace of the Europe-Africa plate boundary, is inferred but not documented.

In the Algerian cross section, the duality of the sedimentary and orogenic series is emphasised by the intervening crustal panel of the Oran Meseta. The Tellian and Bibanais segments show the thrust style of the Numidian series. The Sahara Atlas is dominated by late Triassic and younger buoyant and piercing evaporite; its style is different from the regular fold style shown for the Tunisian Atlas. This, however, may be in part an artifact.

The northeasterly strike direction of the Maghrebide fold-thrust belt in Tunisia still reflects the geometry of the Tethyan collision suture prior to the opening of the Tyrrhenian sea. To varying degrees, the accomplished folds are lengthened along their axes and offset on strike-slip faults. These features of Pliocene and younger age (Castany 1951, Burollet 1967) are consistent with extensional tectonics.

Europe-Africa plate boundary

Near the EGT in Tunisia, global plate models (Le Pichon 1968, Livermore and Smith 1985) suggest a convergence rate between Africa and Europe of 10 to 12 mma^{-1} since the early Miocene, and a finite convergence of about 250 km. The Atlas structure suggests that much or all of this convergence has been absorbed in a detachment at the base of the sediment series. It also suggests that this implied subduction zone should involve a 200 km long NW-dipping, lower plate slab of African crust and lithosphere. Two possible sites for the surface trace of this suture are the Sardinia Channel and the south front of the Algerian Meseta block, but neither site is confirmed by a Benioff zone. There is no documented step in the southward Moho slope which could be interpreted as a suture.

The distribution of relative (residual) P-wave velocities as obtained by the seismic tomography method (Spakman 1990b) is consistent with a north-dipping subduction slab at the site of the Europe–Africa convergent plate boundary. Tomography data also suggest that the plate boundary is somehow linked to the Saharan and Tunisian Atlas and not to the Numidian thrusts and the Sardinia channel (Figure 6-33). These data suggest that the residue of a subducted slab may be represented by a deeper high-velocity body detached from the lithosphere. Extensional tectonics, suggested by crustal structure in the Sardinia Channel, is coeval with Atlasian compression and with early stages of the Tyrrhenian opening. By its setting in the upper plate of a subduction zone, this extension would qualify as a site of backarc spreading.

6.7 RECENT TECTONICS OF THE MEDITERRANEAN

D. Roeder and P. Scandone

What are the driving forces of the swirling field of orogenic belts and centres of localised extension so typical of the Mediterranean? Much of the pattern originates from the convergence between Africa and Europe (Argand 1924), evidence of which is available to us as dated plate motion paths at a resolution of 10–100 km (Dewey *et al.* 1989, Roeder 1989a). However, the topographic, paleogeographic, and geodynamic evidence suggests a more complex story (Smith 1971). Its understanding requires a look at the dynamics of compressional belts with extending and collapsing hinterlands. It requires a look at ridge push, slab pull, and sublithospheric convection patterns, and it requires a scrutiny of planetary mantle flow patterns, the subject of much of Chapter 7.

Fundamental to an understanding, however, is a knowledge of the evidence. Mediterranean tectonics is at present producing local ocean basins deeper than 3 km. It is also producing active volcanoes, seismic zones, and countless human tragedies, such as the destruction of the Minoan culture (1450 BC), of Pompeii and Herculaneum (AD79), Messina (1906), Friuli (1976), the Napoli area (1983), and Kalamata (1986) to name just a few. The system is geologically set for future disasters, although in critical areas anywhere between Budapest, Algiers, and Kurdistan we cannot easily predict their location.

A majority of orogenic belts in the Mediterranean segment of the Tethyan suture zone display a combination of convex-outward curvature, thrusting on the external side, and radial extension on the internal side (see Figure 7-6). This Mediterranean geometry is known in other parts of our planet as divergent arc (Dewey 1980) or subduction with backarc spreading (Le Pichon *et al.* 1973).

6.7.1 MEDITERRANEAN PLATE TECTONICS AND LITHOSPHERIC DYNAMICS

Our kinematic understanding of the Mediterranean became quantitative after the Mesozoic opening of the Atlantic was charted with magnetic surveys and JOIDES drilling results (Pitman and Talwani 1972). The restoration by Eulerian vector addition follows the techniques proposed by Wilson (1965) and by McKenzie and Morgan (1969). Smith (1971) showed that Mediterranean kinematics requires local plate vectors in addition to the Europe–Africa convergence. The first plate tectonic basin study of the Mediterranean (Dewey *et al.* 1973) used the global Europe–Africa convergence vectors, restorations of extensional tectonics, and available estimates of bulk strain in the fold-thrust belts. Newer attempts use improved databases. References can be found in Ziegler (1988), and improved numerical plate paths have been presented by Dewey *et al.* (1989).

Our dynamic understanding of the Mediterranean lithosphere follows the work by Le Pichon (1983) and his students, and it uses the concepts of local indentation and lateral escape flow (Pavoni 1961, Tapponnier 1977). We explain the fields of extension behind orogens as topographically determined plateau collapse (Dalmayrac and Molnar 1981, Dewey *et al.* 1986, Molnar and Lyon-Caen 1988). Advanced stages after orogenic collapse are more abundant and more obvious in the Mediterranean and are driven by the stress fields of lateral density changes between crust, lithosphere and asthenosphere (Bott and Kuznir 1979), including ridge push (Solomon and Sleep 1974, Turcotte 1983) and slab pull.

In slow or decaying convergence, subducted slabs are affected by a kinematic process called hinge retreat (Molnar and Atwater 1978) which leads to mushrooming asthenosphere and backarc spreading. The growth of these local spreading sites is governed by the geometry of the subduction zone and by points of indentation between continents. An account of areas lost and gained in convergence and backarc spreading does not reveal a direct relationship to the convergence rate. Rather, it suggests that the overall supply of fresh asthenosphere is controlled by the Europe–Africa convergence. The overall polarity of the Mediterranean orogens, their local convergence vectors, the dip direction of their subduction slabs, and their age succession perhaps suggests the effect of a global pattern, but there is no clear evidence for this.

Continental convergence

Figure 6-34 shows a collection of published plate paths of Alpine convergence since early Oligocene obtained by vector addition through time between the three plates of North America, Europe and Africa, and by vector addition between the Adriatic microplate and Africa. To illustrate the earlier effects of the Atlantic opening, Figure 6-34 also shows a plate path of Iberia relative to Africa since the Dogger age of the earliest mid-Atlantic ocean floor.

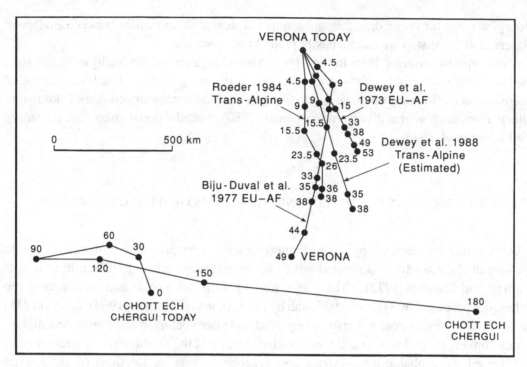

Figure 6-34. Collection of trans-Alpine plate paths projected into a vector through Verona just east of the EGT trace, redrawn after Roeder (1989a). Also shown is a plate path of Iberia relative to Africa redrawn after Roeder (1990). This path shows plate movements associated with the opening of the mid-Atlantic since the Middle Jurassic. Numbers are ages in Ma.

The trans-Alpine paths, just east of the EGT, are oriented northerly or northwesterly, and they show between 340 and 500 km of finite convergence, or between 9 and 13 mma⁻¹ since collision. The convergence pattern has the shape of a spherical pie wedge. Finite and present rates increase eastward from the Eulerian pole at the Azores triple junction to nearly 50 mma⁻¹ in southeast Asia (Le Pichon 1968). On a great circle through the Aegean–Hellenic system, the present rate is given as 7–10 mma⁻¹ (Le Pichon 1983).

Local vectors of convergence and extension

Convergence parallel to the continental path is suggested by thrust fronts on both flanks of the Alps, on the Carpathian north front, in mountainous North Africa, and on a short segment of the Hellenide and Tauride south front. At more than half of the frontage of Mediterranean orogens, however, the thrust vectors deviate by more than 45° from the continental path. In several areas, the fold-thrust belts are accompanied by backarc extension, often at higher rates than the continental convergence. At the Hellenic trench, for example, the plate vector approaches zero (Le Pichon 1983). The vector addition becomes even more complex where the foreland of the compound orogenic belt moves independently, such as in the Apennines (Patacca and Scandone 1989).

Data for plate reconstructions in the western Mediterranean are scarce and lead to subjective and widely differing results. For individual events of backarc spreading, finite microplate paths are based on interpretively dated oceanic magnetic anomalies and on stratigraphic data at the basin flanks (Cohen 1980) as illustrated in Figure 6-35. This type of local microplate path is defined relative to the continental plate on the backside of the spreading site. The associated convergence vector at the front of the spreading site can be measured only by structural analysis of the fold-thrust belt.

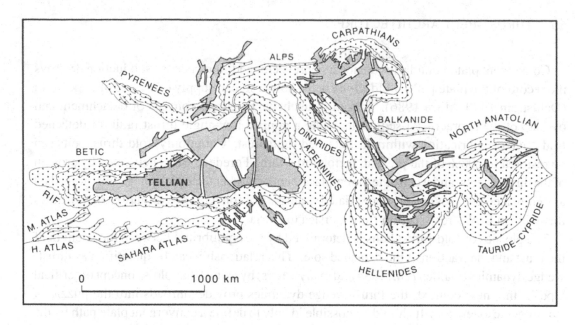

Figure 6-35. Map of Mediterranean orogens (stippled) in Recent plate configuration, showing Neogene sites of extension and/or backarc spreading (dark shading), largely based on Boccaletti et al. (1982). Coastlines are not shown.

Path of Adriatic microplate

The common foreland of Apennines, Southern Alps, and Dinarides, measuring roughly 1000 km by 700 km, the Adriatic microplate is a cratonic fragment of the early Mesozoic passive extensional margin of the Tethys embayment between Europe and Africa. Since the Oligocene, this fragment became partially and increasingly wedged between the converging continents. Since the Miocene it has been indenting Europe in a hard collision and has, we assume, been transform-sliding along, or converging across, Ionian oceanic crust connected to the African plate. Its plate path is critical to its role of indenter and to the dynamics of Apennine–Tyrrhenian backarc spreading. However, its plate path cannot be determined because it has only convergent and transforming plate boundaries. Any record of early divergence from African plate terranes is now obliterated by burial beneath the thick Ionian sediments.

Quantitative attempts at determining the paths of the Adriatic microplate and other fragments have been based on paleomagnetic data suggesting counterclockwise rotations against the global magnetic field, of the Apennine foreland in SE Italy and of the island chain of Corsica and Sardinia (Vandenberg and Zijderveld 1982). Livermore and Smith (1985) have obtained a path by determining the rotation required to restore the Apennine foreland from its present known location and orientation to its assumed initial location and orientation. Anticlockwise rotation of west Mediterranean terranes around a pole near the indenter point of the Adriatic terrane has left behind a trail of wedge-shaped backarc basins. In this conceptual model, the pivot and indenter point are assumed to show some movement relative to Africa and to move northwestward, almost along the continental path and almost at the Europe-Africa convergence rate, relative to Europe.

6.7.2 THRUST-BELT ARCHITECTURE

Convergent plate boundaries do not support vector addition because subduction destroys the record of the plate path. Fold-thrust belts, however, can supply estimates of bulk strain (Dahlstrom 1970, Mitra 1986). In fold-thrust belts, the strain consists of detachment and overlap of the supracrustal sediments along a ramped low-angle thrust fault, of detached folds, and of imbrication within the detached supracrust. Commonly, fold-thrust belts can be restored to 200% of the present, tectonised width. Foredeep fill is generated by erosion of the thrust sheets, and it can date the emplacement of marker tectonic units. Bulk-strain data and emplacement ages can produce a dated strain path of convergence. Figure 6-36 illustrates this with an E–W cross section from Corsica to Adria.

The physics of fold-thrust belts is determined by the equilibrium between the strength of the crust and the traction at the detached sole. This relationship can be quantified as thrust-wedge dynamics (Dahlen *et al*. 1984 and many others) by using Chapple's concept of 'critical taper'. In a new context, the thrust-wedge dynamics provides inroads into the puzzle of synorogenic extension. It should be possible ideally to define a convergent plate path by the strain path obtained from a fold-thrust belt. However, two additional tectonic processes limit this use of fold-thrust belts.

First, additional subduction zones may dislocate the basal detachment of a fully developed fold-thrust belt. This has been postulated in the north-central Apennine (Royden and Karner 1984) by the flexural geometry of the Pliocene foredeep fill. The implied crustal break is also visible in seismic data, and it extends southwards into the Tyrrhenian Benioff zone. It adds an unknown distance to the plate path.

Secondly, the lithosphere may detach in the weak middle crust and may create plate convergence without supracrustal expression. An intracrustal blind thrust is suggested by seismic data and seismicity data along the Alpine north front (Mueller *et al*. 1980). It may also be developed at the contact between the Adriatic microplate and the Ionian oceanic crust SE of Sicily. Intracrustal detachment is also needed to explain the effect of Tyrrhenian extension in the Tunisian Atlas.

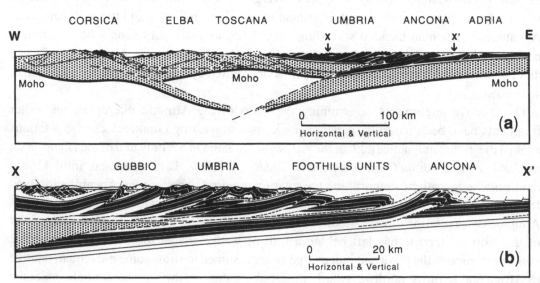

Figure 6-36. (a) Structure cross section of the Umbrian segment of the Apennines showing a fold-thrust belt detaching and imbricating a sediment series of Triassic to Miocene age. Redrawn after Roeder (1991).

(b) Detail of cross section (a) between points X and X'.

6.7.3 PLATEAU COLLAPSE AND ASTHENOSPHERE PUSH

Fold-thrust belts moving under conditions of critical taper accumulate hinterland terrains coalescing into a large, high, and extending orogenic plateau. In the model of Molnar and Lyon-Caen (1988), extensional tectonics limits the height of the plateau to 4–5 km, depending on the rheology, the rate of convergence, and the amount of erosion. In a steady-state process, the orogenic plateau grows laterally as a field of extensional shards, and its marginal fold-thrust belt advances over its foreland. The growing crustal root leads to isostatic uplift, to more top extension, and to thermal softening of the deeply buried felsic root terranes.

Figure 3-36 is a cross section of the Umbrian segment of the Apennines east of Corsica. Flexure of the foreland basement is documented in offshore seismic data and in the geometry of the Pliocene foredeep fill. It suggests an intracrustal Benioff zone piggybacking the internal (west) part of the fold-thrust belt and its crustal substratum. This subduction is also documented in a seismically mapped Moho offset and in the Calabrian Benioff zone.

Plateau collapse is driven by crustal convergence and controlled by the density contrast between felsic or supracrustal rocks and air, that is, by the same mechanism that controls the critical taper of fold-thrust wedges. This mechanism does not work in the Mediterranean sites of extension shown in Figure 6-35, because the sites are topographically lower than the forelands of their fringing fold-thrust belts. However, at sites of thin lithosphere, there is a lateral density gradient of the high-risen mantle asthenosphere adjacent to thicker lithosphere and crust with felsic and supracrustal orogenic rocks. The compressive stress exerted by this gradient easily exceeds the basal traction required to move a fold-thrust belt and build up or maintain its topography (Bott and Kuznir 1979, Turcotte 1983, Le Pichon 1983, Chapter 7.1). Because this force also exists in the flanks of oceanic spreading ridges, it is sometimes referred to as ridge push (Solomon and Sleep 1974).

Despite their fundamental difference, both forms of synorogenic extension occur in the same geological domain. At least some of the Mediterranean extensional sites may have started as collapsing orogenic plateaus. For example, the largely compressional and high-rising Western Alps grade continually into the low and largely extensional Pannonian basin. Where is the switch from deep mantle and plateau collapse to shallow mantle and ridge push?

This problem is not solved, but it leads to considering two more aspects of the lithosphere, namely metamorphic core complexes and indentation or slip-line tectonics.

Metamorphic core complexes

Extension of lithosphere is accompanied by thinning and by buoyant rise of its layers. Sharply localised fields of extension and tectonic uplift have been described in much detail from the North American cordillera as metamorphic core complexes (Crittenden *et al*. 1980, Coney 1980). Metamorphic haloes accompanied by extensional tectonics are known from several Mediterranean sites of orogeny-related extension, such as in the northern Apennines (Carmignani and Kligfield 1990), in the Pelagonian massif in the Hellenides (Le Pichon 1983), in the Menderes massif in the Taurides of Turkey (Dewey *et al*. 1986), and in the Pannonian site (Horvath, pers. comm.). There may be more core complexes to discover.

The metamorphism in some of the Mediterranean sites has been interpreted as burial beneath excessive overburden (Carmignani and Kligfield 1990) or as unspecified orogenic metamorphism (Jacobshagen 1986). In its modern form (Roeder 1989a,c and others), however, the Cordilleran model directly relates the metamorphism to the extension. As

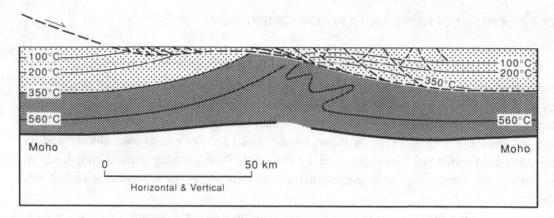

Figure 6-37. Schematic cross section of a metamorphic core complex showing the thermal effects of large-scale extension by trans-crustal low-angle normal faulting. The thermal structure, with selected isotherms, is shown ideally at the instant of accomplished extension. Stipple: elastically reacting upper crust, shading: viscous lower crust. The hangingwall is extended and warped. The footwall is flexed elastically by buoyant upwarp. Its hot basal area is in contact with shallow parts of the hangingwall. Redrawn after Roeder (1989a).

shown in Figure 6-37, the setting for this model includes a three-layer lithosphere, an average or less than Barrovian geothermal gradient, and major extensional bulk strain. Dip slip in the order of 100 km along a trans-crustal or trans-lithospheric low-angle detachment is accompanied by smaller-scale extension in the hangingwall and by buoyant and viscous uplift of the footwall. The buoyant rise of the footwall compensates for the load of the tectonically removed overburden. The upper-crustal or lithospheric parts of the footwall are upbent elastically. The viscous underpinning rises in the shape of a pillow. This setting can achieve amphibolite-grade metamorphism in the thinned hangingwall at the surface. It can also generate conduits for granitic melts into the hangingwall, but it cannot migmatise hangingwall rocks.

The temperature at the top of the buoyant viscous pillow is that of the lower crust cooled during the removal of the hangingwall. If the viscous uplift is 15 km or more, and if its strain rate exceeds the cooling rate, amphibolite-grade metamorphism and granitic wet-melt conditions (Dallmeyer *et al.* 1986, Snoke and Miller 1987, and others) can be generated in the shallow and extended hangingwall rocks. At an average dip of 15° on the detachment, this uplift requires 60 km of extension, and one-dimensional thermal modelling suggests that the extension must take place at rates well above 10 mma[-1].

Depending on the size of the extension and on its duration, the lithospheric effects of extension by rising viscous pillows may include rise of the Moho, flattening of orogenic Moho roots, rise of the thermal top of the asthenosphere, and incipient sea-floor spreading. The sites of extension in the lower lithosphere may be located directly below the stretched upper crust or may be found elsewhere in the system transferred laterally along detachments in the viscous lower crust.

Indentation and escape

During a collision, crustal edges with promontories and embayments will generate regional orogenic complications (Dewey and Burke 1973). During the subduction of a jagged crustal edge, the lithosphere beneath crustal promontories will sink more slowly than the lithosphere beneath the embayments. Unless the lithosphere detaches from the crust, it

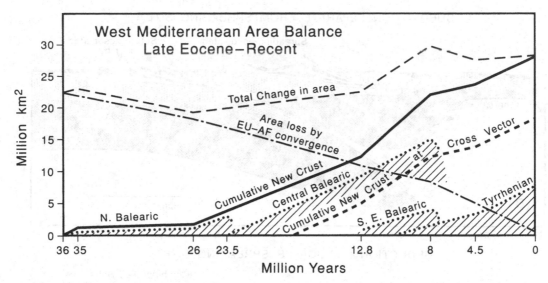

Figure 6-38. Graph of area in km² changing through time, depicting gains and loss of area during the Oligocene and Neogene evolution of the western Mediterranean.

will react with hinge advance, and it will localise compressional belts. In slow convergence, the crustal embayments will sink faster, perform extensional hinge retreat, and serve to localise the sites of backarc spreading.

Tapponnier (1977) has used an analogy from metal shaping by indenting or extrusion. This mechanism explains how material is pushed away from sites of indentation, and how it fills the voids opening at the sites of backarc spreading. As previously indicated by Pavoni (1961), the escaping material moves along slip lines, that is, faults of predictable orientation and strain. Tapponnier sees Mediterranean tectonics as a field of escape sites between the indenting Adriatic microplate and Asia minor. Material from the compressed Alps is escaping into the opening Pannonian basin accompanied by strike slip on the Insubric line. The tectonic boundary between Alps and Apennines may be a slip line. The southern edges of the west Mediterranean backarc basins require strike slip in their strain geometry. All tectonic mechanisms discussed serve to generate thickness variations in the Mediterranean lithosphere: subduction and convergent stacking, extension and viscous pillowing, indentation, and escape. Although thermal thickness variations are ephemeral, compositional thickness variations will survive the final freezing of the convergence. In the following chapter, these ideas are given quantitative expression.

6.7.4 EXPANDING BACKARC BASINS

The west Mediterranean series of backarc basins suggests that continental convergence displaces asthenosphere at depth, as depicted in Figure 4-19, and induces it to ascend in a stringer of mantle diapirs or viscous pillows. This can be derived from a west Mediterranean plate model since the early Oligocene (Cohen 1980, Livermore and Smith 1985).

Figure 6-38 is a graph showing cumulative changes in area through time. It shows the convergence as a loss in area of 0.3×10^6 km² in 26 Ma. It also shows the growth of new sea floor along spreading directions parallel to the convergence vector, and it shows the growth of backarc basins spreading normal to the convergence. The areas compared are confined to the western Mediterranean; the backarc basins do not include the Pannonian, Aegean, and

SUBDUCTION SEPARATE FROM SPREADING SITE

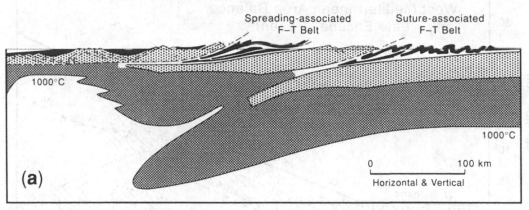

SUBDUCTION DISLOCATES SPREADING SITE

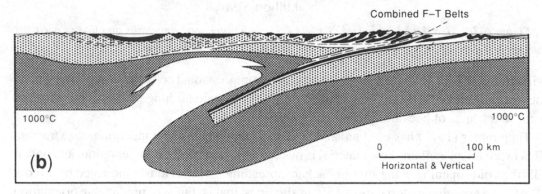

ALLOCHTHONOUS SPREADING SITE

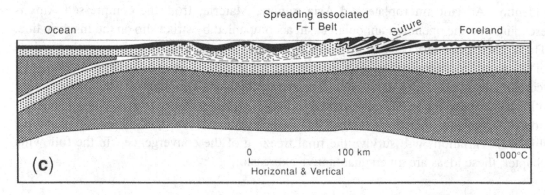

Figure 6-39. Sketch cross sections showing three varieties of orogenic belts common in the Mediterranean.

(a) Two fold-thrust belts geographically separate. The internal one is associated with an extensional site and has a shallowing basal detachment at the base of severely thinned crust, possibly overlying a lithospheric thermal high. The external one is the surface expression of a plate boundary. Its hangingwall block is a high and inclined crustal slab in outcrop or at shallow depth, referred to as median crustal slab. Its footwall block is an elastically deflected segment of crust with pre-Alpine thinning. This variety is best realised in the Tello-Rifian and Atlasian systems of Algeria.

(b) Two convergent crustal contacts with supracrustal fold-thrust belts. The external contact dislocates the internal detachment, piggybacking its fold-thrust belt. The site of subduction may not be evident in the thrust architecture, but it can be recognised by the geometry of the foredeep fill, by Moho offsets, and by Benioff zones. At the internal fold-thrust belt, extended crust is thrust

Alboran sites of extension. The graph suggests that in the western Mediterranean, extension exceeded convergence by about 50%. About half of the excess spreading is absorbed in west-to-east subduction at the Sardo–Corsican front and at the Calabro–Panormide front. One third of the remaining excess spreading is absorbed by Apennine thrusting and possible subduction within the Adriatic microplate. The remaining fifth is absorbed in the Dinaride foothills. Hence the net area change by post-collisional tectonics in the western Mediterranean is about zero. The four individual spreading events follow each other like successive explosions from the NE to the SW, building up to a climax during the Balearic event and declining through the Tyrrhenian event.

6.7.5 MEDITERRANEAN OROGENIC CYCLES

For the study of Mediterranean tectonics with cross sections, it is appropriate to separate fold-thrust belts at the edges of extensional sites from fold-thrust belts in the forelands of lithospheric subduction zones. This cannot be done everywhere. In the Apennines, both systems overlap but have been separated through elastic-load studies (Royden and Karner 1984). The Hellenide site is dominated by subduction of the retreating hinge, and the extension-related system is carried in piggyback fashion (Le Pichon 1983). In the Atlas of North Africa, the subduction system generates the Sahara Atlas and the High Atlas; it is geographically well separated from the Tellorifian fold-thrust belt along the edge of the Alboran and Balearic extension sites. In a series of sketches (Figure 6-39a,b,c) we show the possible geometric relationships between the two associated types of compressional belts. Of the three illustrated combinations, two are common and well documented. A third type (Figure 6-39c) is theoretically possible and was suggested a long time ago (Andrieux *et al.* 1971), but it is not yet acceptably documented. Transitional forms between the sketched types are also possible and in part documented.

Alpine–Mediterranean orogenic belts undergo a life cycle of subduction, collision, topographic buildup, extensional collapse, and backarc spreading. The sequence of stages is determined by the tectonics of the converging lithosphere and the flowing asthenosphere, but the time spent in any of the stages is not fixed. We have identified three dynamic settings in which the lithospheric tectonics shapes the orogenic cycle: collision, decay of convergence, and hinge retreat.

over the median crustal slab. Incipient backarc spreading is suggested by crustal extension and by a lithospheric thermal high. This tectonic configuration is common, such as in the northern Apennines, Hellenides, and possibly in the Dinarides, Taurides, and northern Carpathians.

(c) Shallow internal thrust system which has overridden and covered the external subduction system over a distance of several hundred km. The detachment is most likely located in the viscous lower crust of the hangingwall slab. In the footwall, the crustal type is probably thinned-continental or oceanic; it may display a Moho at a normal continental depth and may mask the thrust overlap. The existence of this variety is suggested by paleogeographic reconstructions and by mismatches between subsidence and apparent crustal thickness; it implies complex map-view geometry and highly efficient mechanisms of indentation and escape. An example of this type is the Alboran Sea. In this diagram, the Alboran Sea would be seen from the north, with the open Atlantic on the right edge. Another possible example is the Pannonian basin with the Balkanide Benioff zone.

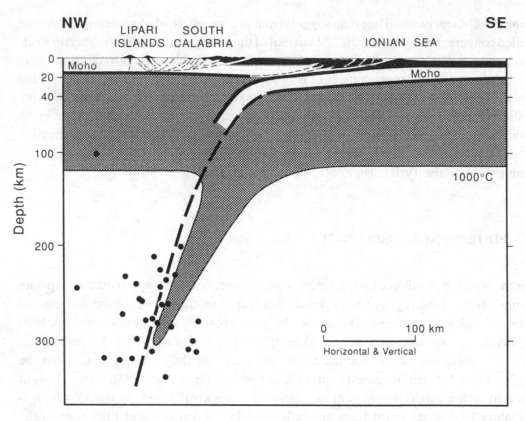

Figure 6-40. Cross section of Calabrian arc and Tyrrhenian Benioff zone, redesigned after Giese et al. (1982), with seismic foci from Ogniben (1969). Shaded area: lithosphere cooler than 1000°C.

Collision or indentation are needed to build the topographic elevation which initiates extension. Hinge advance in a subduction slab can also set an orogen into a compressional mode, but there are no Mediterranean examples of this situation.

Decay of convergence and/or hinge retreat are needed to change the setting from plateau buildup to crustal thinning and eventual sea-floor spreading. In the Neogene to Recent Mediterranean tectonic setting, both of these factors are present. Figure 6-40 shows a cross section of the Calabrian arc and Tyrrhenian Benioff zone. The sharp bend in the Ionian-Sea slab is constrained by the elastic-flexure parameter of the crust in the Ionian Sea and by the location of the Benioff zone. This cross section across the Calabrian arc and Tyrrhenian Sea serves to document the Mediterranean type of piggyback thrust belt with hinge retreat.In several Mediterranean models (Alvarez 1976, Biju-Duval *et al.* 1977, Livermore and Smith 1985), the age of the crust being subducted beneath the Apennines and the Aegean arc increases toward the trapped oceanic crust of the Ionic sea. Figure 6-41 illustrates the piggyback structure of the entire Apennines fold-thrust belt. Its location precludes that it was generated by the Tyrrhenian subduction. We interpret it as generated by Tyrrhenian backarc spreading.

In the Mediterranean, the predominance of west-dipping subduction slabs, of east-vergent fold-thrust belts, and of eastward migration of spreading sites is remarkable. Possible explanations can be based on global or on more local, specifically Mediterranean arguments but, based on available data, neither global nor local explanations are conclusive. Doglioni and collaborators suggest a westward directed toroidal shear motion between the lithosphere and deeper mantle realms (Doglioni 1990, Ricard *et al.* 1991). Some of their supportive geological observations had been used earlier to support an easterly mantle flow (Nelson and

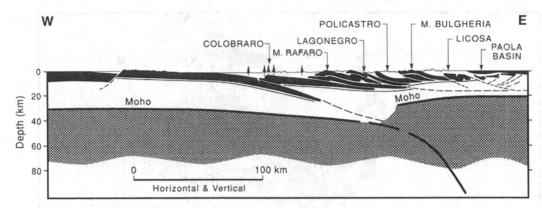

Figure 6-41. Structure cross section of southern Apennines just north of Calabria, simplified after Roeder (1984), and projected into lithospheric and Benioff geometry of Calabrian–Tyrrhenian arc as in Figure 6-40.

Temple 1972). The global hotspot configuration (reviewed by Duncan and Richards 1991) suggests a pattern of mantle convection unaffected by toroidal and global shear. It also suggests a slow toroidal displacement between the core-inducted magnetic field and the hotspot reference system; this torus is oriented roughly normal to the Earth's spin axis. A more local interpretation of Mediterranean mantle tectonics is imposed by a convecting mantle without toroidal shear. Local tectonic elements can still be explained by shallow mantle convection and lithospheric thickening (Channell and Mareschal 1989). Regional elements include the eastward increasing age of Tethyan oceanic crust. In maintaining the eastward hinge retreat, this shallow feature has predetermined much of the Mediterranean tectonic pattern. The density gradient between shallow mantle and deep continental crust (Bott and Kuznir 1979, Turcotte 1983, Le Pichon 1983) can also generate a tectonic polarity, with Tethyan mantle predominant to the east, and deep continental crust to the northwest and the south. This point is addressed again in Chapter 7.2.4.

6.7.6 VARISCAN ELEMENTS IN MEDITERRANEAN TECTONICS

Mediterranean tectonics clearly shows that orogenic loops and sites of extension and compression originate together and depend on each other. It also shows that some tectonic features outlast their orogenic cycle and help to predetermine the plate pattern of the subsequent cycle. It is therefore worth briefly re-examining Late Paleozoic orogens in the western Mediterranean area and their remnants of extensional basin fill, which possibly signal post-collisional plateau collapse. In the modern Mediterranean configuration, there are outward-vergent loops around sites of extension and inward-vergent loops around subducted or overrridden terranes. Based on our present understanding, extension at outward-vergent loops is synkinematic with orogeny. Extension may soon reach the stage of sea-floor spreading and may disperse the orogenic fringe toward distant shores. At inward-vergent loops, extension can only start as plateau collapse after terminal collision of the orogenic fringes, helped by the subducted foreland terrane being converted into a soft, buoyant orogenic root. If plateau collapse extends to the base of the crust and is not blocked by intraplate compression, the root can serve as the focus for the next generation of extension. Figure 6-42 shows the late-orogenic pattern of Atlantic Variscan belts as prepared by Ziegler (1988) and projected onto the pre-Atlantic plate configuration suggested by Livermore and

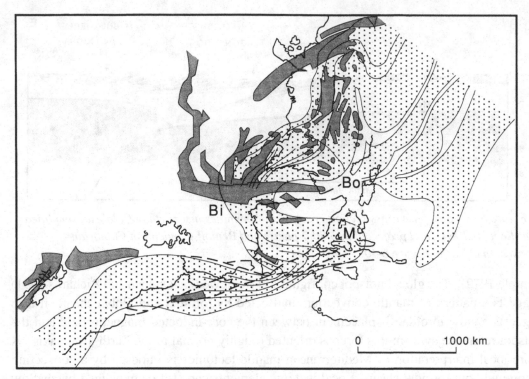

Figure 6-42. Sketch map showing the Atlantic Variscan architecture (Ziegler 1988) as projected into the plate restoration at 180 Ma by Livermore and Smith (1985). Stipple: main orogenic belts, M: Meseta loop, Bi: Biscay–Asturia loop, Bo: Bohemian loop. Dark shading: late- and post-orogenic rift sediments, possibly indicating plateau collapse.

Smith (1985). Figure 6-42 also shows Stephanian and Permian clastic units (Ziegler 1988) which are candidates for plateau collapse sediments. The Appalachian–Variscan system is shown to be bivergent. Its loops have a dual polarity and are therefore all capable of evolving towards plateau collapse. On the African–European side, the east-convex Meseta loop has no documented Late Paleozoic debris but forms the site of the Atlantic–Tethys group of transform faults of Jurassic age. The west-convex Biscay–Asturia loop is inward-vergent viewed from the east and contains considerable Stephanian and Rotliegend debris. It also serves as a transform system terminating the Faroe–Rockall rift (Ziegler 1988). The Bohemian loop is outward-vergent and its plateau collapse may have been part of the extensional field covering the Variscan heartland. As well as the well-mapped Rotliegend troughs, late Paleozoic metamorphic core complexes could still be discovered there. However, this part of the Variscan system did maintain its cratonic coherence after the Variscan orogeny. Additional Variscan areas with possible plateau collapse may have predetermined the pattern of Alpine–Penninic basins.

7 Geodynamics of Europe

D. BLUNDELL, ST. MUELLER AND K. MENGEL

7.1 HOW DOES GEOLOGY WORK?

In this book we have endeavoured to present a coherent view of the lithosphere of Europe, brought into focus through the European Geotraverse which provides continuity across the continent on a scale that encompasses the whole lithosphere. We have made use of this to reconstruct the geological evolution of the crustal units that now make up the continent of Europe, within a framework of plate tectonic processes taking place throughout the past 2 Ga of Earth history. But what of the mechanisms? What forces drive these processes? And how do some areas of the continent remain stable over long periods of time whilst others are remobilised again and again? What are the processes that split continents, fragment them into microplates, and what are the processes of terrane accretion and continental growth? Within a continent, deformation and alteration take place across extensive regions so that plate boundaries are not definable as lines of activity separating rigid undeformable plates as is the case in the oceanic regime. This is understandable from the discussion in Chapter 4 because the composition and physical conditions of continental crust are such that the rheology creates a weak zone in the middle crust that gives continental lithosphere a 'soft centre'. In contrast, the thin crust and basaltic composition of the oceanic lithosphere provide a more uniform and stronger rheological character. But if the continental lithosphere does have a 'soft centre' and is more readily deformable, why does it not deform more extensively than it does? Why, for example, is the Alpine collision zone so narrow? Many of these questions can be answered by looking at the geodynamics of Europe where, as a result of EGT, we have for the first time sufficient evidence available.

7.2 WHAT DRIVES TECTONIC PROCESSES?

7.2.1 GEODYNAMIC MODELLING

We should not get into too detailed a discussion of the forces that drive plate tectonics, but refer the reader to modern textbooks such as the one written by Fowler (1990). Suffice to say that the driving forces are generally regarded as arising from thermal instabilities in the

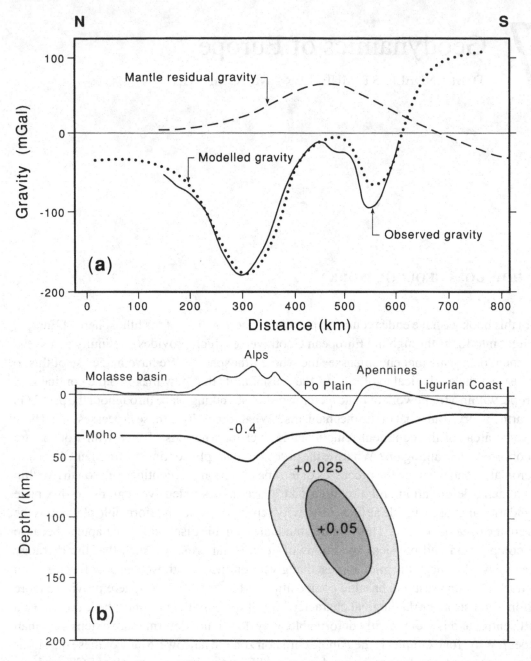

Figure 7-1. Gravity model of Werner (1985) for the Alps along the line of EGT (adapted from Mueller 1989).

(a) Observed Bouguer gravity anomaly corrected for near-surface effects (full line) and modelled gravity variations (dotted line) computed from density contrast of -0.4 gcm⁻³ of the crust at the Moho and a lithospheric root with density contrast of +0.05 gcm⁻³ (+0.025 gcm⁻³ around the periphery) relative to the upper mantle, as shown in (b). The positive residual gravity variation due to the mantle lithosphere root alone is shown by the dashed line.

(b) Density model cross section showing contrasts in gcm⁻³ between crust and upper mantle and a high density region within the upper mantle. The background upper mantle density relative to which the contrasts are expressed could be taken to be 3.3 gcm⁻³.

mantle which produce mechanical movements, particularly within the asthenosphere, its most mobile region. The global dynamics of this system interacts with the lithosphere to create the primary stresses driving plate movements. We have seen in Chapters 2 and 6 how the geology of Europe has been dominated by the relative movements of the major plates, carrying large continents, with smaller microplates inexorably caught up within the larger system. However, as is clear from Chapters 3, 4 and 5, the continental lithosphere is far from uniform, varying in thickness, composition, physical properties and behaviour. Internal stresses are created from body forces derived from density contrasts both within the lithosphere, for example at the Moho, and between the lithosphere and asthenosphere. In two key papers, Fleitout and Froidevaux (1982, 1983) recognised the dynamic role of lithosphere heterogeneities. They set up a theoretical basis for working out the stress fields due to particular density contrast configurations and then determined the velocities of mass movements in response to these stresses by means of the Navier–Stokes equation and the principle of conserving mass. As one example of their numerical treatment, they considered the mechanics of continent–continent collision leading to thickening of both crust and lithosphere as a whole. They recognised that thickening of lower density crust which displaces higher density upper mantle results in buoyancy forces that create uplift and extension in the upper crust. Such forces serve to readjust the crust to its former thickness. Thickening of the lower lithosphere, however, has the opposite effect. The lithosphere root, being slightly denser than the surrounding asthenosphere, creates downward directed forces, leading to subsidence and lateral compression. Because the lithosphere is in a metastable state, once thickening has started the body forces act to reinforce the process by positive feedback. Of course, treating the system in a purely mechanical way ignores the very important thermal controls which act to stabilise the situation, since the lithosphere base is as much a thermal boundary as a mechanical one. A full understanding of the process clearly requires a comprehensive thermo-mechanical treatment. None the less, Fleitout and Froidevaux demonstrated quantitatively that lithosphere heterogeneities could give rise to significant tectonic stresses in localised and self-supporting systems and, in the case of a mountain belt resulting from continent–continent collision, one that could sustain the mountain building process.

In parallel with these theoretical and modelling studies Kissling *et al.* (1983) and Werner and Kissling (1985) began to apply the same basic principles to the Alps, as we shall see in the following section, and to match their models against observed gravity (Figure 7-1).

The geodynamic modelling of mountain belts was taken a stage further by Bott (1990) using finite element analysis to follow the consequences of continent collision and lithospheric thickening, based on the theory developed by Fleitout and Froidevaux. His models required some 1000 iterations computed for 500 year intervals to arrive at a dynamic equilibrium after 0.5 Ma. The models assume that the lithosphere is made up of an elastic upper crustal layer 20 km thick above a viscoelastic layer, consistent with the rheological behaviour discussed in Chapter 4.2. He incorporated a density contrast of -0.4 gcm^{-3} at the Moho between crust and mantle and a density contrast of +0.05 gcm^{-3} in the centre of the lithospheric root, relative to the surrounding asthenosphere, diminishing to +0.025 gcm^{-3} towards the edge of the root, to match Werner's (1985) model for the Alps. The topography of the crustal root is modelled to be in Airy isostatic equilibrium with the surface topography of the mountains. The mass excess of the lithosphere root gives rise to a residual gravity anomaly of +78 mGal, comparable with that observed (Figure 7-1). Bott considered first the forces developed by just the topographic load of the mountains and the buoyancy of the crustal root. Large horizontal deviatoric stress is concentrated within the elastic upper crust, amounting to tension of over 100 MPa, as shown in Figure 7-2a, equivalent to forces of over 4×10^{12} Nm^{-1}. Tensional stress is also maintained in the lower crust which continues to creep slowly so as to thin (by 90 mma^{-1}) and widen (at 350 mma^{-1}) and thus reduce and spread the crustal

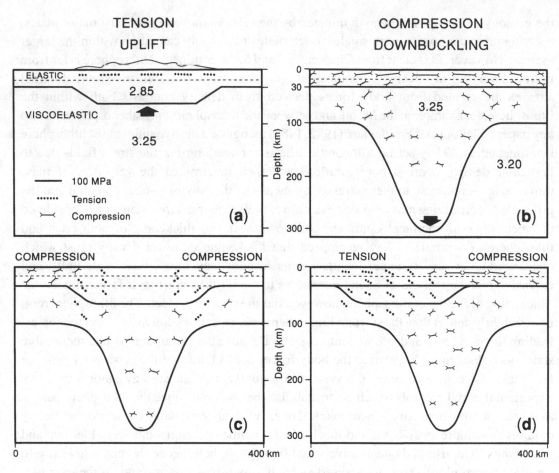

Figure 7-2. Sketches of stress distributions in models of the lithosphere with an elastic upper layer overlying a visco-elastic sub-stratum due to thickening of crust and lithosphere; simplified after Bott (1990):

(a) Tensional stresses due to a crustal root .

(b) Compressional stresses due to a lithospheric root.

(c) Stresses due to symmetrically disposed crust and lithosphere roots.

(d) Stresses due to asymmetrically disposed crust and lithosphere roots.

root. Next, Bott considered the forces developed solely from the load of the lithospheric root, Figure 7-2b. Their effect is to produce large horizontal compressional deviatoric stresses in the upper crust of up to 112 MPa which extend laterally beyond the collision zone. There is an associated downwarping of the crust by over 1 km. A significant proportion of the stress remains, however, within the lower crust, the lithosphere root itself and the nearby mantle lithosphere, and is inclined to the horizontal. These stresses maintain creep movements as the root continues to sink and to deform internally. Their effect is to cause the crustal root to thicken (at 50 mma^{-1}) and narrow (by 240 mma^{-1}), in opposition to the effect of crustal forces to diminish it, and for the upper part of the mantle lithosphere root to narrow laterally and stretch vertically as it sinks. Finally, Bott brought together the loads induced from the crustal root and the lithosphere root, combining the two models of Figure 7-2a and b to create two further models, shown in Figure 7-2c and d. One places the lithosphere root directly beneath the crustal root whilst the other has the lithosphere root displaced laterally by 200 km from the crustal root. In the symmetrical model, Figure 7-2c, the tensional stress from the crustal root and the compressional stress from the lithosphere root more or less cancel out leaving a negligible horizontal deviatoric stress in the upper crust in the region of the mountain belt itself, but moderate horizontal compressional stress is present on the flanks and bordering regions. Significant stresses of 20–30 MPa are also present in the lower crust and the mantle lithosphere which would result in a slow dissipation of the crustal root together with a narrowing of the upper part of the lithosphere root. In the asymmetrical model shown in Figure 7-2d, the lithosphere root underlies the right edge of the mountain range. Because of its lateral displacement, the compressional stress from the lithosphere root does not cancel the tensional stress from the crustal root so that quite large horizontal deviatioric stresses are present in the upper crustal layer, 60 MPa tensional on the left side and some 90 MPa compressional on the right, in juxtaposition. At depth, the crustal root is dissipating by creep on the left but is being maintained on the right, leading to an asymmetric Moho shape. Downbuckling above the lithosphere root of about 1 km could lead to sedimentary fill of the basin which would give an additional load and an overall basin thickness of 4–5 km. In contrast, the crustal root will provide crustal uplift commensurate with the mass loss resulting from the erosion and denudation of the mountain range. Bott has calculated that the overall force involved in the sinking of the lithosphere root, which effectively pulls the two plates together, is around 1.2×10^{12} Nm^{-1}. This is significantly greater than the force developed in the continental lithosphere from the push developed at an oceanic ridge plate boundary, of around 0.6×10^{12} Nm^{-1} which would give rise to a horizontal deviatoric compressive stress of some 6 MPa in the upper crust. Once formed, the collision zone is a self-sustaining system in which the localised forces dominate. The lithosphere root not only pulls down but also holds the zone together and keeps it narrow. Initiation of the system is most readily explained if a subducted slab of oceanic lithosphere is present at the onset of continental collision, as is likely. Eventually the lithosphere root will narrow to a point where it detaches and becomes a spent force so that the crustal forces remain to dominate. At that stage, uplift and extension result in the collapse of the orogen, bringing metamorphic core complexes to the surface and initiate the creation of extensional basins.

7.2.2 THE GEODYNAMICS OF THE ALPS

The discovery of a lithospheric root to the Alps came as no great surprise to Panza and Mueller (1978) because Austrian and Swiss geologists had understood from the turn of the century that the Alps had formed by continental collision, and Ampferer had postulated in 1906 that the crustal shortening must be accompanied by some form of subduction process of mantle material. As evidence about the lithosphere root accumulated during the early 1980s, the geophysical group at Zurich led by Mueller was able to build up a geodynamic model for the Alps (cf. Mueller 1989). The lithosphere root is characterised, as Kissling *et al.* (1983) pointed out, by relatively high seismic velocities, high density and low temperatures. The first step in building a model was to relate the gravity and seismic evidence. Seismic velocities within the root range between 7.8 and 8.4 kms^{-1} for V_p and between 4.3 and 4.6 kms^{-1} for V_s but range between 7.5 and 7.8 kms^{-1} for V_p and between 4.1 and 4.3 kms^{-1} for V_s within the surrounding asthenosphere. When the observed Bouguer gravity is modelled by accounting for density variations within the crust and the varying thickness of the crust, there remains, as shown in Figure 7-1, a broad residual gravity anomaly of some + 80 mGal centred just to the south of the Alps. Since the crustal effects have been eliminated, and in view of the breadth of the anomaly, it is clear that its source resides within the upper mantle. A density contrast of +0.05 gcm^{-3} dipping southwards within the lithospheric root can account for the magnitude of the residual anomaly and its asymmetrical form, and is consistent with the seismic evidence from P-wave tomography presented in Figure 3-19.

The next step was to consider whether the increased seismic velocities and density within the root were due simply to reduced temperature as cool, subducted lithosphere is transported downwards. Assuming uniform mantle material, whose density depends only on temperature, Kissling *et al.* (1983) developed a kinematic model for crustal thickening and uplift, and lower lithosphere thickening and subsidence based on crustal shortening of the Alps during the past 40 Ma, to calculate the transport of heat resulting from the mass movements. They solved the classic thermal diffusion equation by a finite difference method to calculate the temperature distribution step by step through the 40 Ma period, and end up with temperatures within the lithosphere root reduced by as much as 450 °C at 150 km depth, relative to the surrounding lithosphere.

Using a linear relationship between density change and temperature change governed simply by the coefficient of thermal expansion (taken to be $3 \times 10^{-5} °C^{-1}$), Kissling *et al.* (1983) calculated the gravity anomaly resulting from their computed temperature reduction and found it to be consistent with the observed residual gravity high. The mass excess of the lithospheric root could thus be explained simply as a thermal effect except that it would to be distributed across a larger volume than is observed. The density contrast obtained from gravity modelling must therefore owe something to compositional variation as well as to temperature. Moreover, the kinematic model predicted a present day subsidence rate of 1 mma^{-1} of material in the lithosphere root, consistent with an average uplift rate of the surface topography of 0.5 mma^{-1} (also equal to the erosion rate of the Alps).

This kinematic model was checked by Werner and Kissling (1985) with a dynamic model treating the mantle material as behaving like a Newtonian viscous fluid, to see if the downward force of the additional mass of the lithosphere root (its extra weight) could drive subduction at a rate of 1 mma^{-1}. Their calculations were based on solving the Navier–Stokes equation in two dimensions, following an approach similar to Fleitout and Froidevaux (1982), taking into account the topographic load, the buoyancy forces due to the density contrast at the Moho as well as the weight of the lithosphere root. They tried various models for the variation of viscosity with depth, to show that this rate of 1 mma^{-1} was, indeed, re-alistic. Subsequently, they went on to calculate the crustal uplift/subsidence rates and the amount of near-surface horizontal stress along the profile (Werner and Kissling 1985). An uplift rate for the Alps of 1.4 mma^{-1} is predicted, along with subsidence of up to 1.5 mma^{-1} across the Po basin, the northern Apennines and the Ligurian Sea to the south. Horizontal stress is shown to be compressional up to 15 MPa across the area of thrusting in the northern foreland of the Alps and across the Po basin to the south, but extensional (*ca.* 10 MPa) across the Alpine mountain chain and across the Ligurian Sea. Recently Werner and Gudmundsson (1992) have refined the earlier 1988 rheological model taking into account the results of more recent repeated precise levelling profiles across the Alps. The new model displayed in Figure 7-3c now reproduces correctly the subsidence in the Po basin (Figure 7-3b) as well as the pronounced uplift in the Central Alps and the very moderate uplift in the northern Apennines (Figure 7-3b). The horizontal stress (σ_x) in the upper and middle crust exhibits compression north of the Alps, extension in the central part and compression again beneath the Southern Alps and the Po basin (Figure 7-3a).

In order to arrive at the agreement described it has been necessary to modify the lithosphere-asthenosphere system considerably. The dominant deep-reaching lithospheric root beneath the Po basin (*ca.* 300 km depth) represents the remnant of an earlier subduction episode and may well have been broken up into 'blobs' during the course of time (cf. Figure

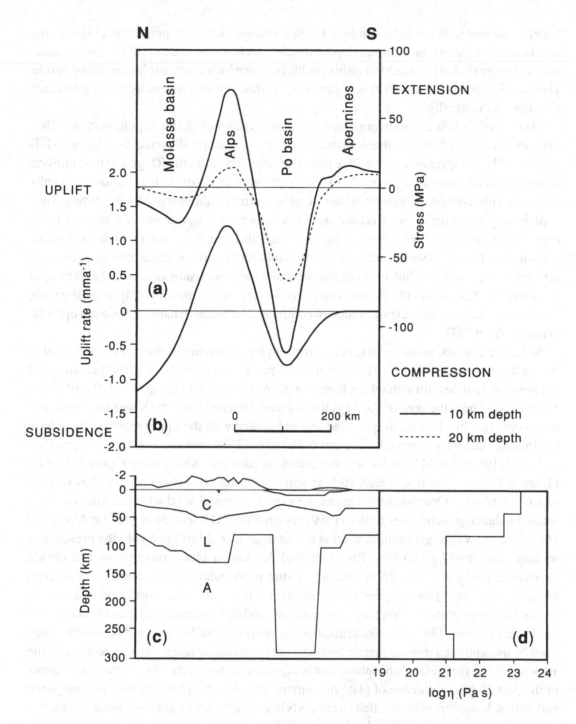

Figure 7-3. Geodynamic model of the Alps along the line of EGT prepared by Werner and Gudmundsson (1992) to predict rates of uplift or subsidence and horizontal stress variations.

(a) Horizontal stress (σ_x) computed from model shown in (c) and (d) at 10 km depth (full line) and at 20 km depth (dashed line).

(b) Rates of vertical movement, with uplift positive and subsidence negative, computed for the model shown in (c) and (d).

(c) Model cross section giving densities in gcm^{-3} for the crust (C), mantle lithosphere (L) and asthenosphere (A). This model matches the observed gravity profile shown in Figure 7-1a.

(d) Viscosity–depth profile used in conjunction with the model shown in (c) to compute the stress field and vertical movements. Viscosity η is in Pas.

3-19). This root now has to be combined with a younger, less deep penetrating lithospheric root beneath the Alps (*ca.* 130 km depth) which can clearly be seen in all the tomographic sections by Spakman (1990a,b). It is this double feature which seems to play the decisive role. The two-dimensional stress field in Figure 7-4 illustrates the heterogeneity of the stress both laterally and vertically.

These models bring a number of insights to the dynamics of the Alpine system. They explain why the observed stress distribution across Europe depicted in Figure 5-1 is dominated by compression due to ridge push from the North Atlantic Ridge and is so uniform across much of Europe to the north of the Alps, whilst the stress distribution is radially disposed within the Alpine system in the direction of crustal shortening (Mueller 1984). They explain why the collision zone is narrow. They demonstrate how deviatoric stresses in the upper crust of up to 100 MPa can be generated that are powerful enough to produce earthquakes (Scholz 1990), and how tensile and compressive stresses can be present in juxtaposition. They also link uplift of the mountain range with subsidence on the flanks just as observed. They also indicate the likely time sequence of Alpine evolution. Altogether, they have given us a much clearer vision of continental collision dynamics, founded upon the evidence from EGT.

So far the discussion has dealt specifically with the mechanics of the system, but what of the rocks that are involved? How do they respond to the forces, and the pressure and temperature regimes through which they move? As discussed in Chapter 6.4, the EGT has demonstrated how the core of the Alps has become detached from the lower crust and how the lower crust has become fragmented and imbricated with the upper mantle. As Adrian Pfiffner has elegantly sketched in Figures 6-26 and 6-27, delamination of the lower crust (cf. Frei *et al.* 1989) should have led to considerable subduction. Comparing Figure 6-27 with Figure 3-11, it is clear that at least 100 km length of lower crust is not observed as such by seismic methods. Opinion differs about the extent of crustal subduction. Using classical section balancing techniques Butler (1990) restored the EGT section across the Alps to 20 Ma ago to show that, given thinned crust at that time, it could balance with the present day section, shortened by 160 km. But it implied that earlier plate convergence and crustal shortening could not be accommodated. Using mass balance considerations, however, Ménard *et al.* (1991) have argued that if the crust was initially much stretched and only 15 km in thickness, plate convergence can be accommodated without significant loss of crust. Laubscher (1988a, 1989) has also argued on the basis of mass balance and shows that crust must be incorporated into the mantle and 'lost' and that lithosphere must also be lost into the asthenosphere, the present lithosphere root being able to account, at most, for the convergence of the past 20 Ma. Evidence of plate movements reconstructed from ocean crust magnetic anomalies, however, indicates that overall NNW convergence might have been as much as 800 km during the past 80 Ma (Laubscher 1988b).

These arguments of mass balance, and the calculations made to support them, are made all the more difficult to sustain by the clear evidence of the three-dimensional geometry of the Alpine lithosphere. Not only are the Alps strongly arcuate in plan view, evidence of the geometries of three Moho horizons beneath the Alps (Figure 3-13a) points to rotational movements (anticlockwise about vertical axes) being just as important as overall plate convergence. None the less, the balance of evidence does suggest, as revealed in Figures 6-26 and 6-27 that there has been a history of loss of lower crust and its 'absorption' into the geophysically defined upper mantle. If that is the case, then the process may be one of conversion of granulite to eclogite (Laubscher 1988a, Austrheim 1991) which has the same seismic velocity properties as peridotite (Figure 4-3) which makes it virtually invisible in the

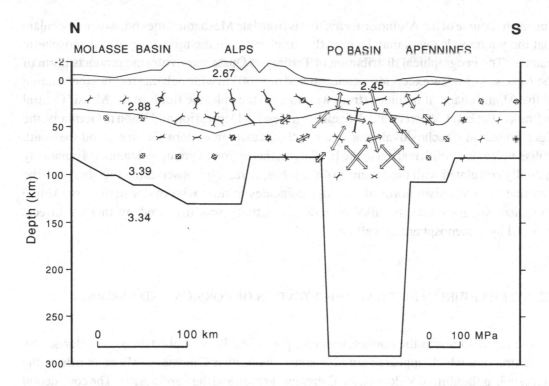

Figure 7-4. Model cross section prepared by Werner and Gudmundsson (1992) along the EGT line across the Alps, as shown in Figure 7-3, with principal stress field pattern within the plane of section. Numbers are densities in gcm⁻³.

uppermost mantle on seismic tomography sections, although the density is greater. Mengel argued in Chapter 4.3 that this was happening during the Variscan plate convergence in the region of the North Hessian depression.

Figure 3.21, supported by the P-wave tomography of Spakman (1991), shows that the lithosphere root beneath the Po basin is oval in plan view and, situated near the centre of curvature of the Alpine arc, represents the result of radial convergence of lithosphere. It is insufficient to account for the total mass of lithosphere subducted from the plate convergence over the past 20 Ma, let alone over the preceding 60 Ma. But reference to the P-wave velocity structure of the upper mantle given in Figure 3.19 indicates relatively high values (equivalent to lower temperatures) throughout a volume reaching to below 800 km beneath the Southern Alps and northern Apennines. It would seem that older lithosphere from the earlier eo-Alpine convergence has detached and sunk well into the mantle and is still slowly being absorbed.

It is interesting to note that north of the Alpine belt the depth of the lithosphere base is in a shallow position over a distance of some 500 km. This thinning of the lithosphere is accompanied by upwelling of asthenospheric mantle material. This probably has a dramatic effect on the thermal state of the lithospheric mantle and is probably reflected by the increase in surface heat flow from the northern end of the Variscan belt to the Alpine belt: as the lithosphere thins out, surface heat flow increases. It is widely believed now that the asthenospheric upper mantle is a chemically and isotopically well mixed layer with a fairly constant temperature of around 1400–1500°C. A temperature of 1400°C at depths of 60–100 km is close to the solidus temperature of anhydrous peridotite and is well above the solidus temperature of metasomatised (OH-bearing) lithospheric peridotite (Mengel and Green 1989).

If we assume that the present distribution of the depth of the lithosphere has developed during the course of the Alpine orogeny, that is from late Mesozoic times on, we can speculate that the young volcanism may be directly correlated with the upwelling of asthenospheric mantle. The geographical distribution of Tertiary to Quaternary volcanic products north of the Alps is, with a few exceptions, concentrated in a 500 km wide belt that follows the bending of the Alpine chain. It includes (from west to east) the volcanic fields of the Massif Central in France, the Eifel, Westerwald, Vogelsberg, Hessian Depression and Rhön in Germany, the Eger graben in Czechoslovakia, as well as the Lausitz in eastern Germany and the south Polish volcanics. With respect to the EGT trace, this region of young volcanism is obviously spatially correlated with the thinning of the lithosphere. This observation together with the fact that the volcanism north of the Alps coincides in time with major periods of Alpine tectonism suggests that the alkaline volcanic activity was triggered by thermal effects induced by asthenosphere upwelling.

7.2.3   THE LIGURIAN RIFT AND THE ROTATION OF CORSICA AND SARDINIA

The Ligurian Sea as the northeasternmost part of the Provençal basin is characterised by a rift structure which appeared simultaneously with other Oligocene–Miocene rifts in the area, such as the Gulf of Valencia, the Camargue graben and the Sardinia rift. The continental rifting episode began about 30 Ma ago and ended 24 Ma ago. According to palaeomagnetic results the anticlockwise rotation of Corsica and Sardinia did not start before 21 Ma. Several authors have therefore concluded that, to reach their present N–S positions, ocean spreading began with a translational phase of the two islands as a single block that took place between 24 and 21 Ma. This was followed by a 30° rotation of Sardinia and a 10° rotation of Corsica, which occurred in the relatively short time span between 21 and 19 Ma (Burrus 1984). These movements are shown in Figure 7-5 within the region of the Provençal basin oceanic crust.

The deep structure of the Ligurian Sea which is displaying all the features of an asymmetric progressing rift system changes gradually in a northeasterly direction from a young oceanic crust to an intermediate-type crust. The average spreading rate of 10 mma^{-1} is relatively slow. Between Corsica and Provence a 3 km thick, two-layer oceanic crust exists which does not show any thickening in the centre (Le Douaran *et al.* 1984). The distinct differences which are observed between the structures on the Provence and Corsica sides disappear when entering the Gulf of Genova (see Figure 3-14). Specifically, the two-layer crust in the centre of the rift is characterised by an increased thickness and lower P-wave velocities (5.2 and 6.0 kms^{-1}) than the ones found for the young oceanic basement in the Ligurian Sea. The velocities at the top of the mantle are also quite low (7.6–7.9 kms^{-1}). It has been suggested that this intermediate-type of crust, which can also be seen in the sections of the EGT profile (Figure 3-15) is an indicator of differential incipient rifting responsible for the anticlockwise rotation of the Corsica–Sardinia block.

The question as to which depth the continental block of the two islands is detached ('decoupled') from the underlying medium such that it can undergo substantial rotations can now be answered as a result of the EGT surface wave and deep seismic sounding results. Figure 3-15 shows that the Corsica–Sardinia block has continental crust of normal thickness and seismic velocities, underlain by upper mantle with properties consistent with normal continental lithosphere of about 70 km thickness (Figures 3-20 and 3-21). There is no evidence of a low-velocity zone at any level within this lithosphere. The detachment level must therefore be sought at the base of the lithosphere in the upper mantle at depths between

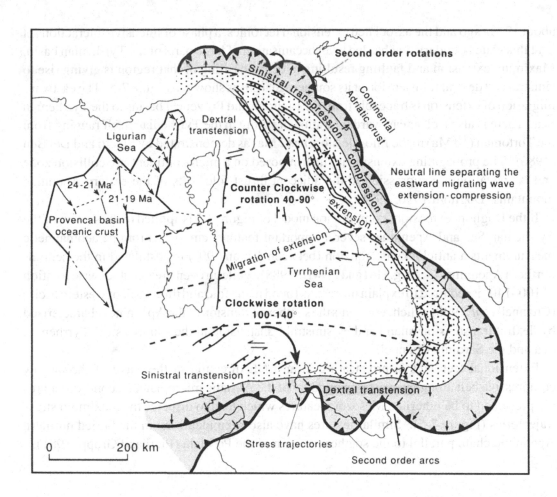

Figure 7-5. Kinematic model by Doglioni (1991) to show how the migration and rotations of stress patterns have fashioned the present shape and state of the Tyrrhenian Sea tectonic system. The translation of Corsica and Sardinia from 24 to 21 Ma ago and their rotations between 21 and 19 Ma ago as the Provençal basin opened are shown, after Burrus (1984).

about 60 and 70 km. This block has a thicker crust and lithosphere than the oceanic lithosphere around it and is therefore in the situation depicted in Bott's symmetrical model shown in Figure 7-3c. We can therefore expect Corsica and Sardinia to have the same local tectonic stress system, as is evident in Figure 7-5.

This region illustrates well how tectonic forces are localised within a self-contained stress system and how such a system evolves in time, as elaborated on in Chapter 6.7. Thus we find extensional tectonics propagating in time so that currently two thirds of the Apennine peninsula is in a state of horizontal tension, as shown in Figure 5-7. This evolving stress system can be analysed by using the modelling procedures described in Sections 7.2.1 and 7.2.2. It is through this approach that we can now begin to quantify the processes of continental fragmentation, drift of microplates, terrane accretion and continental growth.

7.2.4 CORSICA–SARDINIA CHANNEL

The rotation of Corsica and Sardinia came to an end when this block collided with the western margin of the Adriatic (Apulian) microplate. With the opening of the Tyrrhenian Sea about 12 Ma ago and the associated extensional tectonics, a phase of intensive interaction set in at this plate boundary that has led to an oceanisation of large parts of the Tyrrhenian basin. Maximum extension and faulting resulting in oceanic rifting of that region is giving rise to sinistral and dextral transtension at its southern margin as shown in Figure 7-5. The eastward migration of extension is traceable from the Ligurian and Provençal basins to the Tyrrhenian Sea. There is also a clear eastward migration of rifting in the Tyrrhenian itself ranging from the Tortonian (10 Ma) to the Plio-Pleistocene (2 Ma) as demonstrated by Finetti and Del Ben (1986). The propagating extension is superimposed on a microcontinent-arc collision zone and is coeval with compression in the adjacent thrust-fold belts of the eastern Apennine mountains (Doglioni 1990, 1991).

If the Doglioni extension–compression model of Figure 7-5 is applied to the system of the Tyrrhenian Sea and Apennines several important features emerge. From palaeomagnetic measurements a anticlockwise rotation (between 40° and 90°) is postulated in the central-northern Apennines (see e.g. Hirt and Lowrie 1988), while an even greater clockwise rotation of 100–140° is required to explain the results found in the fold and thrust belt of western Sicily (Channell *et al.* 1990) which represents the southern extension of the Apennines characterised by dextral transtension inland and by sinistral transtension in the southwestern Tyrrhenian Sea and the Sardinia Channel.

Extensional tectonics is propagating in an ENE direction in the wake of the narrow compressive belt along the western boundary of the Adriatic microplate. Second-order arcs are presumed to be inherited Mesozoic features which tend to disperse the maximum stress trajectories (Figure 7-5). Similar features have also been identified in the buried northern Apenninic chain parallel to the southern margin of the Po Plain (Pieri and Groppi 1981).

7.2.5 FENNOSCANDIAN UPLIFT

Whilst tectonic processes are active across the southern segment of EGT, so, too, are they active in the north, but from a very different cause. In Chapter 5.3 we discussed the observational evidence of uplift following the removal of the ice loading of Scandinavia at 13000 BP. Information from several lines of evidence, reviewed in a set of papers published in the November 1991 issue of *Terra Nova*, points to a dual mechanism for this postglacial rebound which adds further to our understanding of the geodynamic processes that apply to continental lithosphere. The two components of this dual mechanism are:

(a) lateral movement of relatively low-viscosity material in the asthenosphere radially inward towards the centre of uplift,

(b) vertical uplift in the centre and subsidence around the periphery due to the pressure relief consequent upon the shifting load, in effect a decompression of the mantle below where the load is removed.

Simple calculations of the ratio of the rate of gravity change to rate of uplift from these two mechanisms place limits on their values. For the first, a 'Bouguer' model is appropriate, and yields $g/h^{\cdot} = -0.17$ μGal mm^{-1}. For the second, a 'Free Air' model gives $g/h^{\cdot} = -0.31$ μGal mm^{-1}. Careful repeat measurements of gravity along a profile at 63°N latitude across the centre of the uplift area (Ekman 1991) give a measured value of the ratio as $g/h^{\cdot} = -0.24 \pm 0.03$ μGal mm^{-1}. Because the standard error limits calculated may underestimate the experimental error it is not possible to discount either extreme as the sole mechanism. However the measured value of the ratio suggests that a combination of the two is more probable. Modelling the shape of the uplift contours (Figure 5-11) Fjeldskaar and Cathles (1991) come to the same conclusion. From harmonic analysis they found that short-wavelength features decay faster than longer wavelength ones. From three-dimensional modelling they showed that if the rebound were solely due to lateral flow in the asthenosphere the uplift area would be of far greater extent than is observed. On the other hand, if there were no low-viscosity asthenosphere material, the short-wavelength variations in uplift could not be relaxed quickly enough.

Examining a range of models in which the two effects are combined, their best fit with the observed uplift contours is given by a model made up of lithosphere having a flexural rigidity of no more than 10^{24} Nm overlying asthenosphere of 75 km thickness and a viscosity of 1.8 x 10^{19} Pas above 'mesosphere' mantle with a viscosity of 1.2 x 10^{21} Pas. Although there is some possible trade-off between asthenosphere thickness and viscosity, their models clearly indicate that there has to be a relatively thin asthenosphere underlying the Baltic Shield lithosphere in which there is a relatively low viscosity to allow lateral channel flow to occur. Observed tilting of old shorelines requires a relatively low flexural rigidity, equivalent to an effective elastic thickness (EET) of 56 km, which matches the values determined by Cloetingh and Banda (see Chapter 4.2 and Figure 4-12). Mörner (1990) also argues for a dual mechanism, based on his analysis of the changing uplift pattern since 13000 BP which can be followed by tracing old shorelines, well dated at 9300, 7700 and 2300 BP across the uplift area and its surrounding peripheral trough. He has observed a sequence of lateral movement radially outwards between 25000 and 19000 BP due to the ice loading, lateral movement radially inwards between 13000 and 4500 BP after the ice had melted, followed from 4500 BP to the present by vertical movement due to pressure relief from the unloading. The time scales match the viscosities of the asthenosphere for the radial movements and the mantle as a whole for the vertical movements. Mörner points out that the latter also include the effects of pressure-induced phase changes as mantle material rises through the 670 km discontinuity (see Chapter 3.3). Mörner also argues that the estimate of flexural rigidity for the lithosphere of 10^{24} Nm is based on observations near the coast of Norway within the Caledonian realm where seismic evidence shows the crust and lithosphere thicknesses to be less than beneath the Baltic Shield and he estimates that in the centre of the shield the lithosphere may have a flexural rigidity ranging between 7×10^{24} Nm in the south and 9×10^{25} Nm further north where lithosphere thickness is greater. Cloetingh and Banda also argue in Chapter 4.2 (see Figure 4-12) for lateral variation in flexural rigidity and effective elastic thickness.

Although, as Balling and Banda have explained in Chapter 5.3, it is difficult to be too precise with these figures, the essential outcome of this analysis is that the mechanical responses to changing loads in the lithosphere come from channel flow within a relatively thin, low-viscosity asthenosphere and the vertical compression or decompression effects of the mantle as a whole (with substantially higher viscosity) in response to loading or off-loading of the lithosphere. These mechanical responses are likely to apply equally when the loading and off-loading are tectonic in origin rather than simply the passive load of an ice cap. However, thermal considerations must also always be borne in mind. Returning to the seismic tomography section, Figure 3-19, we can see the approximate upper and lower boundaries of the asthenosphere, as depicted in Figure 7-6, and note the variations in its thickness which any modelling of isostatic response to crustal movement will need to take into account.

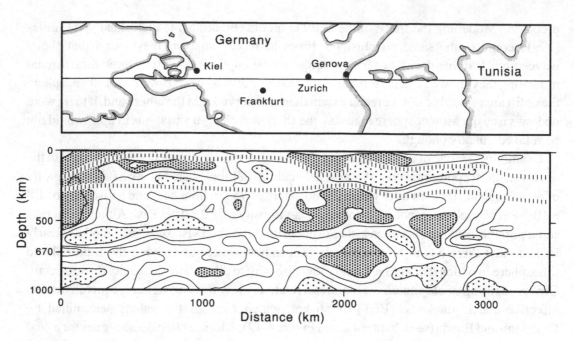

Figure 7-6. P-wave tomographic cross section of the upper mantle along EGT (Spakman, 1991), shown in Figure 3-19, with interpretation superimposed of the upper and lower boundaries of the asthenosphere, marked by the vertical shading. The 670 km discontinuity is marked by a dashed line.

7.3   GEODYNAMIC PROCESSES IN THE PAST?

An understanding of the mechanisms of modern tectonic processes involving continental lithosphere and the appreciation that plate tectonic processes have been taking place for at least the past 2 Ga, give us the opportunity to consider the geodynamic processes of past geological events and how they have come to fashion the continent of Europe as it exists at present.

7.3.1   THE STABILITY OF THE BALTIC SHIELD

EGT has brought out the contrasts between the Baltic Shield and western Europe. For the shield, both crust and lithosphere are substantially thicker and have remained essentially stable and undeformed for a long period of time during which western Europe has been through a succession of tectonic events which have modified it in many ways. Even today, there are distinct contrasts in the levels of tectonic activity, as have been discussed in Chapter 5. And yet the Baltic Shield appears to have developed through plate tectonic processes much the same as in western Europe. The seismic reflection image of a collision zone of 1.9 Ga age beneath the Gulf of Bothnia (Figure 3-5) is remarkably similar to that of the Alps (Figure 3-11), formed within the past 90 Ma. The geological evolution of the Baltic Shield has been shown by Berthelsen in Chapter 2 and by Windley in Chapter 6.1 to be wholly consistent with plate tectonics. So, why should these contrasts exist? If the mechanisms are the same and the driving forces have the same root cause, have the material properties altered? It can be seen from the geodynamics of the Alps that ocean subduction followed by continental collision leads to the formation of a lithosphere root. By virtue of the localised forces that develop, the collision zone operates as a self-sustaining system and stays as a relatively narrow region of deformation, leaving adjacent continental units relatively undisturbed. Once the subsiding mantle of the lithosphere root detaches and sinks into the asthenosphere, however, there is an opportunity for buoyancy forces due to the locally thickened crust to dominate so that the root thins and spreads and post-orogenic crustal extension takes over. Somehow, this latter stage was never reached in the Baltic Shield. The crustal thickening of the Bothnian collision zone is beautifully preserved. The lithosphere is uniformly as thick as the Alpine root, with no subsidence and detachment. Why? If the mechanisms of plate tectonics were the same, perhaps the conditions were different from the Alps. The clues are in Windley's account of Baltic evolution in Chapter 6.1. In the far north, collision tectonics in the Early Proterozoic involved the amalgamation of Archaean terranes to form the Kola–Karelian orogen which, even by that time (2.0–1.9 Ga), were of considerable antiquity and substance. However, to the south, the Sveco-Fennian orogen contained no Archaean terranes but 'developed by the growth and collision of 2.0–1.8 Ga juvenile arcs and by extensive crustal melting in the period 1.8–1.55 Ga'. The juvenile arcs would have been separated by juvenile oceanic lithosphere. In such circumstances all the young lithosphere would have been thin and relatively strong. Subduction of oceanic lithosphere would have been at a low angle and the extensive melting would have led to crustal underplating across a wide region. Consequently, neither crustal roots nor lithospheric roots would have formed as in the Alpine scenario and the localised body forces could not have developed. A closer modern analogy, though by no means an exact one, is the subduction of the Juan de Fuca plate beneath the Cordillera of western Canada. There, evidence from the Lithosphere Geotransect (Monger *et al.* 1985) and the deep reflection profiling efforts of Clowes *et al.* (1987) and Cook *et al.* (1988) in the Lithoprobe Programme have shown how the Cordillera has no crustal root and is extensively underthrust and underlain by the oceanic lithosphere of the Juan de Fuca plate. Consequently, it becomes less difficult to envisage a situation in which the Sveco-Fennian orogen was created from units of juvenile lithosphere that welded together into a sizeable province with laterally uniform properties, at least on a lithospheric scale. The crust was thickened, relatively uniformly, through melt processes and is currently preserved as the layer shown in Figure 3-3 with P-wave velocities greater than 7.0 kms^{-1}. The thick lithosphere is preserved because of its initial lateral extent of at least 1000 km and perhaps its crude layering (Figure 3-18) is a remnant of successive low-angle subductions of thin oceanic layering. It would seem that the later Gothian and Sveco-Norwegian orogens formed peripherally to the Sveco-Fennian as, indeed, have the Norwegian Caledonides, without materially affecting the core of the Baltic Shield lithosphere which had by then become firmly established as a stable block. Once such a substantial block of lithosphere becomes established as a shield with a keel, as is apparent elsewhere around the world, it takes powerful forces to move or disrupt it. The influence of the stability of the Baltic Shield on the geological evolution of western Europe, particularly through the Phanerozoic, is evident in Chapter 6.

The contrasts between Baltic Shield and western European lithosphere do not require different plate tectonic mechanisms but can be explained as a consequence simply of differing material properties resulting from the ages and characteristics of the elements they contained at the time of their amalgamation. None the less, we should continue to bear in mind that although the mechanisms during the Proterozoic may have been essentially the same as those prevailing at the present, radioactivity and temperatures within the Earth would have been greater then than now, so that the physical parameters and boundary conditions of geodynamic processes would have been different then and may have slowly evolved with time.

7.3.2 DISAPPEARING VARISCAN ROOTS

The Variscan units crossed by the central segment of EGT show all the evidence of a tectonic collage brought together by collision processes. Yet they have a crust of remarkably uniform 30 km thickness, with no lithosphere thickening of the kind found in the Alps or the Baltic Shield. In Chapter 6.3.1, Franke has explained that the main continental Rheno-Hercynian, Saxo-Thuringian and Moldanubian terranes were separated during the Lower Palaeozoic by relatively narrow oceans which had closed by the Early Carboniferous, to be followed by collision of the continental units, island arcs and other unsubducted oceanic features. An estimate of at least several hundred kilometres N–S crustal shortening across this tectonic collage is an indication of the significant impact of the collision and the extent of the accompanying crust and lithosphere thickening. Although the original Benioff slabs from oceanic subduction may not have been of any great size, it seems fairly clear that crustal and lithosphere roots must have developed during continental collision. So where are the mountain roots now? The bilateral nature of underthrusting (Rheno-Hercynian and Saxo-Thuringian SE, Moldanubian NW) might have had its cause in the asymmetry of crustal and lithosphere roots (as in Figure 7-2d). At the end of the orogeny, as Franke has pointed out in Chapter 6.3.1, the high heat flow required for the substantial magmatic activity that occurred is incompatible with a thick continental crust and lithosphere. It can be understood if the lithosphere root had eventually dropped down and separated from the rest of the lithosphere, allowing asthenosphere to reach a higher level in the upper mantle and the buoyancy forces at the Moho to degrade the crustal root and thin the crust. Uplift and extension, creating the variety of Permian basins described by Franke, could then be accompanied by magmatic activity, intrusions no doubt taking advantage of pre-existing shear zones to infuse into the lower crust. These processes would all tend to stabilise the Moho at a uniform level. In addition, the circumstances of island arc collision, underplating of basaltic magmas and crustal thickening were conducive to the transformation to eclogite, as explained by Mengel in Chapter 4.3, which underlie the seismically defined Moho.

Thus we can make use of the models of the Alpine stress system which indicate that once the lithospheric upper mantle root detaches and no longer provides a downward force on the lithosphere, the buoyancy force between crust and denser upper mantle (particularly with high density eclogite, see Figure 4-4) produces uplift, crustal extension and the destruction of the crustal root as the crust returns to a uniform thickness. This model for the tectonic mechanism has been applied by Andersen *et al.* (1991) to the Norwegian Caledonides to explain the return of the crust to normal thickness in the aftermath of the Caledonian orogeny, without necessarily requiring to invoke either large-scale lateral flow within the lower crust (Kusznir and Park 1988) or magmatic underplating. The same model can serve equally well for the Variscides, although in this case post-orogenic magmatism has occurred.

7.4   EGT – THE FUTURE?

Over a decade, a series of experiments has been conducted in a carefully directed collaborative research project across Europe, the European Geotraverse. In addition, a comprehensive range of geological and geophysical information has been compiled and mapped in compatible form on a uniform scale over the length and breadth of the EGT swathe. A complete, 4600 km long profile revealing the lithosphere of Europe from northern Norway to central Tunisia has been prepared. This is the first of its kind, anywhere, to provide a unified view of continental geology on a scale that allows the lateral variability to be measured and compared from one geological province to another. EGT coordinated scientific collaboration on a scale unprecedented in the Earth sciences, crossing frontiers of discipline, language and nationality, where it was needed to match the scale of the geological problems that it sought to resolve. Many of the results of this research have been published during the course of the EGT and are already entrenched in the scientific literature. But only in the past year has this evidence been assembled so that comparisons can readily be made. Only now, coincident with this book, have the results been brought together comprehensively and become generally accessible through the publication of the EGT Atlas. The final report of the EGT, presented to its founding body, the European Science Foundation, concluded that after all the experimental work and analysis had been completed, 'in many ways, EGT has only just begun'.

Perhaps the most lasting achievement of the EGT is to have brought so many scientists from such a wide range of disciplines to work together for a common purpose. The workshops and, in particular, the study centres have broadened the vision of many and became the breeding grounds of new ideas, new scientific liaisons and new collaborative experiments which should continue well into the future.

This book represents simply the first thoughts of a few of the many scientists who have been involved in this enterprise. We have attempted to narrate the way that EGT has brought new and vital evidence to bear on our understanding of how the geology of a continent works. We have sought to arouse your interest and to stimulate your criticism of the ideas that we have presented. We expect that we shall be proved wrong on a number of counts but hope, modestly, that a reasonable number of our views are confirmed. We know that we have only scratched the surface of the great wealth of information that has been generated through the EGT. Most of it is contained in the Atlas. May we urge you to make use of the Atlas to carry forward the leading edge of our science, even if you do end up proving us wrong!

We began this book by suggesting that the search for geological truth is much like a detective investigation. We have held to this analogy subconsciously in the way that we have attempted to present this account, discussing each line of evidence in turn and relating one clue with another. We have, we believe, exposed the lithosphere to a more intensive view than has ever been achieved before, but we are equally aware that there remain a good many more bodies in the basement to be unearthed.

References

Abrahamsen, N. and Madirazza, I. 1986. Gravity and magnetic anomalies: the enigma of the Silkeborg gravity and magnetic anomalies, Central Jutland, Denmark. In: J.T. Müller (ed.), Twenty five years of geology in Århus. *Geoskrifter* **24**: 45–59.

Åhäll, K.I. and Daly, J.S. 1989. Age, tectonic setting and provenance of Östfold-Marstrand Belt supracrustals: westward crustal growth of the Baltic Shield at 1760 Ma. *Precambrian Res.* **45**: 45–61.

Ahjos, T. 1990. A look at the Fennoscandian earthquake data. In: R. Freeman and St. Mueller (eds), *Sixth EGT workshop: Data compilations and synoptic interpretation*, pp. 349–55, European Science Foundation, Strasbourg, France.

Ahjos, T. and Uski, M. 1991. Earthquake epicentres in northern Europe. Institute of Seismology, Univ. Helsinki Report **S-25**: 81–86.

Ahjos, T. and Uski, M. 1992. Earthquakes in Northern Europe 1375–1989. *Tectonophysics* **207**: in press.

Ahorner, L. 1970. Seismo-tectonic relations between the graben zones of the Upper and Lower Rhine valley. In: H. Illies and St. Mueller (eds), *Graben problems*, pp. 155–66, Schweizerbart, Stuttgart.

Ahorner, L. 1975. Present-day stress field and seismotectonic block movements along major fault zones in western Europe. *Tectonophysics* **29**: 233–49.

Ahorner, L., Baier, B. and Bonjer, K.-P. 1983. General pattern of seismotectonic dislocation and the earthquake generating stress field in Central Europe between the Alps and the North Sea. In: K. Fuchs *et al.* (eds), *Plateau Uplift*, pp. 187–97, Springer, Berlin, Heidelberg, New York.

Aichroth, B. and Prodehl, C. 1990. EGT Central Segment refraction seismics. In: R. Freeman and St. Mueller (eds), *Sixth EGT workshop: Data compilations and synoptic interpretation*, pp. 187–97, European Science Foundation, Strasbourg, France.

Aichroth, B., Ye, S., Feddersen, J., Maistrello, M. and Pedone, R. 1990. A Compilation of Data from the 1986 European Geotraverse Experiment (Main Line) from Genova to Kiel. Geophysical Institute, University of Karlsruhe Open File Report **90-1**: 115pp.

Aichroth, B., Prodehl, C. and Thybo, H. 1992. Crustal structure along the central segment of the EGT from seismic refraction studies. *Tectonophysics*, in press.

Alvarez, W. 1976. A former continuation of the Alps. *Bull. Geol. Soc. Amer.* **87**: 891–6.

Anderle, H.-J. 1987. The evolution of the South Hunsrück and Taunus Borderzone. *Tectonophysics* **137**: 101–14.

Andersen, T.B., Jamtveit, B., Dewey, J.F. and Swensson, E. 1991. Subduction and eduction of continental crust: major mechanisms during continent–continent collision and orogenic extensional collapse, a model based on the south Norwegian Caledonides. *Terra Nova* **3**: 303–10.

Andersson, U.B. 1991. Granitoid episodes and mafic-felsic magma interaction in the Svecofennian of the Fennoscandian shield, with main emphasis on the ~ 1.8 Ga plutonics. *Precambrian Res.* **51**: 127–49.

Andréasson, P.-G. and Rodhe, A. 1990. Geology of the Protogine zone south of Lake Vättern, southern Sweden: a reinterpretation. *Geol. Fören. Stockh. Förh.* **112**: 107–25.

Andrieux J., Fontbote, J.M. and Mattaeur, M. 1971. Sur un modele explicatif de l'arc de Gibraltar. *Earth Planet. Sci. Lett.* **12**: 191–8.

Argand, E. 1924. La Tectonique de l'Asie. *13th International Geological Congress Comptes-Rendus* **5**: 171–372.

Arndt, N.T. 1983. Role of a thin, komatiite-rich oceanic crust in the Archean plate-tectonic process. *Geology* **11**: 372–75.

Atkin, B.P. and Brewer, T.S. 1990. The tectonic setting of basaltic magmatism in the Kongsberg, Bamble and Telemark sectors, southern Norway. In: C.F. Gower, T. Rivers and B. Ryan (eds), Mid-Proterozoic Laurentia–Baltica. *Geol. Assoc. Can. Spec. Pap.* **38**: 471–83.

Austrheim, H. 1991. Eclogite formation and dynamics of crustal roots under continental collision zone. *Terra Nova* **3**: 492–9.

BABEL Working Group 1990. Evidence for early Proterozoic plate tectonics from seismic reflection profiles in the Baltic Shield. *Nature* **348**: 34–8.

BABEL Working Group 1991. Deep seismic survey images crustal structure of Tornquist Zone beneath southern Baltic Sea. *Geophys. Res. Lett.* **18**: 1091–4.

Bachmann, G.H. and Grosse, S. 1989. Struktur und Entstehung des Norddeutschen Beckens - geologische und geophysikalische Interpretation einer verbesserten Bouguer-Schwerekarte. *Veröff. der Niedersächsischen Akademie der Geowissenschaften* **2**: 24–47.

Bachmann, G.H., Müller, M. and Weggen, K. 1987. Evolution of the Molasse Basin (Germany, Switzerland). *Tectonophysics* **137**: 77–92.

Baker, J.H. and Hellingwerf, R.H. (eds) 1988. The Bergslagen Province, Central Sweden: structure, stratigraphy and ore-forming processes. *Geol. en Mijn. Spec. Iss.* **67**: 121–478.

Balling, N. 1980. The land uplift in Fennoscandia, gravity field anomalies and isostasy. In: N-A. Mörner (ed.), *Earth Rheology, Isostasy and Eustasy*, pp. 297–321, John Wiley & Sons, New York.

Balling, N. 1984. Gravity and isostasy in the Baltic Shield. In: D.A. Galson and St. Mueller (eds), *First EGT Workshop: The Northern Segment*, pp. 53–68, European Science Foundation, Strasbourg.

Balling, N. 1990a. Heat flow and lithospheric temperature along the Northern Segment of the European Geotraverse. In: R. Freeman and St. Mueller (eds), *Sixth EGT workshop: Data compilations and synoptic interpretation*, pp. 405–16, European Science Foundation, Strasbourg, France.

Balling, N. 1990b. Integrated geophysical investigations of the lithosphere beneath southern Scandinavia-objectives and first results. In: R. Freeman and St. Mueller (eds), *Sixth EGT workshop: Data compilations and synoptic interpretation*, pp. 441–57, European Science Foundation, Strasbourg, France.

Balling, N. 1992. Heat flow and lithospheric temperature along the northern segment of the European Geotraverse. *Tectonophysics*, in press.

Bally, A. W., Burbi, L., Cooper, C. and Ghelardoni, R. 1988. Balanced Sections and Seismic Reflection Profiles Across the Central Apennines. *Mem. Soc. Geol. Ital.* **35**: 257–310.

Bamford, D. 1977. Pn velocity anisotropy in a continental upper mantle. *Geophys. J. R. Astr. Soc.* **49**: 29–48.

Barbey, P. and Cuney, M.K. 1982. Rb, Sr, Ba, U and Th geochemistry of the Lapland granulites: LILE fractionating controlling factors. *Contrib. Mineral. Petrol.* **81**: 304–16.

Barbey, P. and Raith, M. 1990. The granulite belt of Lapland. In: D. Vielzeuf and Ph. Vidal (eds), *Granulites and Crustal Evolution*, pp. 111–32, Kluwer, Dordrecht.

Barbey, P., Convert, J., Moreau, B., Capdevila, R. and Hameurt, J. 1984. Petrogenesis and evolution of an early Proterozoic collisional orogenic belt: the Granulite belt of Lapland and the Belomorides (Fennoscandia). *Bull. Geol. Surv. Finl.* **56**: 161–88.

Båth, M., 1956. An earthquake catalogue for Fennoscandia for the years 1891–1950. *Sv. Geol. Unders. Årsbok* **50**: 52 pp.

Behrens, K., Hansen, J., Flüh, E. R., Goldflam, S. and Hirschleber, H. 1986. Seismic investigations in the Skagerrak and Kattegat. *Tectonophysics* **128**: 209–28.

Behrens, K., Goldflam, S., Heikkinen, P., Hirschleber, H., Lindquist, C. and Lund, C.-E. 1989. Reflection seismic measurements across the Granulite Belt of the POLAR Profile in the northern Baltic Shield, Northern Finland. *Tectonophysics* **162**: 101–11.

Behrmann, J., Drozdzewski, G., Heinrichs, T., Huch, M., Meyer, W. and Oncken, O. 1991. Crustal-scale balanced cross-sections through the Variscan fold belt, Germany – the central EGT-segment. *Tectonophysics* **196**: 1–21.

Beiersdorf, H., 1969. Druckspannungsindizien in Karbonatgesteinen Süd-Niedersachsens, Ost-Westfalens und Nord-Hessens. *Geol. Mitt.* **8**: 217–62.

Beloussov, V. V. and Pavlenkova, N. I. 1984. The types of the Earth's crust. *J. Geodyn.* **1**: 167–83.

Bergström, J., 1984. Strike-slip faulting and Cambrian biogeography around the Tornquist Zone. *Geol. Fören. Stockh. Förh.* **106**: 382–3.

Bernard-Griffiths, J., Peucat, J.J., Postaire, B., Vidal, P., Convert, J. and Moreau, B. 1984. Isotopic data (U–Pb, Rb–Sr, Pb–Pb, and Sm–Nd) of mafic granulites from Finnish Lapland. *Precambrian Res.* **23**: 325–48.

Bernoulli, D., Heitzmann, P. and Zingg, A. 1990. Central and southern Alps in southern Switzerland: tectonic evolution and first results of reflection seismics. In: F. Roure, P. Heitzmann and R. Polino (eds), Deep structure of the Alps. *Mém. Soc. géol. France* **156**; *Mem. Soc. Geol. Suisse* **1**; *Vol. Spec. Soc. Geol. Italy* **1**: 289–302.

Berthelsen, A. 1977. Tentative tectonic division of the Sveco-Norwegian belt. In: *Field Guide COMTEC 77 on Tectonics of Precambrian orogens of southern Scandinavia,* Copenhagen.

Berthelsen, A. 1980. Towards a palinspastic tectonic analysis of the Baltic Shield. *Int. Geol. Congr. Coll.* **C6**: 5–21.

Berthelsen, A. 1987. A tectonic model for the evolution of the Baltic Shield. In: J.P. Schaer and J. Rodgers (eds), *The Anatomy of Mountain Ranges,* pp. 31–58, Princeton University Press, New Jersey.

Berthelsen, A. 1990. Tectonic map of the Northern Segment of the EGT. In: R. Freeman and St. Mueller (eds), *Sixth EGT workshop: Data compilations and synoptic interpretation,* pp. 5–7, European Science Foundation, Strasbourg, France.

Berthelsen, A. and Marker, M. 1986a. Tectonics of the Kola collision suture and adjacent Archaean and early Proterozoic terrains in the northeastern region of the Baltic Shield. *Tectonophysics* **126**: 31–55.

Berthelsen, A. and Marker, M. 1986b. 1.9–1.8 Ga old strike-slip megashears in the Baltic Shield and their plate tectonic implications. *Tectonophysics* **128**: 163–81.

Bertotti, G. 1990. The deep structure of the Monte Generoso basin: an extensional basin in the south-Alpine Mesozoic passive continental margin. In: F. Roure, P. Heitzmann and R. Polino (eds), Deep structure of the Alps. *Mém. Soc. géol. France* **156**: 303–308.

Betz, D., Führer, F., Greiner, G. and Plein, E. 1987. Evolution of the Lower Saxony Basin. *Tectonophysics* **137**: 127–70.

Biella, G.C., Gelati, R., Maistrello, M., Mancuso, M., Massiotta, P. and Scarascia, S. 1987. The structure of the upper crust in the Alps–Apennines region deduced from refraction seismic data. *Tectonophysics,* **142**: 72–85.

Biju-Duval B., Dercourt, J. and Le Pichon, X. 1977. From the Tethys Ocean to the Mediterranean Seas: a plate tectonic model of the evolution of the Western Alpine system. In: B. Biju-Duval and L. Montadert (eds), *International symposium on the structural evolution of the Mediterranean basins,* pp. 145–64, Editions Technip, Paris.

BIRPS and ECORS 1986. Deep seismic reflection profiling between England, Ireland and France. *J. Geol. Soc. Lond.* **143**: 15–40.

Bjerhammar, A. 1980. Postglacial uplifts and geopotentials in Fennoscandia. In: Mörner, N. A. (ed.), *Earth Rheology, Isostasy and Eustasy,* pp. 323–6, John Wiley & Sons, New York.

Bjerreskov, M. and Jørgensen, K. 1983: Late Wenlock graptolite-bearing tuffaceous sandstone from Bornholm, Denmark. *Bull. Geol. Soc. Den.* **3l**: 129–49.

Blundell, D.J. and the BABEL Working Group 1992. Seismic reflectivity of the crust in transition from basin to platform regions in Europe. *Mém. Soc. Géol. France,* in press.

Bobier, C., Viguier, C., Chaari, A. and Chine, A. 1991. The post-Triassic sedimentary cover of Tunisia: Seismic sequences and structure. *Tectonophysics* **195**: 371–410.

Boccaletti, M., Conedera, C., Daineli, P. and Gogev, P. 1982. The recent (Miocene–Quaternary) regmatic system of the western Mediterranean region. *J. Petrol. Geol.* **5**: 31–49.

Bohlen, S.R. and Mezger, K. 1988. Origin of granulite terranes and the formation of the lowermost continental crust. *Science* **244**: 326–9.

Bott, M.H.P. 1990. Stress distribution and plate boundary force associated with collision mountain ranges. *Tectonophysics* **182**: 193–209.

Bott, M.H.P. and Kusznir, N. J., 1979. Stress distribution associated with compensated plateau uplift structures with application to the continental splitting mechanism. *Geophys. J. Roy. Astr. Soc.* **56**: 451–9.

Bowes, D.R., Halden, N.M., Koistinen, T.J. and Park. A.F. 1984. Structural features of basement and cover rocks in the eastern Svecokarelides, Finland. In: A. Kröner and R. Greiling (eds), *Precambrian Tectonics Illustrated*, pp.147–71, Schweizerbart'sche, Stuttgart.

Brink, H.-J., Franke, D., Hoffmann, N., Horst, W. and Oncken, O. 1990. Structure and evolution of the North German Basin. In: R. Freeman, P. Giese and St. Mueller (eds), *The European Geotraverse: Integrative studies*, pp. 195–212, European Science Foundation, Strasbourg, France.

Brink, H.-J., Dürschner, H. and Trappe, H. 1992. The late and post-Variscan development of the NW German Basin. *Tectonophysics*, in press.

Brinkmann, R. 1986. *Historische Geologie: Erd- u. Lebensgeschichte*. Neubearbeitet von K.Krömmelbein; 12./13. Auflage, durchgesehen von F.Strauch., Enke.

Britze, P. and Japsen, P. 1991. Geologic map of Denmark 1:400 000, The Danish Basin: 'Top Zechstein' and the Triassic. Geological Survey of Denmark, Map Series No. 31.

Brochwicz-Lewinski, W., Pozaryski, W. and Tomczyk, H. 1981. Mouvements coullisantes de grande ampleur au Paleozoique inferieur le long de la marge sud-ouest de la Platform Est-Europenne. *C. R. Acad.Sci. Paris* Ser. II **293**: 855–8.

Brochwicz-Lewinski, W., Pozaryski, W. and Tomczyk, H. 1984. Sinistral strike-slip movements in Central Europe in the Paleozoic. *Publ. Inst. Geophys. Pol. Acad. Sci.* **A-13**: 3–13.

Bucher-Nurminen, K. 1991. Mantle fragments in the Scandinavian Caledonides. *Tectonophysics* **190**: 173–92.

Buness, H. 1990. *A Compilation of Data from the 1983 European Geotraverse Experiment from the Ligurian Sea to the Southern Alps*. Institute of Geophysics, Free University of Berlin.

Buness, H. and Giese, P. 1990. A crustal section through the Northwestern Adriatic Plate. In: R. Freeman, P. Giese and St. Mueller (eds), *The European Geotraverse: Integrative studies*, pp. 297–304, European Science Foundation, Strasbourg, France.

Bungum, H. 1989. Earthquake occurrence and seismotectonics in Norway and surrounding areas. In: S. Gregersen and P.W. Basham (eds), *Earthquakes at North-Atlantic Passive Margins: Neotectonics and Postglacial Rebound*, pp. 501-519, Kluwer Academic Publishers.

Bungum, H. and Fyen, J. 1980. Hypocentral distribution, focal mechanisms, and tectonic implications of Fennoscandiann earthquakes, 1954–1978. *Geol. Fören. Stockh. Förh.* **101**: 261–71.

Burollet, P. F. 1967. General geology of Tunisia. In: *Guidebook to the geology and history of Tunisia*, pp. 51–8, Petroleum Society of Libya Ninth Ann. Field Conf.

Burollet, P.F. 1991. Structures and tectonics of Tunisia. *Tectonophysics* **195**: 359–69.

Burrus, J. 1984. Contribution to a geodynamic synthesis of the Provençal Basin (North-Western Mediterranean). *Marine Geol.* **55**: 247–69.

Butler, R.W.H. 1990. Notes on crustal balancing along the Alpine segment of the European Geotraverse. In: R. Freeman, P. Giese and St. Mueller (eds), *The European Geotraverse: Integrative studies*, pp. 263–76, European Science Foundation, Strasbourg, France.

Butler R.W.H., Matthews, S.J. and Parish, M. 1986. The NW external Alpine thrust belt and its implications for the geometry of the Western Alpine orogen. In: M.P. Coward, and A.C. Ries (eds), Collision Tectonics. *Geol. Soc. Lond. Spec. Publ.* **19**: 245–60.

Bylinski, R., Glebovitski, V., Dolivo-Dobrovolski, A. and Porotova, G. 1977. The major Belomorian deep-fault zone. In: C. Kortman (ed.), Fault Tectonics in the Eastern Part of the Baltic Shield. *Proc. Finnish-Soviet Symp.*, Helsinki: 49–62.

Calcagnile, G. and Panza, G.F. 1990. Crustal and upper mantle structure of the Mediterranean area derived from surface-wave data. *Phys. Earth . Planet. Int.* **60**: 163–8.

Calcagnile, G., Perri, P., Del Gaudio, V. and Mueller, St. 1990. A two-dimensional velocity model for the upper mantle beneath FENNOLORA from seismic surface waves and body waves. In: R. Freeman, P. Giese, and St. Mueller (eds), *The European Geotraverse: Integrative Studies*, pp. 49–66, European Science Foundation, Strasbourg, France.

Carmichael, R.S. 1989. *Practical handbook of physical properties of rocks and minerals*, CRC Press Inc., Boca Raton, Florida, USA.

Carmignani, L. and Kligfield, R. 1990. Crustal Extension in the Northern Apennines: The transition from Compression to Extension in the Alpe Apuane Core Complex. *Tectonics* **9**: 1275–1303.

Carswell, D.A. 1991. Variscan high P–T metamorphism and uplift history in the Moldanubian Zone of the Bohemian Massif in Lower Austria. *Eur. J. Mineral.* **3**: 323–42.

Carter, N.L. and Tsenn, M.C. 1987. Flow properties of continental lithosphere. *Tectonophysics* **136**: 27–63.

Cartwright, J.A. 1987. Tranverse structural zones in continental rifts – an example from the Danish Sector of the North Sea. In: J.Brooks and K.Glennie (eds), *Petroleum Geology of North West Europe*, pp. 441–52, Graham and Trotman.

Cartwright, J.A. 1990. The structural evolution of the Ringkøbing-Fyn High. In: D.J. Blundell and A.D. Gibbs (eds), *Tectonic evolution of the North Sea Rifts*, pp. 200–16, Oxford University Press.

Cassano, E., Anelli, L., Fichera, R. amd Cappelli, V. 1986. Pianura Padana Interpretazione integrata di dati Geofisici e Geologici. *73 Congresso SGI, AGIP*: 1-28.

Castany, G. 195. Etude geologique de l'Atlas tunisien oriental. *Ann. Mines Geol. (Tunis)* **8**: 1–324.

Castellarin, A., Eva, C., Giglia, G. and Vai, G.-B. 1985. Analisi strutturale del Fronte Appenninico Padano. *Giorn. Geol. Ser.3* **47**: 47–76.

Cathles, L.M. 1975. *The viscosity of the Earth's mantle.* Princeton Univ. Press, Princeton, N.J.

Cêrmák, V. and Bodri, L. 1992. Three-dimensional deep temperature modelling along the European Geotraverse. *Tectonophysics*, in press.

Channell, J.E.T. and Mareschal, J.C. 1989. Delamination and asymmetric lithospheric thickening in the development of the Tyrrhenian Rift. In: M.P. Coward, D. Dietrich and R.G. Parks (eds), Alpine tectonics, *Geol. Soc. Lond. Spec. Publ.* **45**: 285–302

Channell, J.E.T., Oldow, J.S., Catalano, R. and D'Argenio, B. 1990. Palaeomagnetically determined rotations in the western Sicilian fold and thrust belt. *Tectonics* **9**: 641–60.

Chen, W.P. and Molnar, P. 1983. Focal depths of intracontinental and intraplate earthquakes and their implications for the thermal and mechanical properties of the lithosphere. *J. Geophys. Res.* **88**: 4183–214.

Choukroune, P. and ECORS Team 1989. The ECORS Pyrenean deep seismic profile reflection data and the overall structure of an orogenic belt. *Tectonics* **8**: 23–39.

Cloetingh, S., Kooi, H., and Groenewoud, W. 1989. Intraplate stresses and sedimentary basin evolution. *Amer. Geophys. U. Geophys. Mono.* **48**: 1–16.

Cloetingh, S., Reemst, P.A., Kooi, H. and Fanavoll, S. 1992. Intraplate stresses and Post-Cretaceous uplift and the evolution of northern Atlantic margins. *Norsk Geol. Tidsk.*, in press.

Clowes, R. M., Gens-Lenartowicz, E., Demartin, M. and Saxov, S. 1987. Lithospheric structure in southern Sweden – results from FENNOLORA. *Tectonophysics* **142**: 1–14.

Cogné, J. and Wright, A.E. 1980. L'orogène Cadomien. In: *Int. Geol. Congr. Coll.* C6: 29–55.

Cohen, C. R. 1980. Plate tectonic model for the Oligocene–Miocene evolution of the western Mediterranean. *Tectonophysics* **68**: 283–312.

Colley, H. and Westra, L. 1987. The volcano-tectonic and mineralization of the early Proterozoic Kemiö–Orijärvi–Lohja belt, SW

Finland. In: T.C. Pharaoh, R.D. Beckinsale and D. Rickard (eds), Geochemistry and Mineralization of Proterozoic Volcanic Suites. *Geol. Soc. Lond. Spec. Publ.* **33**: 95–107.

Coney, P. J. 1980. Cordilleran metamorphic core complexes: An overview. In: M. D. Crittenden, P. J. Coney, and G. H. Davis (eds), Cordilleran Metamorphic Core Complexes. *Geol. Soc. Amer. Mem.* **153**: 7–34.

Cook, F., Green, A., Simony, P., Price, R., Parrish, R., Milkereit, B., Gordy, P., Brown, R., Coflin, K. and Patenaude, C. 1988. LITHOPROBE seismic reflection structure of the southeastern Canadian Cordillera: initial results. *Tectonics* **7**: 157–80.

Crawford, M.B. and Windley, B.F. 1990. Leucogranites of the Himalaya/Karakoram: implications for magmatic evolution within collisional belts and the study of collision-related leucogranite petrogenesis. *J. Volc. Geotherm. Res.* **44**: 1–19.

Dadlez, R. 1990. Tectonics of the southern Baltic (in Polish with English summary and captions to Figs). *Kvartalnik Geol.* **34**:1–20.

Dahl-Jensen, T., Dyrelius, D. and Palm, H. 1991. Deep crustal seismic reflection profiling across two major tectonic zones in southern Sweden. *Tectonophysics* **195**: 209–40.

Dahlen, F.A., Supppe, J. and Davis, D. 1984. Mechanics of Foldand Thrust Belts and Accretionary Wedges: Cohesive Coulomb Theory. *J. Geophys. Res.* **89**: 10087–101.

Dahlstrom, C.D.A. 1970. Structural geology in the eastern margin of the Canadian Rocky Mountains. *Can. Bull. Petrol. Geol.* **18**: 332–402.

Dallmeyer, R.D., Snoke, A.W. and McKee, E.H. 1986. The Mesozoic–Cenozoic Tectonothermal Evolution of the Ruby Mountains, East Humboldt Range, Nevada: A Cordilleran Metamorphic Core Complex. *Tectonics* **5**: 931–54.

Dallmeyer, R.D., Franke, W. and Weber, K. (eds) 1992. *Tectonostratigraphic evolution of the Central and East European Palaeozoic orogens.* Springer, Berlin, Heidelberg, New York.

Dalmayrac, B. and Molnar, P. 1981. Parallel thrust and normal faulting in Peru and constraints on the state of stress. *Earth Planet. Sci. Lett.* **55**: 473–81.

De Geer, G. 1888/1890. Om Skandinaviens niv öf örandringar under Quarterperioden. *Geol. Fören. Stockh. Förh.* **10**: 366–79, **12**: 61–110.

De Mets, C., Gordon, R.G., Argus, D.F. and Stein, S. 1990. Current plate motions. *Geophys. J. Int.* **101**: 425–78.

Decker, K., Faupl, P. and Möller, A. 1987. Synorogenic Sedimentation on the Northern Calcareous Alps during the Early Cretaceous. In: H.W. Flögel, and P. Faupl (eds), *Geodynamics of the Eastern Alps*, pp. 126–41, Deuticke, Vienna.

Deichmann, N. 1992. Structural and rheological implications of lower crustal earthquakes below northern Switzerland. *Phys. Earth Planet. Int*, in press.

Deichmann, N. and Baer, M. 1990. Earthquake focal depths below the Alps and northern Alpine foreland of Switzerland. In: Freeman, R., Giese, P., and Mueller, St. (eds.), *The European Geotraverse: Integrative Studies*, pp. 277–88, European Science Foundation, Strasbourg.

Deichmann, N. and Rybach, L. 1989. Earthquakes and temperatures in the lower crust below the northern Alpine foreland of Switzerland. In: R.F. Mereu, St. Mueller and D.M. Fountain (eds), Properties and Processes of the Earth's Lower Crust. *Amer. Geophys. Union Geophys. Monogr.* **51**: 197–213.

Deichmann, N., Ansorge, J. and Mueller, St. 1986. Crustal structure of the Southern Alps beneath the intersection with the European Geotraverse. *Tectonophysics* **126**: 57–83.

DEKORP Research Group 1985. First results and preliminary interpretation of deep-reflection seismic recordings along profile DEKORP 2-S. *J. Geophys.* **57**: 137–63.

DEKORP Research Group 1988. Results of the DEKORP 4/KTB Oberpfalz deep seismic reflection investigations. *J. Geophys.* **62**: 69–101.

DEKORP Research Group 1990. Crustal structure of the Rhenish Massif: results of deep seismic reflection lines DEKORP 2-N and 2-N-Q. *Geol. Rundschau* **79**: 523–66.

DEKORP Research Group 1991. Results of the DEKORP 1 (BELCORP–DEKORP) deep seismic reflection studies in the western part of the Rhenish Massif. *Geophys. J. Int.* **106**: 203–27.

Della Vedova, B., Lucazeau, F., Pellis, G. and Pasquale, V. 1990. Heat flow and tectonics along the EGT Southern Segment. In: R. Freeman and St. Mueller (eds), *Sixth EGT workshop: Data compilations and synoptic interpretation*, pp. 431–40, European Science Foundation, Strasbourg, France.

Della Vedova, B., Lucazeau, F., Pasquale, V., Pellis, G., and Verdoya, M. 1992. Heat-flow along the EGT southern segment and adjacent areas. *Tectonophysics*, in press.

Denton, G.H. and Hughes, T.J. 1981. *The Last Great Ice Sheets*. Wiley, Chichester.

Dercourt, J., Zonenshain, L.P., Ricou, L.-E., Kazmin, V.G., Le Pichon, X., Knipper, A.L., Grandjacquet, C., Sbortshikov, I.M., Geyssant, J., Lepvrier, C., Pechersky, D.H., Boulin, J., Sibouet, J.-C., Savostin, L.A., Sorokhtin, O., Westphal, M., Bazhenov, M., Lauer, J.P. and Biju-Duval, B. 1986. Geological evolution of the Tethys belt from the Atlantic to the Pamirs since the Lias. *Tectonophysics* 123: 241–315.

Dewey, J.F. 1980. Episodicity, sequence, and style at convergent plate boundaries. In: D.W. Strangway (ed.), The continental crust and its mineral deposits, *Geol. Assoc. Can. Spec. Pap.* 20: 553–73.

Dewey, J.F. and Burke, K.C. 1973. Tibetan, Variscan, and Precambrian basement reactivation: products of continental collision. *J. Geol.* 81: 683–92.

Dewey, J.F. and Windley, B.F. 1988. Palaeocene–Eocene tectonics of NW Europe. In: A.C. Morton and C.M. Parson (eds), Early Tertiary volcanism and the opening of the NE Atlantic. *Geol. Soc. Lond. Spec. Publ.* 39: 25–31.

Dewey, J.F., Pittman III, W.C., Ryan, W B.F. and Bonnin, J. 1973. Plate tectonics and the evolution of the Alpine system. *Bull. Geol. Soc. Amer.* 84: 3137–80.

Dewey, J.F., Hempton, M.R., Kidd, W.F S., Saroglu, F. and Sengor, A.M.C. 1986. Shortening of continental lithosphere: the neotectonics of Eastern Aanatolia – a young collision zone. In: M.P. Coward and A.C. Ries (eds), Collision Tectonics. *Geol. Soc. Lond. Spec. Publ.* 19: 3–36.

Dewey, J.F., Helman, M.L., Turco, E., Hutton D.H.W. and Knott, S.D. 1989. Kinematics of the western Mediterranean. In: M.P. Coward, D. Dietrich and R.G. Park (eds), Alpine Tec-

tonics. *Geol. Soc. Lond. Spec. Publ.* 45: 265–83.

Doglioni, C. 1990. The Global Tectonic Pattern. *J. Geodyn.* 12: 21–38.

Doglioni, C. 1991. A proposal for the kinematic modelling of W-dipping subduction – possible applications to the Tyrrhennian–Apennines system. *Terra Nova* 3: 423–34.

Doglioni, C. and Bosellini, A. 1987. Eoalpine and mesoalpine tectonics in the Southern Alps. *Geol. Rundschau* 76: 735–54.

Dohr, G., Lukic, P. and Bachmann, G. 1983. Deep crustal reflections in the Northwest German Basin. In: A. W. Bally (ed.), Seismic Expression of Structural Styles, a Picture and Work Atlas. *Amer. Assoc. Petrol. Geol. Studies in Geol. Ser.* 15: 1.5.1–5.

Dorr, W., Kramm, U., Francke, W. and von Gehlen. 1991. U–Pb systematics of detrital zircons from the Sazothuringian belt: constraints on the tectonic development. *Terra Abstr.* 3: 206–7.

Downes, H., Dupuy, C., Leyreloup, A. 1990. Crustal evolution of the Hercynian belt of Western Europe: Evidence from lower crustal granulitic xenoliths. *Chem. Geol.* 68: 291–303.

Drozdzewski, G. 1988. Die Wurzel der Osning-Überschiebung und der Mechanismus herzynischer Inversionsstörungen in Mitteleuropa. *Geol. Rundschau* 77: 127–41.

Duncan, R.A., and Richards, M.A., 1991. Hotspots, Mantle Plumes, Flood Basalts, and True Polar Wander. *Rev. Geophys.* 29: 31–50.

Edel, J.-B. and Wickert, F. 1991. Paleoposition of the Saxothuringian (Northern Vosges, Pfalz, Odenwald, Spessart) in Variscan times: paleomagnetic investigation. *Earth Planet. Sci. Lett.* 103: 10–26.

Edelman, N. and Jaanus-Järkkälä, M. 1983. A plate tectonic interpretation of the Precambrian of the archipelago of SW Finland. *Bull. Geol. Surv. Finl.* 325: 1–33.

Egger, A. 1990. *A Comprehensive Compilation of Seismic Refraction Data along the Southern Segment of the European Geotraverse from the Northern Apennines to the Sardinia Channel (1979–1985)*. Open File Report, Institute of Geophysics, ETH Zürich, Switzerland.

Egger, A. 1992. Lithospheric structure along a traverse from the Northern Apennines to Tunisia derived from seismic refraction data. PhD Thesis, Swiss Fed. Inst. Technology, Zürich.

Ehlers, C. and Lindros, A. 1990. Early Proterozoic Sveco-Fennian volcanism and associated plutonism in Enklinge, SW Finland. *Precambrian Res.* **47**: 307–18.

Ekman, M. 1991. Gravity change, geoid change and remaining postglacial uplift of Fennoscandia. *Terra Nova* **3**: 390–2.

Elming, S.A., Pesonen, L.J., Leino, M., Khramov, A.N., Michailova, N.P., Krasnova, A.F., Mertanen, S., Bylund, G. and Terho, M. 1992. The continental drift of Fennoscandia and Ukraina during the Precambrian. *Tectonophysics*, in press.

Elo, S., Lanne, E., Ruotoistenmäki, T. and Sindre, A. 1989. Interpretation of gravity anomalies along the POLAR Profile in the northern Baltic Shield. *Tectonophysics* **162**: 135–50.

Emmermann, R. 1977. A petrogenetic model for the origin and evolution of the Hercynian granite series of the Schwarzwald. *N. Jb. Geol. Miner. Abh.* **128**: 219–53.

Emmermann, R. and Wohlenberg, J. (eds), 1989. *The German Continental Deep Drilling Program (KTB)*. Springer, Berlin, Heidelberg, New York.

Emmert, U. 1981. Perm nördlich der Alpen. In: Bayerisches Geologisches Landesamt (ed.), Erläuterungen zur Geologischen Karte von Bayern, pp. 34-40.

Emslie, R.F. 1991. Granitoids of rapakivi-anorthosite and related associations. *Precambrian Res.* **51**: 173–92.

Engel, W. 1984. Migration of folding and flysch sedimentation on the southern flank of the Variscan Belt (Montagne Noire, Mouthoumet Massif, Pyrenees). *Z. Deutsch Geol. Ges.* **135**: 279–92.

Engel, W. and Franke, W. 1983. Flysch-sedimentation: its relation to tectonism in the European Variscides. In: H. Martin and F.W. Eder (eds), *Intracontinental Fold Belts*, pp. 290–321, Springer, Berlin, Heidelberg, New York.

Engel, W., Franke, W., Grote, C., Weber, K., Ahrendt, H. and Eder, F.W. 1983. Nappe tectonics in the southeastern part of the Rheinisches Schiefergebirge. In: H. Martin and F.W. Eder (eds), *Intracontinental Fold Belts*, pp. 267-287, Springer, Berlin, Heidelberg, New York.

ERCEUGT Group 1992. An electrical resistivity crustal section from the Alps to the Baltic Sea (central segment of the EGT). *Tectonophysics* **207**, in press.

EREGT Group 1990. An electrical resistivity transect from the Alps to the Baltic Sea (Central Segment of the EGT). In: R. Freeman and St. Mueller (eds), *Sixth EGT workshop: Data compilations and synoptic interpretation*, pp. 299–313, European Science Foundation, Strasbourg, France.

ETH Working Group on Deep Seismic Profiling 1991. Integrated analysis of seismic normal incidence and wide-angle reflection measurements across the eastern Swiss Alps. In: R. Meissner, L. Brown, H.J. Dürbaum, W. Franke, K. Fuchs and F. Seifert (eds), Continental Lithosphere: Deep Seismic Reflections. *Amer. Geophys. Union Geodyn. Ser.* **22**: 195–205.

EUGEMI Working Group 1990. The European Geotraverse seismic refraction experiment of 1986 from Genova, Italy, to Kiel, Germany. *Tectonophysics* **176**: 43–57.

EUGENO-S Working Group 1988. Crustal structure and tectonic evolution of the transition between the Baltic Shield and the North German Caledonides (the EUGENO-S Project). *Tectonophysics* **150**: 253–348.

Eva, C., Augliera, P., Cattaneo, M., and Giglia, G., 1990. Some considerations on seismotectonics of northwestern Italy. In: Freeman, R., Giese, P., and Mueller, St. (eds.), *The European Geotraverse: Integrative Studies*, pp. 289–96, European Science Foundation, Strasbourg.

Falkum, T. and Petersen, J.S. 1980. The Sveco-Norwegian orogenic belt, a case of late Proterozoic plate collision. *Geol. Rundschau* **69**: 622–47.

Faupl, P., Pober, E. and Wagreich, M. 1987. Facies Development of the Gosau Group of the eastern parts of the Northern Calcareous Alps during the Cretaceous and Paleogene. In: H.W. Flögel and P. Faupl (eds), *Geodynamics of the Eastern Alps*, pp. 142–55, Deuticke, Vienna.

Finetti, I. and Del Ben, A. 1986. Geophysical study of the Tyrrhenian opening. *Boll. Geofis. Teor. Appl.* **28**: 75–155.

Finger, F. and Steyrer, H.P. 1990. I-type granitoids as indicators of a late Paleozoic convergent ocean–continent margin along the southern flank of the central European Variscan orogen. *Geology* **18**: 1207–10.

Fjeldskaar, W. and Cathles, L. 1991. The present rate of uplift of Fennoscandia implies a low-viscosity asthenosphere. *Terra Nova* **3**: 393–400.

Fleitout, L. and Froidevaux, C. 1982. Tectonics and topography for a lithosphere containing density heterogeneities. *Tectonics* **1**: 21–56.

Fleitout, L. and Froidevaux, C. 1983. Tectonic stresses in the lithosphere. *Tectonics* **2**: 315–24.

Flick, H. 1987. Geotektonische Verknüpfung von Plutonismus und Vulkanismus im südwestdeutschen Variscikum. *Geol. Rundschau* **76**: 699–707.

Flöttmann, T. 1988. *Strukturentwicklung, P-T-Pfade und Deformationsprozesse im zentralschwarzwälder Gneiskomplex.* Frankfurter Geowiss. Arb., Ser. A 6.

Flöttmann, T. and Kleinschmidt, G. 1989. Structural and Basement Evolution in the Central Schwarzwald Gneiss Complex. In: R. Emmermann, and J. Wohlenberg (eds), *The German Continental Deep Drilling Program (KTB)*, pp. 265–76, Springer, Berlin, Heidelberg, New York.

Floyd, P.A. 1984. Geochemical characteristics and comparison of the basic rocks of the Lizard Complex and the basaltic lavas within the Hercynian troughs of SW England. *J. Geol. Soc. Lond.* **141**: 61–70.

Floyd, P.A., Leveridge, B.E., Franke, W., Shail R. and Dörr, W. 1990. Provenance and depositional environment of Rheno-Hercynian synorogenic greywackes from the Giessen Nappe, Germany. *Geol. Rundschau* **79**: 611–26.

Fowler, C.M.R. 1990. *The Solid Earth.* Cambridge University Press.

Frank, W., Gansser, A. and Trommsdorff, V. 1977. Geological observations in the Ladakh area (Himalayas). *Schwiz. Mineral. Petrogr. Mitt.* **57**: 89–113.

Frank, W., Hoinkes, G., Purtscheller, F. and Thöni, M. 1987. The Austroalpine unit west of the Hohe Tauern: the Ötztal-Stubai Complex as an example for the Eoalpine Metamor-

phic Evolution. In: H.W. Flögel and P. Faupl (eds), *Geodynamics of the Eastern Alps*, pp. 179–225, Deuticke, Vienna.

Franke, D. 1991. Der präpermische Untergrund der Mitteleuropäischen Senke–Fakten und Hypothesen. *Nied. Akad. Geowiss.* **4**: 19–76.

Franke, D., Hoffman, N. and Kamps, J. 1989. Alter und struktureller Bau des Grundgebirges im Nordteil der DDR. *Z. Angew. Geol. Berlin* **35**: 289–96.

Franke, W. 1984a. Variszischer Deckenbau im Raume der Münchberger Gneismasse; abgeleitet aus der Fazies, Deformation und Metamorphose im umgebenden Paläozoikum. *Geotekton. Forsch.* **68**: 1–253.

Franke, W. 1984b. Late events in the tectonic history of the Saxothuringian Zone. In: D.H.W. Hutton and D.J. Sanderson (eds), *Variscan tectonics of the North Atlantic Region*, pp. 33–45, Blackwell Scientific Publications.

Franke, W. 1989a. Tectonostratigraphic units in the Variscan belt of central Europe. *Geol. Soc. Amer. Spec. Paper* **230**: 67–90.

Franke, W. 1989b. The geological framework of the KTB drill site, Oberpfalz. In R. Emmermann and J. Wohlenberg. (eds), 1989. *The German Continental Deep Drilling Program (KTB)*, pp. 37–54, Springer, Berlin, Heidelberg, New York.

Franke, W. 1989c. Variscan plate tectonics in Central Europe – current ideas and open questions. *Tectonophysics* **169**: 221–8.

Franke, W. and Oncken, O. 1990. Geodynamic evolution of the North-Central Variscides – a Comic Strip. In: R. Freeman, P. Giese and St. Mueller (eds), *The European Geotraverse: Integrative studies*, 187–94, European Science Foundation, Strasbourg, France.

Franke, W., Bortfeld, R. K., Brix, M., Drozdzewski, G., Dürbaum, H. J., Giese, P., Janoth, W., Jödicke, H., Reichert, Ch., Scherp, A., Schmoll, J., Thomas, R., Thünker, M., Weber, K., Wiesner, M. G. and Wong, H. K. 1990a. Crustal structure of the Rhenish Massif: results of deep seismic reflection lines DEKORP 2-North and 2-North-Q. *Geol. Rundschau* **79**: 523–66.

Franke, W., Giese, P., Grosse, S., Haak, V., Kern, H., Mengel, K. and Oncken, O. 1990b. Geophysical imagery of geological structures along the central segment of the EGT. In: R. Freeman, P. Giese and St. Mueller (eds), *The European Geotraverse: Integrative studies*, pp. 177–86, European Science Foundation, Strasbourg, France.

Freeman, R., Von Knorring, M., Korhonen, H., Lund, C. and Mueller, St. (eds) 1989. The European Geotraverse, Part 5: The POLAR Profile. *Tectonophysics* **162**: 1–171.

Frei, W., Heitzmann, P., Lehner, P., Mueller, St., Olivier, R., Pfiffner, O. A., Steck, A. and Valasek, P. 1989. Geotraverses across the Swiss Alps. *Nature* **340**: 544–8.

Friend, R.L., Nutman, A.P. and McGregor, V.R. 1988. Late Archaean terrane accretion in the Godthåb region, southern West Greenland. *Nature* **335**: 535–8.

Froitzheim, N. 1988. Synsedimentary and synorogenic normal faults within a thrust sheet of the Eastern Alps (Ortler zone, GraubÅnden, Switzerland). *Ecologae Geol. Helv.* **81**: 593–610.

Frost, R.T.C., Fitch, F.J. and Miller, J.A. 1981. The age and nature of the crystalline basement of the North Sea Basin. In: L.V.Illing and G.C. Hobson (eds), *Petroleum Geology of the Continental Shelf of North-West Europe*, pp. 43–57, Heyden, London.

Fuchs, K. 1983. Recently formed elastic anisotropy and petrological models for the continental subcrustal lithosphere in southern Germany. *Phys. Earth Planet. Inter.* **31**: 93–118.

Fuchs, K., von Gehlen, K., Mälzer, H., Murawski, H., and Semmel, A. (eds) 1983. *Plateau Uplift: The Rhenish Shield – A Case History*. Springer, Berlin, Heidelberg, New York.

Fuchs, K., Bonjer, K.-P., Gajewski, D., Lüschen, E., Prodehl, C., Sandmeier, K.-J., Wenzel, F. and Wilhelm, H. 1987. Crustal evolution of the Rhinegraben area. 1: Exploring the lower crust in the Rhinegraben rift by unified geophysical experiments. *Tectonophysics* **141**: 261–75.

Funk, H.P. 1985. Mesozoische Subsidenzgeschichte im Helvetischen Schelf der Ostschweiz. *Ecologae Geol. Helv.* **78**: 249–72.

Gaál, G. 1986. 2200 million years of crustal evolution: the Baltic Shield. *Bull. Geol. Soc. Finl.* **58**: 149–58.

Gaál, G. 1990. Tectonic styles of early Proterozoic ore deposition in the Baltic Shield. *Precambrian Res.* **46**: 83–114.

Gaál, G. and Gorbatschev, R. 1987. An outline of the Precambrian evolution of the Baltic Shield. *Precambrian Res.* **35**: 15–52.

Gaál, G., Berthelsen, A., Gorbatschev, R., Kesola, R., Lehtonen, M.I., Marker, M. and Raase, P. 1989. Structure and composition of the Precambrian crust along the POLAR Profile in the northern Baltic Shield. *Tectonophysics* **162**: 1–25.

Gajewski, D., Stangl, R., Fuchs, K., and Sandmeier, K.J. 1990. A new constraint on the composition of the topmost continental mantle – anomalously different depth increases of P and S velocity. *Geophys. J. Int.* **103**: 497–507.

Gall, H., Hüttner, R. and Müller, D. 1977. Erläuterungen zur Geologischen Karte des Rieses 1:50000. *Geol. Bavarica* **76**: 1–171.

Gast, R.E. 1988. Rifting im Rotliegenden Niedersachsens. *Die Geowiss.* 1988/4: 115–22.

Gebauer, D., Williams, I.S., Compston, W. and Grünenfelder, M. 1989. The development of the Central European continental crust since the Early Archaean based on conventional and ion-microprobe dating of up to 3.84 b.y. old detrital zircons. *Tectonophysics* **157**: 81–96.

Giannini, E., and Lazzarotto, A. 1975. Tectonic evolution of the Northern Apennines. In: C. H. Squyres, (ed.), Geology of Italy. *Earth Sci. Soc. Libyan Arab Rep.* **2**: 237–87.

Giese, P. 1968. Die Struktur der Erdkruste im Bereich der Ivrea-Zone. *Schweiz. Mineral. Petrogr. Mitt.* **48**: 261–84.

Giese, P., Prodehl, C. and Stein, A. (eds) 1976. *Explosion Seismology in Central Europe*, Springer, Berlin, Heidelberg, New York.

Giese, P., Reutter, K.J., Jacbshagen, V., and Nicolich, R. 1982. Explosion-seismic crustal studies in the Alpine Mediterranean region and their implications to tectonic processes. In: H. Berckhemer and K. Hsu (eds), Alpine–Mediterranean Geodynamics, *Amer. Geophys. Union Geodyn. Ser.* **7**: 39–74.

Giese, P., Ibbeken, S., Baier, B. and Schulze-Frerichs, K. 1990. Accompanying seismic refraction investigations along the profile DEKORP *North. Geol. Rundschau* **79**: 567–79.

Giger, M. and Hurford, A.J. 1989. Tertiary intrusives of the Central Alps, their Tertiary uplift, erosion, redeposition and burial in the south-alpine foreland. *Ecologae Geol. Helv.* **82**: 857–66.

Glahn A., Sachs P.M. and Achauer U. 1992. A teleseismic and petrologlcal study of the crust and upper mantle beneath the geothermal anomaly of Urach/SW Germany. *Phys. Earth Planet. Int.* **69**: 176–206.

Goetze, C. and Evans, B. 1979. Stress and temperature in the bending lithosphere as constrained by experimental rock mechanics. *Geophys. J. Roy. Astr. Soc.* **59**: 463–78.

Gorbatschev, R. and Gaál, G. 1987. The Precambrian history of the Baltic Shield. In: A. Kröner (ed.), Proterozoic Lithospheric Evolution. *Amer. Geophys. Union*, 149–59.

Gorbatschev, R.,Lindh, A., Solyom, Z., Laitakari, I., Aro, K., Lobach-Zhuchenko, S.B., Markov, M.S., Ivliev, A.I. and Bryhni, I. 1987. Mafic Dyke Swarms of the Baltic Shield. In: H.C. Halls and W.F. Fahrig (eds), Mafic Dyke Swarms. *Geol. Assoc. Can. Spec. Paper* **34**: 361–72.

Govers, R., Wortel, M.J.R., Cloetingh, S.A.P.L. and Stein, C.A. 1992. Stress magnitude estimates from earthquakes in oceanic plate interiors. *J. Geophys Res.*, in press.

Gower, C.F.1990. Mid-Proterozoic evolution of the eastern Grenville Province, Canada. *Geol. Fören. Stockh. Förh.* **112**: 127–39.

Grauert, B., Mecklenburg, S., Vinx, R. and Vocke, R.D. 1987. Geochronology and petrogenesis of the Harzburg gabbro and Brocken intrusive complex. *Terra Cognita* **7**: 335.

Green, C.M., Stuart, G.W., Lund, C-E. and Roberts. R.G. 1988. P-wave crustal structure of the Lake Vänern area, Sweden: EUGENO-S profile 6. *Tectonophysics* **150**: 349–61.

Gregersen, S., Flüh, E.R., Möller, C. and Hirschleber, H. 1987. *Seismic Data of the EUGENO-S Project,* Department of Seismology of the Danish Geodetic Institute Charlottenlund (Denmark).

Gregersen, S. 1979. Earthquakes in the Skagerrak recorded at small distances. *Bull. Geol. Soc. Den.* **28**: 5–9.

Gregersen, S., Kohonen, H., and Husebye, E.S. 1991. Fennoscandian dynamics: Present-day earthquake activity. In: S. Björnsson, S.

Gregersen, E.S. Husebye, H. Korhonen and C.-E. Lund (eds), Imaging and Understanding the Lithosphere of Scandinavia and Iceland. *Tectonophysics* **189**: 333–44.

Griffin, W.L. andKreston, P. 1987. Scandinavia– the cabonatite connection. In: P.H. Nixon (ed.), *Mantle xenoliths*, pp. 103–6, John Wiley and Sons, Chichester, England.

Growes, D.I., Ho, S.E., Rock, N.M.S., Barley, M.E. and Muggeridge, T.1987. Archaean cratons, diamond and platinum: Evidence for coupled long-lived crust-mantle systems. *Geology* **15**: 801–5.

Gubler, E. 1990. Compilation of recent crustal movement data for the European Geotraverse project. In: R. Freeman, P. Giese and St. Mueller (eds), *The European Geotraverse: Integrative studies*, pp. 371–9, European Science Foundation, Strasbourg, France.

Guggisberg, B. 1986. Eine zweidimensionale refraktionsseismische Interpretation der Geschwindigkeits-Tiefenstruktur des oberen Mantels unter dem fennoskandischen Schild (Projekt FENNOLORA). PhD Thesis, Swiss Fed. Inst. Technology Zürich.

Guggisberg, B. and Berthelsen, A. 1987. A two-dimensional velocity model for the lithosphere beneath the Baltic Shield and its possible tectonic significance. *Terra Cognita* **7**: 631–8.

Guggisberg, B., Kaminski, W. and Prodehl, C. 1991. Crustal structure of the Fennoscandian Shield: a travel time interpretation of the long-range FENNOLORA seismic refraction profile. *Tectonophysics* **195**: 105–37.

Gunzenhauser, B.A. 1985. Zur Sedimentologie und Paleogeographie der oligo-miocenen Gonfolite Lombarda zwischen Lago Maggiore und Brianza (Südtessin, Lombardei). *Beitr. Geol. K. Schweiz* **159**: 1–114.

Guterch, A., Grad, M., Materzok, R. and Perchuc, E.1986. Deep structure of the earth's crust in the contact zone of the Palaeozoic and Precambrian platforms in Poland (Tornquist-Teisseyre Zone). *Tectonophysics* **128**: 251–79.

Haapala, I. and Rämö, T. 1990. Petrogenesis of the Proterozoic rapakivi granites of Finland. *Geol. Soc. Amer. Spec. Pap.* **246**: 275–86.

Haenel, R. 1983. Geothermal investigations in the Rhenish Massif. In: K. Fuchs *et al.* (eds),

Plateau uplift, the Rhenish Massif – a case history, pp. 228–46, Springer, Berlin, Heidelberg, New York.

Hageskov, B.1980. The Sveconorwegian structures of the Norwegian part of the Kongsberg–Bamble–Østfold basement complex, SE Norway. *Geol. Fören. Stockh. Förh.* 102: 150–5.

Hageskov, B.1985. Constrictional deformation of Koster dyke swarm in a ductile sinistral shear zone, Koster islands, SW Sweden. *Bull. Geol. Soc. Den.* 34: 151–97.

Haggerty, S.E. 1986. Diamond genesis in a multiply-constrained mantle. *Nature* 320: 34–8.

Hahn, A. and Wonik, T. 1990. Interpretation of aeromagnetic aniomalies. In: R. Freeman and St. Mueller (eds), *Sixth EGT workshop: Data compilations and synoptic interpretation*, pp. 225–36, European Science Foundation, Strasbourg, France.

Hamann, N.E. 1989. Bornholms Mesozoicum. In: Bornholms Geologi IV: Mesozoicum, *VARV No. 3 (1989)*, 75–104.

Hambrey, M.J. and Harland, W.B. (eds.) 1982. *Earth's pre-Pleistocene glacial record*, Cambridge University Press.

Handy, M.R. 1987. The structure, age and kinematics of the Pogallo Fault Zone; Southern Alps, northwestern Italy. *Ecologae Geol. Helv.* 80: 593–632.

Hansen, B.T. Teufel, S. and Ahrendt, H. 1989. Geochronology of the Moldanubian-Saxothuringian Transition Zone, Northeast Bavaria. In: R. Emmermann and J. Wohlenberg (eds), *The German Continental Deep Drilling Program (KTB)*, pp.55–65, Springer, Berlin, Heidelberg, New York.

Haskell, N.A. 1935. The motion of a viscous fluid under a surface load. *Physics* 6: 265–69.

Hauser, F. 1989. Die Struktur der Erdkruste in Südschweden, abgeleitet aus P- und S-Wellenbeobachtungen des FENNOLORA-Profiles. Diploma Thesis, University of Karlsruhe.

Hauser, F., Prodehl, C. and Schimmel, M. 1990. *A compilation of data from the Fennolora Seismic Refraction Experiment 1979*. Geophysical Institute, University of Karlsruhe (Federal Republic of Germany), Open File Report

Haverkamp ,J., Kramm, U. and Walter, R. 1991. U–Pb isotope variations of detrital zircons in sediments of the Rheno-Hercynian and their significance for the Palaeozoic geotectonic development of NW Central Europe. *Terra Abstracts* 3: 207.

Heller, F., Lowrie, W. and Hirt, A.M. 1989. A review of palaeomagnetic and magnetic anisotropy results from the Alps. In M.P. Coward, D. Dietrich and R.G. Park (eds), Alpine Tectonics. *Geol. Soc. Lond. Spec. Publ.* 45: 399–420.

Henes-Klaiber, U., Holl, A. and Altherr, R. 1989. The Odenwald: more evidence for Hercynian arc magmatism. *Terra Abstracts* 1: 281.

Henkel, H. 1991. Magnetic crustal structures in northern Fennoscandia. *Tectonophysics* 192: 57–79.

Henkel, H., Lee, M. K., Lund, C.-E. and Rasmussen, T. 1990. An integrated geophysical interpretation of the 2000 km FENNOLORA section of the Baltic Shield. In: R. Freeman, P. Giese and St. Mueller (eds), *The European Geotraverse: Integrative studies*, pp. 1–47, European Science Foundation, Strasbourg, France.

Hfaiedh, M., Chadi, M. and Allouche, M. 1985. Seismicity of Tunisia and neighbouring areas. In: D.A. Galson and St. Mueller (eds), *European Geotraverse (EGT) Project, the Southern Segment*, pp.261–7, European Science Foundation, Strasbourg, France.

Hietanen, A. 1975. Generation of potassium-poor magmas in the northern Sierra Nevada and the Sveco-Fennian of Finland. *J. Res. U.S. Geol. Surv.* 3: 631–45.

Hirschmann, G. and Okrusch, M. 1988. Spessart-Kristallin und Ruhlaer Kristallin als Bestandteile der Mitteldeutschen Kristallinzone – ein Vergleich. *N. Jb. Geol. Paläont. Abh.* 177: 1-39.

Hirt, A.M. and Lowrie, W. 1988. Palaeomagnetism of the Umbrian–Marches orogenic belt. *Tectonophysics* 146: 91–103.

Hjelt, S.E. 1990. Electromagnetic studies on the EGT northern segment. Summary of results. In: R. Freeman and St. Mueller (eds), *Sixth EGT workshop: Data compilations and synoptic interpretation*, pp. 287–97, European Science Foundation, Strasbourg, France.

Hjelt, S.E. 1991. Geoelectric studies and conductivity structures of the eastern and northern parts of the Baltic Shield. *Tectonophysics* **189**: 249–60.

Hoegen, J.von, Kramm, U. and Walter, R. 1990. The Brabant Massif as part of Armorica/Gondwana: U–Pb isotopic evidence from detrital zircons. *Tectonophysics* **185**: 37–50.

Hoffman, P.F. 1989. Precambrian geology and tectonic history of North America. In: The Geology of North America, vol. A, The Geology of North America-an overview. *Geol. Soc. Amer.*, 447–512.

Holl, A. and Altherr, R., 1987. I-type granitoids of northern Vosges: documents of increasing arc maturity. *Terra Cognita* **7**: 174.

Holliger, K. and Kissling, E. 1991. Ray theoretical depth migration: Methodology and application to deep seismic reflection data across the eastern and southern Swiss Alps. *Ecologae Geol. Helv.* **84**: 369–402.

Holliger, K. and Kissling, E. 1992. Gravity interpretation of a unified 2D-acoustic image of the central alpine collision zone. *Geophys. J. Int.*, in press.

Hörmann, P.K., Raith, M., Raase, P., Ackermand, D. and Seifert, F. 1980. The granulite complex of Finnish Lapland: petrology and metamorphic conditions in the Ivalojoki-Inarijärvi area. *Bull. Geol. Surv. Finl.* **308**: 1–95.

Hossain, Md. A. 1989. Seismic refraction studies in the Baltic Shield along the Fennolora profile. PhD Thesis, University of Uppsala, 133pp.

Huhma, H. 1987. Provenance of early Proterozoic and Archaean metasediments in Finland: a Sm–Nd study. *Precambrian Res.* **35**: 127–43.

Hunziker, J.C., Desmons, J. and Martinotti, G. 1989. Alpine thermal evolution in the Central and the Western Alps. In M.P. Coward, D. Dietrich and R.G. Park (eds), Alpine Tectonics. *Geol. Soc. Lond. Spec. Publ.* **45**::353–67.

Hurford, A.J., Flisch, M. and Jéger, E. 1989. Unravelling the thermo-tectonic evolution of the Alps: a contribution from fission track analysis and mica dating. In M.P. Coward, D. Dietrich and R.G. Park (eds), Alpine Tectonics. *Geol. Soc. Lond. Spec. Publ.* **45**:369–98.

Husebye, E.S., Bungum, H., Fyen, J., and Gjöystdal, H. 1978. Earthquake activity in Fennoscandia between 1497 and 1975 and intraplate tectonics. *Nor. Geol. Tidsskr.* **58**: 51–8.

Jacobshagen, V. 1986. *Geologie von Griechenland: Beitr. Region,* pp. 1–363, Geologie der Erde, 19, Borntraeger, Berlin.

Jahn, B.M., Vidal, Ph. and Kröner, A. 1984. Multi-chronometric ages and origin of Archean tonalitic gneisses in Finnish Lapland: a case for long crustal residence time. *Contrib. Mineral. Petrol.* **86**: 398–408.

Japsen, P.1992. Landhøvningerne i Sen Kridt of Tertiar i det nordlige Danmark (with English abstract). *Dan. Geol Foren.*, in press.

Jeffreys, H. 1970. *The Earth.* Cambridge University Press.

Jeffreys, H., 1975. The Fenno-Scandian uplift. *J. Geol. Soc. Lond.* **131**: 32–5.

Jödicke, H. 1990. *Zonen hoher elektrischer Krustenleit-föhigkeit im Rhenoherzynikum und seinem nördlichen Vorland.* Hochschulschriften Bd. 24, Lit Verlag Münster and Hamburg.

Johansson, Å. and Larsen, O. 1989. Radiometric age determinations and Precambrian geochronology of Blekinge, southern Sweden. *Geol. Fören. Stockh. Förh.* **111**: 35–90.

Jones, A.G. 1981. Geomagnetic induction studies in Scandinavia II: geomagnetic sounding, induction vectors and coast effects. *J. Geophys.* **50**: 23–36.

Kähkönen, Y., Huhma, H. and Aro, K. 1989. U–Pb zircon ages and Rb–Sr whole-rock isotope studies of early Proterozoic volcanic and plutonic rocks near Tampere, southern Finland. *Precambrian Res.* **45**: 27–43.

Kamoun, Y. and Hfaiedh, M. 1985. Neotectonics of Tunisia; a synthesis. In: D.A. Galson and St. Mueller (eds), *European Geotraverse (EGT) Project, the Southern Segment,* pp.255–59, European Science Foundation, Strasbourg, France.

Kelts, K. 1981. A comparison of some aspects of sedimentation and translational tectonics from the Gulf of California and the Mesozoic Tethys, northern Penninic margin. *Ecologae Geol. Helv.* **74**: 317–38.

Kern, H. 1982. P- and S-wave velocities in crustal and mantle rocks under simultaneous action of high confining pressure and high temperature and the effect of rock microstructure. In.: W. Schreyer (ed.) *High Pressure Research in Geosciences,* pp.15–45, Schweizerbart, Stuttgart.

Kern, H. and Schenk V. 1985. Elastic wave velocities from a lower crustal section in Southern Calabria (Italy). *Phys. Earth Planet. Int.* **40**: 147–60.

Kent, D. V. and Van der Voo, R. 1990. Palaeozoic palaeogeography from palaeomagnetism of the Atlantic-bordering continents. In: M.W. McKerrow and C.R.Scotese (eds), Palaeozoic Palaeogeography and Biogeography. *Geol. Soc. Lond. Mem.* **12**:49–56.

Kerrich, R., Beckinsdale, R.D. and Durham, J.J. 1977. The transition between deformation regimes dominated by intercrystalline diffusion and intercrystalline creep evaluated by oxygen isotope thermometry. *Tectonophysics* **38**: 241–57.

Khattach, D. 1989. Paléomagnetisme des formations Paléozoiques du Maroc. *Mém. Centre Amoricain Étud. Struct. Socle (Rennes)* **30**: 1–207.

Kinck, J.J., Husebye, E.S. and Lund, C.-E. 1991. The South Scandinavian crust: Structural complexities from seismic reflection and refraction profiling.. *Tectonophysics* **189**: 117–33.

Kissling, E., Mueller, St. and Werner, D. 1983. Gravity anomalies, seismic structure and geothermal history of the central Alps. *Ann. Geophys.* **1**: 37–46.

Klemperer, S.L. and Hurich, C.A. 1990. Lithospheric structure of the North Sea from deep seismic reflection profiling. In: D.J. Blundell and A.D. Gibbs (eds), *Tectonic evolution of the North Sea Rifts*, pp. 37–63, Oxford University Press.

Klingelé, E., Lahmeyer, B. and Freeman, R. 1990a. The EGT Bouguer gravity compilation. In: R. Freeman and St. Mueller (eds) *Sixth EGT workshop: Data compilations and synoptic interpretation*, pp. 247–54, European Science Foundation, Strasbourg, France.

Klingelé, E., Lahmeyer, B., Marson, I. and Schartz, G. 1990b. A 2-D gravity model of the seismic refraction profile of the EGT Southern Segment. In: R. Freeman and St. Mueller (eds) *Sixth EGT workshop: Data compilations and synoptic interpretation*, pp. 271–8, European Science Foundation, Strasbourg, France.

Knudsen, H.J., Balling, N. and Jacobsen, B.H. 1991. Seismic modelling of the Norwegian–Danish Basin along a refraction profile in northern Jutland. *Tectonophysics* **189**: 209–18.

Kontinen, A. 1987. An early Proterozoic ophiolite – the Jormua mafic-ultramafic complex, northeastern Finland. *Precambrian Res.* **35**: 313–42.

Köppel, V., Günthert, A. and Grünenfelder, M. 1980. Patterns of U–Pb zircon and monazite ages in polymetamorphic units of the Swiss Central Alps. *Schweiz. Mineral. Petrogr. Mitt.* **61**: 97–119.

Korja, T. 1992. Electrical conductivity distribution in the lithosphere in the central Fennoscandian Shield. *Precambrian Res.*, in press.

Korja, T., Hjelt, S.-E., Kaikkonen, P., Koivukoski, K., Rasmussen, T.M. and Roberts, R.G. 1989. The geoelectric model of the POLAR profile, Northern Findland. *Tectonophysics* **162**: 113–33.

Korsch, R.J. and Schäfer, A. 1991. Geological interpretation of DEKORP deep seismic reflection profiles 1C and 9N across the Variscan Saar–Nahe Basin, southwest Germany. *Tectonophysics* **191**: 127–46.

Korsman, K. (ed.) 1988. Tectono-metamorphic evolution of the Raahe–Ladoga zone. *Bull. Geol. Surv. Finl.* **343**: 1–96.

Kossmat, F. 1927. Gliederung des varistischen Gebirgsbaues. *Abh. Sächs. Geol. L.-Amt* **1**: 1–39.

Kozlovsky, Y.A. (ed.) 1987. *The superdeep well of the Kola Peninsula*, Springer, Berlin, Heidelberg, New York.

Kramer, W. 1988. *Magmengenetische Aspekte der Lithosphärenentwicklung*. Ser. in Geol. Sci. 23, Akademie-Verlag, Berlin.

Kratz, K.O., Glebovitsky, V.A. and Bylinsky, R.V. 1978. *The Earth's Crust in the Eastern Baltic Shield*, Nauka, Leningrad (in Russian).

Krebs, W. 1976. Zur geotektonischen Position der Bohrung Saar 1. *Geol. Jb. A* **27**: 489–98.

Kreuzer, H., Seidel, E., Schüssler, U., Okrusch, M., Lenz, K.-L. and Raschka, H. 1989. K–Ar geochronology of different tectonic units at the northwestern margin of the Bohemian Massif. *Tectonophysics* **157**: 149–78.

Krill, K.O. 1985. Svecokarelian thrusting with thermal inversion in the Karasjok-Levajok area of the northern Baltic Shield. *Nor. Geol. Unders.* **403**: 89–101.

Kröner, A. (ed.) 1981. *Precambrian plate tectonics*, Elsevier, Amsterdam.

Kröner, A. 1991. Tectonic evolution in the Archaean and Proterozoic. *Tectonophysics* **187**: 393–410.

Kröner, A., Puustinen, K. and Hickman, M. 1981. Geochronology of an Archaean tonalitic gneiss dome in northern Finland and its relation with an unusual overlying volcanic conglomerate and komatiitic greenstone. *Contrib. Mineral. Petrol.* **76**: 33–41.

Kullinger, B. and Lund, C.-E. 1986. A preliminary interpretation of S-wave traveltimes from FENNOLORA data. *Tectonophysics* **126**: 375–88.

Kunze, Th., Leydecker, G. and Schneider, G. 1986. Seismicity and seismotectonics in Central Europe. In: R. Freeman, St. Mueller and P. Giese (eds), *European Geotraverse (EGT) Project, The Central Segment*, pp.255–60, European Science Foundation, Strasbourg, France.

Kusznir, N.J. and Park, R.G., 1984. Intraplate lithosphere deformation and the strength of the lithosphere. *Geophys. J. R. Astr. Soc.* **79**: 513–38.

Kusznir, N.J. and Park, R.G. 1988. The extensional strength of the continental lithosphere: its dependence on geothermal gradient, and crustal composition and thickness. In: M.P. Coward, J.F. Dewey and P.L. Hancock (eds), Continental extensional tectonics. *Geol. Soc. Lond. Spec. Publ.* **28**: 35–52.

Lagabrielle, Y., Polino, R., Auzende, J.M., Blanchet, R., Caby, R., Fudral, S., Lemoine, M., Mevel, C., Ohnenstetter, M., Robert, D. and Tricart, P. 1984. Les témoins d'une tectonique intraocéanique dans le domaine téthysien: analyse des rapports entre les ophiolites et leurs couvertures métasédimentaires dans la zone piémontaise des Alpes franco italiennes. *Ofioliti* **9**: 67–88.

Lambeck, K., Johnston, P. and Nakada, M. 1990. Holocene glacial rebound and sea-level change in NW Europe. *Geophys. J. Int.* **103**: 451–68.

Langer, H. 1990. Seismicity along the Central Segment of the EGT. In: R. Freeman, P. Giese and St. Mueller (eds), *The European Geotraverse: Integrative Studies*, pp. 121–9, European Science Foundation, Strasbourg, France.

Larson, S.Å., Berglund, J., Stigh, J. and Tullborg, E-L. 1990. The Protogine zone, southwest Sweden: a new model – an old issue. In: C.F. Gower, T. Rivers, and B. Ryan (eds), Mid-Proterozoic Laurentia–Baltica. *Geol. Assoc. Can. Spec. Pap.* **38**: 317–33.

Laubscher, H. 1987. Die tektonische Entwicklung der Nordschweiz. *Ecologae Geol. Helv.* **80**: 287–303.

Laubscher, H.P. 1988a. Material balance in Alpine orogeny. *Bull. Geol. Soc. Amer.* **100**: 1313–28.

Laubscher, H.P. 1988b. The arcs of the Western Alps and the Northern Apennines: an updated view. *Tectonophysics*, **146**: 67–78.

Laubscher, H.P. 1989. The tectonics of the southern Alps and the Austro–Alpine nappes: a comparison. In M.P. Coward, D. Dietrich and R.G. Park (eds), Alpine Tectonics. *Geol. Soc. Lond. Spec. Publ.* **45**: 229–41.

Le Douaran, S., Burrus, J. and Avedik, F. 1984. Deep structure of the northwestern Mediterranean Basin: results of a two-ship survey. *Marine Geol.* **55**: 325–45.

Le Pichon, X. 1968. Sea-floor spreading and continental drift. *J. Geophys. Res.* **73**: 3661–97.

Le Pichon, X. 1983. Land-Locked Oceanic Basins and Continental Collision. In: K. J. Hsu (ed.), *Mountain Building Processes*, pp.201–12, Academic Press.

Le Pichon, X. and Barbier, F. 1987. Passive margin formation by low-angle faulting within the upper crust: the northern Bay of Biscay margin. *Tectonics* **6**: 133–50.

Le Pichon, X., Francheteau, J. and Bonnin, J. 1973. *Plate Tectonics: Developments in Geotectonics 6*, Elsevier, Amsterdam.

Lee, M.K., Pharaoh, T.C. and Soper, N.J. 1990. Structural trends in central Britain from images of gravity and aeromagnetic fields. *J. Geol. Soc. Lond.* **147**: 241–58.

Lemoine, M. 1980. Serpentinites gabbros and ophicalcites in the western Alps: possible indicators of oceanic fracture zones and associated serpentinite protrusions in the Jurassic–Cretaceous Tethys. *Arch. Sci. Genäve* **33**: 103–16.

Lemoine, M., Dardeau, G., Delpech, P.-Y., Dumont, T., De Graciansky, P.C., Graham, R., Jolivet, L., Roberts, D. and Tricart, P.

1989. Extension synrift et failles transformantes jurassiques dans les Alpes Occidentales. *C. R. Acad. Sci. Sér. II* **309**: 1711–16.

Leydecker, G. 1980. Erdbebenkatakog für die Bundesrepublik Deutschland mit angrenzenden Gebieten. *Geol. Jb. E* **36**:1–83.

Liboriussen, J., Ashton, P. and Tygesen, T. 1987. The tectonic evolution of the Fennoscandian Border Zone in Denmark. *Tectonophysics* **137**: 219.

Lidén, R., 1938. Den senkvartära strandförskjutningens förloppoch kronologi i Ångermanland. *Geol. Fören. Stockh. Förh.* **60**: 397–404.

Lie, J.E. and Husebye, E. 1991. *Results from the MOBIL Search experiment in the Skagerak.* Lecture presented at 1st MONA LISA Workshop, Nov.1st, 1991, Copenhagen.

Lippolt, H.J. 1983. Distribution of volcanic activity in space and time. In: K. Fuchs *et al.* (eds), *Plateau Uplift: The Rhenish shield – a case history*, pp.112–20, Springer, Berlin, Heidelberg, New York.

Livermore, R. A. and Smith, A. G. 1985. Some Boundary Conditions for the Evolution of the Mediterranean Region. In: D.J. Stanley and F.C. Wezel (eds), *Geological Evolution of the Mediterranean Basin*, pp.83–100, Springer, Berlin, Heidelberg, New York.

Lobach-Zhuchenko, S.B. 1989. Review of Archean geology of the eastern part of the Baltic Shield. *Terra Abstracts* **1**: 1.

Lobach-Zhuchenko, S.B., Levchenkov, O.A., Chekulaev, V.P. and Krylov, I.N. 1986. Geological evolution of the Karelian granite-greenstone terrain. *Precambrian Res.* **33**: 45–65.

Loock, G., Seck, H.A. and Stosch, H.-G. 1990. Granulite facies lower crustal xenoliths from the Eifel, West Germany: Petrological and chemical aspects. *Contrib. Mineral Petrol.* **105**: 25–41.

Lorenz, V. and Nicholls, I.A. 1984. Plate and intraplate processes of Hercynian Europe during the late Paleozoic. *Tectonophysics* **107**: 25–56.

Lucazeau, F., Vasseur, G., Foucher, J.P. and Mongelli, F. 1985. Heat Flow along the Southern Segment of the EGT. In: D.A. Galson and St. Mueller (eds), *European Geotraverse (EGT) Project, the Southern Segment*, pp.59-63, European Science Foundation, Strasbourg, France.

Lund, C.-E. 1983. Fennoscandian Long Range Project 1979 (FENNOLORA). In: E. Bisztricsany and G.Y. Szeidovitz (eds), *Developments in Solid Earth Geophysics* **15**: 511–5, Proc. of the 17th Assembly of the European Seismological Comm.1980, Budapest.

Lund, C.-E. 1990. Summary of the results from the FENNOLORA profile. In: R. Freeman and St. Mueller (eds), *Sixth EGT Workshop: Data Compilations and Synoptic Interpretation*, pp. 65–70, European Science Foundation, Strasbourg, France.

Lundqvist, J. and Lagerbeck, R. 1976. The Pärve Fault: A late-glacial fault in the Precambrian of Swedish Lapland. *Geol. Fören. Stockh. Förh.* **98**: 45–51.

Luosto, U. and Korhonen, H. 1986. Crustal structure of the Baltic Shield based on off-FENNOLORA refraction data. *Tectonophysics* **128**: 183-208.

Luosto, U. and Lindblom, P. 1990. *Seismic Refraction Data of the EGT POLAR Profile.* Report S-23, Institute of Seismology, University of Helsinki, Finland.

Luosto, U., Flüh, E. R., Lund, C.-E. and Working Group 1989. The crustal structure along the POLAR Profile from seismic refraction investigations. *Tectonophysics* **162**: 51–85.

Lykke-Andersen, H. 1987. Thickness of Quaternary deposits and their relation to the pre-Quaternary in the Fennoscandian border zone in Kattegat and Vendsyssel. *Boreas* **16**: 369–71.

Lyon-Caen, H. and Molnar, P. 1989. Constraints on the deep structure and dynamic processes beneath the Alps and adjacent regions from an analysis of gravity anomalies. *Geophys. J. Int.* **99**: 19–32.

Madirazza, I., Jacobsen, B.H. and Abrahamsen, N.1990. Late Triassic tectonic evolution in northwest Jutland, Denmark. *Bull. Geol. Soc. Denmark.*

Maistrello, M., Scarascia, S., Corsi, A., Egger, A. and Thouvenot, F. 1990. *EGT-S 1985: Compilation of data from seismic refraction experiments in Tunisia and the Pelagian Sea*, Istituto per la Geofisica della Litosfera C.N.R., Milano, Italy.

Maistrello, M., Scarascia, S., Ye, S. and Hirn, A. 1991. *EGT-S 1986: Compilation of data from DSS experiments in NW Italy (Ligurian Sea to the Southern Alps)*, Istituto per la Geofisica della Litosfera C.N.R., Milano, Italy.

Mäkinen, J., Ekman, M., Mudtsundstad, A., and Remmer, O. 1986. The Fennoscandian land uplift gravity lines 1966–84. *Rep. Finn. Geod. Inst.* No. 85.

Mälzer, H. 1986. Recent kinematics in central Europe. In: Freeman, R., Mueller, St. and Giese, P. (eds), *European Geotraverse Project: the Central Segment*, pp. 249–54, European Science Foundation, Strasbourg.

Mancktelow, N. 1985. The Simplon Line: a major displacement zone in the western Lepontine Alps. *Ecologae Geol. Helv.* **78**: 73–96.

Marker, M. 1985. Early Proterozoic (ca. 2000-1900 Ma) crustal structure of the northeastern Baltic Shield: tectonic division and tectogenesis. *Nor. Geol. Unders.* **403**: 55–74.

Marker, M. 1990. Tectonic interpretation and new crustal modelling along the POLAR Profile, northern Baltic Shield. In: R. Freeman and St. Mueller (eds), *Sixth EGT workshop: Data compilations and synoptic interpretation*, pp. 9–22, European Science Foundation, Strasbourg, France.

Marker, M., Henkel, H. and Lee, M.K. 1990. Combined gravity and magnetic modelling of the Tanaelv and Lapland Granulite Belts, POLAR profile, northern Baltic Shield. In: R. Freeman, P. Giese, and St. Mueller (eds), *The European Geotraverse: Integrative Studies*, pp. 67–76, European Science Foundation, Strasbourg, France.

Marquart, G. 1989. Isostatic topography and crustal depth corrections for the Fennoscandian geoid. *Tectonophysics* **169**: 67–77.

Marquart, G. and Lelgemann, D. 1990. An approach towards an interpretation of geoid anomalies along the EGT profiles. In: R. Freeman, P. Giese, and St. Mueller (eds), *The European Geotraverse: Integrative Studies*, pp. 359–69, European Science Foundation, Strasbourg, France.

Marquart, G. and Lelgemann, D. 1992. On the interpretation of geoid anomalies in Europe with special regard to the EGT profiles. *Tectonophysics* **207**: in press.

Martin, H. and Eder, F.W. (eds) 1983. *Intracontinental fold belts: Case studies in the Variscan belt of Europe and the Damara belt in Namibia*, Springer, Berlin, Heidelberg, New York.

Martin, H. and Franke, W., 1985. Sonderforschungsbereich 'Entwicklung, Bestand und Eigenschaften der Erdkruste, insbesondere der Geosynklinalräume' (48), Universität Göttingen: Vom Meeresbecken zum Hochgebirge. In: *Deutsche Forschungsgemeinschaft (DFG), Sonderforschungsbereiche 1969–84*, pp. 275–88, VCH Verlagsgesellschaft, Weinheim.

Matte, P. 1986. Tectonic and plate tectonics model for the Variscan belt of Europe. *Tectonophysics* **126**: 329–74.

Matte, P. (ed.) 1990. Terranes in the Variscan Belt of Europe and circum-Atlantic Palaeozoic orogens. *Tectonophysics* **177**: 1–323.

Matte, P. 1991. Accretionary history and crustal evolution of the Variscam belt in Europe. *Tectonophysics* **196**: 309–37.

Matte, P. and Zwart, H.J. (eds) 1989. Palaeozoic plate tectonics with emphasis on the European Caledonian and Variscan belts. *Tectonophysics* **169**: 221–352.

Matte, P., Maluski, H., Rajlich, P. and Franke, W. 1990. Terrane boundaries in the Bohemian Massif: Result of large-scale Variscan shearing. *Tectonophysics* **177**: 151–70.

Matthews, D.H. and the BIRPS Group 1990. Progress in BIRPS deep seismic reflection profiling around the British Isles. *Tectonophysics* **173**: 387–96.

Matthews, S.C. 1978. Caledonian connexions of Variscan tectonism. *Z. Deut. Geol. Ges.* **129**: 423–8.

Mauritsch, H.J. and Becke, M. 1987. Paleomagnetic investigations in the Eastern Alps and the Southern Border Zone. In: Flögel, H.W. and Faupl, P. (eds), *Geodynamics of the Eastern Alps*, pp. 282–308. Deuticke, Vienna.

McKenzie, D. and Morgan, W. J. 1969. Evolution of triple junctions. *Nature* **224**: 125–33.

McKenzie, D. and Weiss, N. 1975. Speculations on the thermal and tectonic history of the Earth. *Geophy. J. R. Astr. Soc.* **42**: 131–74.

McNutt, M., Diament, M. and Kogan, M.G. 1988. Variations of elastic plate thickness at continental thrust belts. *J. Geophys. Res.* **93**: 8825–38.

Mechie J., Prodehl, C. and Fuchs, K. 1983. The long-range seismic refraction experiment in the Rhenish Massif. In: K. Fuchs *et al.* (eds), *Plateau Uplift, the Rhenish Massif: a case history*, pp. 336–42, Springer, Berlin, Heidelberg, New York.

Meier, L. and Eisbacher, G.H. 1991. Crustal kinematics and deep structure of the northern Rhine Graben, Germany. *Tectonics* **10**: 621-630.

Meisl, S. 1990. Metavolcanic rocks in the 'Northern Phyllite Zone' at the southern margin of the Rheno-Hercynian belt.. In: W. Franke (ed.), Mid-German Crystalline Rise and Rheinisches Schiefergebirge: Field Guide,pp. 25–42, *Int.Conf. Paleozoic Orogens in Central Europe 1990*, Göttingen–Giessen.

Meissner, R. 1986. *The continental crust,* Academic Press, New York.

Meissner, R. and Bortfeld, R. K. (eds) 1990. *DEKORP-Atlas: results of Deutsches Kontinentales Reflexionsseismisches Tiefbohrprogramme*, Springer, Berlin, Heidelberg, New York.

Meissner, R. and Gebauer, D. 1989. Evolution of the European continental crust: deep drilling, geophysics, geology and geochemistry. *Tectonophysics* **157**: 1–219.

Meissner, R. and Kusznir N.J. 1987. Crustal viscosity and the reflectivity of the lower crust. *Ann. Geophys.* **5**: 365–74.

Meissner, R., Wever, Th. and Flüh, E. R. 1987. The Moho in Europe – implications for crustal development. *Ann. Geophys.* **5**: 357–64.

Meissner, R., Brown, L., Duerbaum, H. J., Franke, W., Fuchs, K. and Seiffert, F. (eds) 1991. Continental Lithosphere: Deep Seismic Reflections. *Amer. Geophys. U. Geodyn. Ser.* **22**.

Ménard, G., Molnar, P. and Platt, J.P. 1991. Budget of crustal shortening and subduction of continental crust in the Alps. *Tectonics* **10**: 231–44.

Mengel K. 1990a. Origin of crustal xenoliths from the Northern Hessian Depression (NW Germany): petrological and chemical evolution. *Contrib. Mineral. Petrol.* **104**: 8–26.

Mengel K. 1990b. *The contribution of xenoliths from the Northern Hessian Depression to problems of the constitution and evolution of the lower continental crust.* Habil. Thesis, Universität Göttingen (in German).

Mengel, K. and Green, D.H. 1989. Stability of amphibole and phlogopite in metasomatized peridotite under water-saturated and water-undersaturated conditions. In: J. Ross (ed.), Kimberlites and related rocks. *Geol. Soc. Austral. Spec. Publ.* **14/1**: 571–81.

Mengel K. and Hoefs J. 1990. Li–δ^{18}O–SiO$_2$ relations in young volcanic rocks and mafic granulites: implications for the origin of mafic lower crustal xenoliths. *Earth Planet. Sci. Lett.* **101**:42–53.

Mengel K. and Kern H. 1992. Evolution of the petrological and seismic Moho – implications for the continental crust–mantle boundary. *Terra Nova* **4**: 109–16.

Mengel K. and van Calsteren, P. 1989. Charakterisierung proterozoischer Unterkruste im Basement NW Deutschlands. *Eur. J. Mineral.* **1**: 1–120.

Mengel, K., Sachs, P.M., Stosch, H.G., Worner, G., and Loock, G.1991. Crustal xenoliths from Cenozoic volcanic fields of west Germany: implications for structure and composition of the continental crust. *Tectonophysics* **195**: 271–89.

Meriläinen, K. 1976. The granulite complex and adjacent rocks in Lapland, northern Finland. *Bull. Geol. Surv. Finl.* **281**: 1–129.

Merle, O., Cobbold, P.R. and Schmid, S. 1989. Tertiary kinematics in the Lepontine dome. In: M.P. Coward, D. Dietrich and R.G. Park, R.G. (eds), Alpine Tectonics. *Geol. Soc. Lond. Spec. Publ.* **45**: 113–34.

Meyer,D.E. 1970. Stratigraphie und Fazies des Paläozoikums im Guldenbachtal/SE-Hunsrück am Südrand des Rheinischen Schiefergebirges. Dissertation Univ. Bonn.

Milnes, A.G. 1974. Structure of the Pennine zone (Central Alps): a new working hypothesis. *Geol. Soc. Amer. Bull.* **85**: 1727–32.

Mitra, S. 1986. Duplex structures and imbricate thrust systems: geometry, structural position, and hydrocarbon potential. *Amer. Assoc. Petrol. Geol. Bull.* **70**: 1087–112.

Molnar, P. and Atwater, T. 1978. Interarc spreading and Cordilleran tectonics as alternates related to the age of subducted oceanic lithosphere. *Earth Planet. Sci. Lett.* **41**: 330–40.

Molnar, P., and Lyon-Caen, H. 1988. Some simple physical aspects of the support, structure, and evolution of mountain belts. In: S.P. Clark Jr., B. C. Burchfiel and J. Suppe (eds), Processes in continental lithospheric deformation. *Geol. Soc. Amer. Spec. Pap.* **218**: 179–208.

Monger, J., Clowes, R.M., Price, R.A., Riddihough, R.P., Simony, P. and Woodsworth, G.J. 1985. Continent–ocean transect B2: Juan de Fuca plate to Alberta plains. *Geol. Soc. Amer. Continent–Ocean Transect* **7**: 1–21.

Morelli, C., Giese, P., Carrozzo, M.T., Colombi, B., Guerra, I., Hirn, A., Letz, H., Nicolich, R., Prodehl, C., Reichert, C., Roewer, P., Sapin, M., Scarascia, S. and Wigger, P. 1977. Crustal and upper mantle structure of the northern Apennines, the Ligurian Sea, and Corsica, derived from seismic and gravimetric data. *Boll. Geofis. Teor. Appl.* **75/76**: 199–260.

Moretti, I. and Royden, L. 1988. Deflection, gravity anomalies and tectonics of doubly subducted continental lithosphere: Adriatic and Ionian Seas. *Tectonics* **7**: 875–93.

Mörner, N.-A. 1977. Past and present uplift in Sweden: glacial isostasy, tectonism and bedrock influence. *Geol. Fören. Stockh. Förh.* **99**: 48–54.'

Mörner, N.-A. 1980. The Fennoscandian uplift: Geological data and their geodynamical implication. In: N.-A. Mörner (ed.), *Earth Rheology, Isostasy and Eustasy*, pp. 251-284, John Wiley and Sons, New York.

Mörner, N.-A. 1990. Glacial isostasy and long-term crustal movements in Fennoscandia with respect to lithospheric and asthenospheric processes and properties. *Tectonophysics* **176**: 13–24.

Morton, L. 1987. Italy: a review of xenolithic occurrences and their comparison with Alpine peridotites. In: P.H. Nixon (ed.), *Mantle xenoliths*, pp. 135–48, John Wiley and Sons, Chichester, England.

Mueller, St. 1984. Dynamic processes in the Alpine arc. *Ann. Geophys.* **2**: 161–4.

Mueller, St. 1989. Deep-reaching geodynamic processes in the Alps. In: M.P. Coward, D. Dietrich and R.G. Park (eds), Alpine Tectonics. *Geol. Soc. Lond. Spec. Publ.* **45**: 30–28.

Mueller, St. and Panza, G. F. 1986. Evidence of a deep reaching lithospheric root under the Alpine arc. In: F.-C. Wezel (ed.), The origin of arcs. *Develop. in Geotectonics* **21**: 93–113.

Mueller, St. and Talwani, M. 1971. A crustal section across the Eastern Alps based on gravity and seismic refraction data. *Pageophys.* **85**: 226–39.

Mueller, St., Ansorge, J., Egloff, R., and Kissling, E. 1980. A crustal cross-section along the Swiss Geotraverse from the Rhinegraben to the Po Plain. *Ecologae Geol. Helv.* **73**: 463–83.

Mugnier, J.-L., Guellec, S., Ménard, G., Roure, F., Tardy, M. and Vialon, P. 1990. A crustal scale balanced cross-section through the external Alps deduced from the ECORS profile. In: F. Roure, P. Heitzmann and R. Polino (eds), Deep structure of the Alps. *Mem. Soc. Géol. France* **156**: 203–16.

Muir Wood, R. 1989. Extraordinary deglaciation reverse faulting in northern Fennoscandia. In: S. Gregersen, and P.W. Basham, (eds), *Earthquakes at North-Atlantic passive margins: neotectonics and postglacial rebound*, pp. 141–73, Kluwer Academic Publishers.

Müller, B., Zoback, M.L., Fuchs, K., Mastin, L., Gregersen, S., Pavoni, N., Stephansson, O. and Ljunggren, Ch. 1992. Regional Petterns of tectonic stress in Europe. *J. Geophys. Res.*, in press.

Nafe, S.E. and Drake, C.L. 1963. Physical properties of marine sediments. In: M.N. Hill (ed.), *The Sea*, 3, pp. 794-819, Wiley Interscience, New York.

Nasir, S., Okrusch, M., Kreuzer, H., Lenz, H. and Höhndorf, A. 1991. Geochronology of the Spessart Crystalline Complex, Mid-German Crystalline Rise. *Mineral. Petrol.* **44**: 39–55.

Nelson, T.H. and Temple, P.G. 1972. Mainstream mantle convection: a geological analysis of plate motion. *Amer. Assoc. Petrol. Geol. Bull.* **56**: 226–46.

Neugebauer, J. 1989. The Iapetus model: a plate tectonic concept for the Variscan belt of Europe. *Tectonophysics* **169**: 229–56.

Nicolich, R. 1985. EGT southern segment: reflection seismics in the offshore areas. In: D.A. Galson and St. Mueller (eds), *Second EGT Workshop: The Southern Segment*, pp. 33–8, European Science Foundation, Strasbourg, France.

Nielsen, S.B. and Balling, N. 1990. Modelling subsidence, heat flow, and hydrocarbon generation in extensional basins. *First Break* **8/1**: 23–31.

Nironen, M. 1989a. Emplacement and structural setting of granitoids in the early Proterozoic Tampere and Savo schist belts, Finland – implications for contrasting crustal evolution. *Bull. Geol. Surv. Finl.* **346**: 1–83.

Nironen, M. 1989b. The Tampere schist belt: sturctural style within an early Proterozoic volcanic arc system in southern Finland. *Precambrian Res.* **43**: 23–40.

Nisbet, E.G. and Fowler, C.M.R. 1983. Model for Archean plate tectonics. *Geology* **11**: 376–9.

Noe-Nygaard, A. 1963. The Precambrian of Denmark. In: K.Rankama (ed.), *Geological Systems – The Precambrian*, 1, pp. 1–25, Interscience Publishers, John Wiley and Sons, London.

Nolet, G. 1990. Partitioned waveform inversion and two-dimensional structure under the Network of Autonomously Recording Seismographs. *J. Geophys. Res.* **95**: 8499–512.

Nolet, G., Dost, B. and Paulssen, H. 1986. Intermediate wavelength seismology and the NARS experiment. *Ann. Geophys.* **4**: 305–14.

Nurmi, P.A. and Haapala, I. 1986. The Proterozoic granitoids of Finland: granite types, metallogeny and relation to crustal evolution. *Bull. Geol. Soc. Finl.* **58**: 203–33.

Oberhänsli, R., Schenker, F. and Mercolli, I. 1988. Indications of Variscan nappe tectonics in the Aar massif. *Schweiz. Mineral. Petrol. Mitt.* **68**: 509–20.

O'Brien, P.J., Carswell, D.A. and Gebauer, D.A. 1990. Eclogite formation and distribution in the European Variscides. In: D.A.Carswell (ed.), *Eclogite facies rocks*, pp. 204-224, Blackie and Son, Glasgow.

Ogniben, L. 1969. Schema introduttiva alla geologia del confine Calabro–Lucano. *Mem. Soc. Geol. Ital.* **8**: 453–763.

Öhlander, B., Skiöld, T., Elming, S.-Å., Claesson, S. and Nisca, D.H. 1992. Delineation and character of the Archaean–Proterozoic boundary in northern Sweden. *Precambrian Res.*, in press.

Oldow, J.S., Bally, A.W., Ave Lallemant, H.G. and Leeman, W.P.1989. Phanerozoic evolution of the North American Cordillera: United States and Canada. In: A.W.Bally and A.R.

Palmer (eds), *The Geology of North America - An overview*, pp. 139–232, Geol. Soc. Amer., Boulder, Colorado.

Oxburgh, E. R. and Turcotte, D. L. 1970. Thermal structure of island arcs. *Geol. Soc. Amer. Bull.* **82**: 1665–88.

Oxburgh, E.R. and O'Nions, R.K. 1987. Helium loss, tectonics and the terrestrial heat budget. *Science* **237**: 1583–8.

Panza, G. F. 1985. Lateral variations in the lithosphere in correspondence of the Southern Segment of EGT. In: D. A. Galson and St. Mueller (eds), *Second EGT Workshop: The Southern Segment*, pp. 47–51, European Science Foundation, Strasbourg, France.

Panza, G. F. and Mueller, St. 1978. The plate boundary between Eurasia and Africa in the Alpine area. *Mem. Sci. Geol. Univ. Padova* **33**: 43–50.

Panza, G. F., Calcagnile, G., Scandone, P. and Mueller, St. 1980a. La struttura profonda dell' area Mediterranea. *Le Scienze* **141**: 60–9.

Panza, G.F., Mueller, St., and Calcagnile, G. 1980b. The gross features of the lithosphere–asthenosphere system in Europe from seismic surface waves and body waves. *Pageophys.* **118**: 1209–13.

Papunen, H. and Gorbunov, G.I. (eds) 1985. Nickel–copper deposits of the Baltic Shield and Scandinavian Caledonides. *Bull. Geol. Surv. Finl.* **333**: 1–393.

Park, A.F. 1984. Nature, affinities and significance of metavolcanic rocks in the Outokumpu assemblage of the Svecokarelides, eastern Finland. *Bull. Geol. Soc. Finl.* **56**: 25–52.

Park, A.F. 1985. Accretion tectonism in the Proterozoic Svecokarelides of the Baltic Shield. *Geology* **13**: 725–9.

Park, A.F. 1991. Continental growth by accretion: a tectonostratigraphic terrane analysis of the evolution of the western and central Baltic Shield, 2.50 to 1.75 Ga. *Geol. Soc. Amer. Bull.* **103**: 522–37.

Park, A.F. and Bowes, D.R. 1983. Basement–cover relationships during polyphase deformation in the Svecokarelides of the Kaavi district, eastern Finland. *Trans. Roy. Soc. Edinb., Earth Sci.* **74**: 95–118.

Pasquale, V., Verdoya, M. and Chiozzi, P. 1991. Lithospheric thermal structure in the Baltic Shield. *Geophys. J. Int.* **106**: 611–20.

Patacca, E., and Scandone, P. 1989. Post-Tortonian Mountain Building in the Apennines: the role of the passive sinking of a relic lithospheric slab. In: A. Boriani, M. Bonafede, G.B. Piccardo and G.B. Vai (eds), *The Lithosphere in Italy*, pp. 157-176, Academia Nazionale dei Lincei, Roma, Italia.

Patchett, P.J., Todt, W. and Gorbatschev, R. 1987. Origin of continental crust of 1.9-1.7 Ga age: Nd isotopes in the Sveco-Fennian orogenic terrains of Sweden. *Precamb. Research* **35**: 145–60.

Paulssen, H. 1988. Evidence for a sharp 670 km discontinuity as inferred from P- to -S converted waves. *J. Geophys. Res.* **93**: 10489–500.

Pavoni, N. 1961. Falting durch Horizontalverschiebung. *Ecologae Geol. Helv.* **54**: 515–34.

Pavoni, N. 1987. Zur seismotectonik Nordschweiz. *Ecologae Geol. Helv.* **80**: 461–72.

Pavoni, N. 1990. Seismicity and fault-plane solutions along the EGT: data selection and representation as illustrated by the seismicity of Switzerland. In: R. Freeman and St. Mueller (eds), *Sixth EGT workshop: Data compilations and synoptic interpretation*, pp. 341–8, European Science Foundation, Strasbourg, France.

Pedersen, T., Larsson, F.R., Lie, J.E. and Husebye, E.S. 1990. Deep and ultra-deep seismic studies in the Skaggerrak (Scandinavia). In: B. Pinet and C. Bois (eds), *The potential of deep seismic profiling for hydrocarbon exploration*, pp. 345–51, Edition Technip, Paris.

Peltier, W.R. 1989. Glacial isostasy in Laurentia and Fennoscandia: New results for the anomalous gravitational field. In: S. Gregersen and P.W. Basham (eds), *Earthquakes at North-Atlantic passive margins: Neotectonics and Postglacial Rebound*, pp. 91–103, Kluwer Academic Publishers.

Peruzza, L., Fischer, G. and Ranieri, G. 1990. Analysis of the Sardinian N–S magnetolelluric profile. In: R. Freeman and St. Mueller (eds), *Sixth EGT workshop: Data compilations and synoptic interpretation*, pp. 329–46, European Science Foundation, Strasbourg, France.

Pesonen, L.J., Torsvik, T.H., Elming, S.-A. and Bylund, G. 1989. Crustal evolution of Fennoscandia – palaeomagnetic constraints.

Tectonophysics **162**: 27–49.

Pfiffner, O.A. 1986. Evolution of the north alpine foreland basin in the Central Alps. In: P.A. Allen, and P. Homewood (eds), Foreland Basins. *Int. Ass. Sediment. Spec. Publ.* **8**: 219–28.

Pfiffner, O.A. 1990. Crustal shortening of the Alps along the EGT profile. In: R. Freeman and St. Mueller (eds), *Sixth EGT workshop: Data compilations and synoptic interpretation*, pp. 255–62, European Science Foundation, Strasbourg, France.

Pfiffner, O.A. 1992. Palinspastic reconstruction of the pre-Triassic basement units in the Alps: the Central Alps. In: J.F. Von Raumer and F. Neubauer (eds), *The pre-Mesozoic Geology in the Alps*, Springer Verlag, in press.

Pfiffner, O.A., Frei, W., Valasek, P., Staeuble, M., Levato, L., Dubois, L., Schmid, S. M. and Smithson, S.B. 1990. Crustal shortening in the Alpine orogen: results from deep seismic reflection profiling in the eastern Swiss Alps, line NFP 20–East. *Tectonics* **9**: 1327–55.

Pharaoh, T.C. and Brewer, T.S. 1990. Spatial and temporal diversity of early Proterozoic volcanic sequences – comparisons between the Baltic and Laurentian shields. *Precambrian Res.* **47**: 169–89.

Pharaoh, T.C., Merriman, R.J., Webb., P.C. and Beckinsale, R.D. 1987. The concealed Caledonides of eastern England: preliminary results of a multidisciplinary study. *Proc. Yorks. Geol. Soc.* **46**: 355–69.

Philip, H. 1987. Plio-Quaternary evolution of the stress field in Mediterranean zones of subduction and collision. *Ann. Geophys.* **3**: 301–20.

Pieri, M. and Groppi, G. 1981. Subsurface geological structure of the Po plain, Italy. *CNR Progetto Finalizzato Geodinamico, Sottoprogetto 'Modello Strucurale'* Pub. **414**: 1–13.

Pitman, W.C. III and Talwani, M. 1972. Seafloor spreading in the North Atlantic. *Geol. Soc. Amer. Bull.* **83**: 619–46.

Platt, J.P. 1986. Dynamics of orogenic wedges and the uplift of high-pressure metamorphic rocks. *Geol. Soc. Amer. Bull.* **97**: 1037–53.

Platt, J.P., Lister, G.S., Cunningham, P., Weston, P., Peel, F., Baudin, T. and Dondey, H. 1989. Thrusting and backthrusting in the

Briançonnais domain of the western Alps. In: M.P. Coward, D. Dietrich and R.G. Park (eds), Alpine Tectonics. *Geol. Soc. Lond. Spec. Publ.* **45**:135–52.

Polino, R., Dal Piaz, G.V. and Gosso, G. 1990. Tectonic erosion at the Adria margin and accretionary processes for the Cretaceous orogeny of the Alps. In: F. Roure, P. Heitzmann and R. Polino (eds), Deep structure of the Alps. *Mem. Soc.Géol. France* **156**: 345–67.

Pollack, H.N. and Chapman, D.S. 1977. On the regional variation of heat flow, geotherms and lithospheric thickness. *Tectonophysics* **38**: 279–96.

Pratsch, J.-C. 1979. Regional structural elements in Northwest Germany. *J. Petrol. Geol.* **2**: 159–80.

Press, S. 1986. Detrital spinels from alpinotype source rocks in the middle Devonian sediments of the Rheinish Massif. *Geol. Rundschau* **75**: 333–40.

Prodehl, C. and Aichroth, B. 1992. Seismic investigations along the European Geotraverse and its surroundings in Central Europe. *Terra Nova* **4**: 14–24.

Prodehl, C. and Kaminski, W. 1984. Crustal structure under the FENNOLORA profile. In: D.A. Galson and St. Mueller (eds), *First EGT Workshop: The Northern Segment*, pp. 43–8, European Science Foundation, Strasbourg, France.

Quadt, A. Von and Gebauer, D. 1988. Sm/Nd, U/Pb and Rb/Sr dating of high-pressure ultramafic to felsic rocks from the Moldanubian area of NE Bavaria (FRG) and the Saxonian Granulite Massif (GDR). *Chem. Geol.* **70**: 15.

Qian, W. and Pedersen, L.B. 1992. Inversion of borehole breakout orientation data. *J. Geophys. Res.* **96**: 20093–107.

Rämö, O.T. 1991. Petrogenesis of the Proterozoic rapakivi granites and related basic rocks of southeastern Fennoscandia: Nd and Pb isotopic and general geochemical constraints. *Bull. Geol. Surv..Finl.* **355**: 1–161.

Ramsay, J.G. 1989. Fold and fault geometry in the western Helvetic nappes of Switzerland and France and its implication for the evolution of the arc of the Western Alps. In: M.P. Coward, D. Dietrich and R.G. Park (eds), Alpine Tectonics. *Geol. Soc. Lond. Spec. Publ.* **45**:33–45.

Rapp, R.H. 1981. The earth's gravity field to degree and order 180 using SEASAT altimeter data, terrestrial gravity data and other data, Rep. No. 322, Ohio State University, Dept. of Geodetic Science.

Rasmussen, T.M., Roberts, R.G. and Pedersen, L.B. 1987. Magnetotellurics along the Baltic Long Range profile. *Geophys. J. R. Astr. Soc.* **89**: 799–820.

Ratschbacher, L. and Neubaur, F. 1989. West-directed décollement of Austro-Alpine cover nappes in the Eastern Alps: geometrical and rheological considerations. In: M.P. Coward, D. Dietrich and R.G. Park (eds), Alpine Tectonics. *Geol. Soc. Lond. Spec. Publ.* **45**:243–62.

Raumer, J.F. and Neubauer, F. (eds) 1992. *The pre-Mesozoic geology of the Alps*, Springer, Berlin, Heidelberg, New York.

Research Group for the Lithospheric Structure in Tunisia 1992. The EGT'85 seismic experiment in Tunisia: A reconnaissance of the deep structures. *Tectonophysics* **207**: in press.

Reys, C. 1988. Chemismus und Phasenpetrologie krustaler Auswürflinge aus Vulkaniten des Laacher-See-Gebietes. Dipl. Thesis, Universität Köln.

Ricard, Y., Doglioni, C. and Sabadini, R. 1991. Differential rotation between lithosphere and mantle: a consequence of lateral mantle viscosity variations. *J. Geophys. Res.* **96**: 8407–15.

Ricou, L.E. and Siddans, A.W.B. 1986. Collision tectonics in the Western Alps. In: M.P. Coward and A.C. Ries (eds), Collision Tectonics. *Geol. Soc. Lond. Spec. Publ.* **19**: 229–44.

Ro, H.E., Stuevold, L.M., Faleide, J.I. and Myhre, A.M. 1990. Skagerrak Graben – the offshore continuation of the Oslo Graben. *Tectonophysics* **178**: 1–10.

Robardet, M. and Doré, F. 1988. The late Ordovician diamictite formations from southwestern Europe: North-Gondwana glaciomarine deposits. *Palaeogeog. Palaeoclim. Palaeoecol.* **66**: 19–31.

Roeder, D. 1984. Tectonic evolution of the Apennines. *Amer. Ass. Petrol. Geol. Bull.* **68**: 798.

Roeder, D. 1989a. South-Alpine thrusting and trans-Alpine convergence. In: M. P. Coward, D. Dietrich and R. G. Park (eds), Alpine

Tectonics. *Geol. Soc. Lond. Spec. Pub.* **45**: 211–27.

Roeder, D. 1989b. Thrust belt of central Nevada, Mesozoic compressional events, and the implications for petroleum prospecting. In: L.J. Garside and D.R. Shaddrick (eds), Compressional and extensional structural styles in the northern Basin and Range, Nevada. *Geol. Soc. Nevada Seminar Proc.*, 21-29.

Roeder, D. 1990. Tectonics of south-Alpine crust and cover (Italy). In: J. Letouzey (ed.), *Petroleum and tectonics in mobile belts,* pp. 1–14, Edition Technip, Paris.

Roeder, D. 1991. Western Sahara Atlas of Algeria: salt diapirism, orogenic folding and hydrocarbon potential. *Amer. Assoc. Petrol. Geol. Bull.* **75**: 1420.

Romer, R.L. 1991. The Late Archaean to Early Proterozoic lead isotopic evolution of the northern Baltic Shield of Norway, Sweden and Finland. *Precambrian Res.* **49**: 73–95.

Roth, Ph., Pavoni, N., and Deichmann, N. 1992. Seismotectonics of the eastern Swiss Alps and evidence for precipitation induced variations of seismic activity. *Tectonophysics* **207**: in press.

Roure, F., Polino, R. and Nicolich, R. 1990. Early Neogene deformation beneath the Po plain: constraints on the post-collisional Alpine evolution. In: M. P. Coward, D. Dietrich and R. G. Park (eds), Alpine Tectonics. *Geol. Soc. Lond. Spec. Pub.* **45**: 309–21.

Royden, L. and Karner, G.D. 1984. Flexure of lithosphere beneath Apennine and Carpathian foredeep basins: evidence for an insufficient topographic load. *Amer. Assoc. Petrol. Geol. Bull.* **68**: 704–12.

Rudnick, R.L. and Goldstein, S.L. 1990. The Pb isotopic compositions of lower crustal xenoliths and the evolution of lower crust. *Earth Planet. Sci. Lett.* **98**: 192–207

Rundqvist, D.M. and Mitrofanov, F.T. (eds) 1991. *Precambrian Geology of the USSR.* Elsevier, Amsterdam.

Sachs P.M 1988. Untersuchungen zum Stoffbestand der tieferen Llthosphäre an Xenolithen südwestdeutscher Vulkane. Ph.D. Thesis, Universität Stuttgart.

Sadler, P. 1974. Trilobites from the Gorran Quartzites, Ordovician of S-Cornwall. *Palaeontology* **17**: 71–93.

Sahlström, K.E. 1930. A seismological map of northern Europe. *Sver. Geol. Unders.*, Årsbok 24.

Sandiford, M. and Powell, R. 1991. Some remarks on high-temperature–low-pressure metamorphism in convergent orogens. *J. Metamorphic Geol.* **9**: 333–40.

Sartori, M. 1987. Blocs basculés briançonnais en relation avec leur socle originel dans la nappe de Siviez-Mischabel (Valais Suisse). *C. R. Acad. Sci. Paris* **305**: 999–1005.

Saverikko, M. 1987. The Lapland greenstone belt: stratigraphic and depositional features in northern Finland. *Bull. Geol. Soc. Finl.* **59**: 109–15.

Savostin, L.A., Sibouet, J.-C., Zonenshain, L.P., Le Pichon, X. and Roulet, M.-J. 1986. Kinematic evolution of the Tethys belt from the Atlantic Ocean to the Pamirs since the Triassic. *Tectonophysics* **123**:1–35.

Schäfer, A. 1989. Variscan molasse in the Saar-Nahe Basin (W-Germany), Upper Carboniferous and Lower Permian. *Geol. Rundschau* **78**: 499–524.

Schmid, S.M., Aebli, H.R., Heller, F. and Zingg, A. 1989. The role of the Periadriatic Line in the tectonic evolution of the Alps. In: M. P. Coward, D. Dietrich and R. G. Park (eds), Alpine Tectonics. *Geol. Soc. Lond. Spec. Pub.* **45**: 153–71.

Schmid, S.M., Röck, P. and Schreurs, G. 1990. The significance of the Schams nappes for the reconstruction of the paleotectonic and orogenic evolution of the Penninic zone along the NFP-20 East traverse (Grisons, eastern Switzerland). In: In: F. Roure, P. Heitzmann and R. Polino (eds), Deep structure of the Alps. *Mem. Soc. Géol. France* **156**: 263–87.

Schmucker A. 1989. Zur Petrographie und Geochemie der Krustenxenolithe aus Vulkanaschen der Westeifel. Dipl. Thesis, Universität Köln.

Scholz, C.H. 1990. *The mechanics of earthquakes and faulting*, Cambridge University Press.

Schulmann, K., Ledru, P., Autran, A., Melka, R., Lardeaux, J.M., Urban, M. and Lobkowicz, M. 1991. Evolution of nappes in the eastern margin of the Bohemian Massif: a kinematic interpretation. *Geol. Rundschau* **80**: 73–92,.

Schulz R. 1990. Subsurface temperature and heat

flow density maps for the central segment of the EGT. In: R. Freeman and St. Mueller (eds), *Sixth EGT workshop: Data compilations and synoptic interpretation,* pp. 417–22, European Science Foundation, Strasbourg, France.

Selverstone, J. 1988. Evidence for east–west crustal extension in the Eastern Alps: implications for the unroofing history of the Tauern Window. *Tectonics* 7: 87–105.

Sengör, D.M.C., Burke, K. and Dewey, J.F. 1978. Rifts at high angles to orogenic belts: test for their origin and the Upper Rhine graben as an example. *Amer. J. Sci.* **278**: 24–40.

Serri, G. 1990. Neogene–Quaternary magmatism of the Tyrrhenian region: characterization of the magma sources and geodynamic implications. *Mem. Soc. Geol. It.* **41**: 219–42.

Shudofsky, G.N., Cloetingh, S., Stein, S. and Wortel, R. 1987. Unusually deep earthquakes in east Africa: constraints on the thermo-mechanical structure of a continental rift system. *Geophys. Res. Lett.* **14**: 741–4.

Shurkin, K.A., Mitrofanov, F.P. and Shemyakin, V.M. 1980. *Early Precambrian igneous associations in the USSR,* vols. 1-3, Nedra, Moscow (In Russian).

Siegenthaler, C.H. 1974. Die nordhelvetische Flysch-Gruppe im Sernftal (Kt. Glarus). PhD thesis, Univ. Zürich.

Simkin, T., Tilling, R., Taggart, J. Jones, W. and Spall, H. 1989. *This dynamic planet: world map of volacanoes, earthquakes, and plate tectonics,* US Geological Survey/Smithsonian Institute, USA.

Simonen, A. 1980. The Precambrian in Finland. *Bull. Geol. Surv. Finl.* **304**: 1–58.

Sinclair, H.D., Coakley, B.J., Allen, P.A., and Watts, A.B. 1991. Simulation of foreland basin stratigraphy using a diffusion model of mountain uplift and erosion: an example from the Central Alps, Switzerland. *Tectonics,* **10**: 599–620.

Sivhed, U. 1991. A pre-Quaternary, post-Palaeozoic erosional channel deformed by strike-slip faulting, Scania, southern Sweden. *Geol. Fören. Stockh. Förh.* **113**: 139–43.

Sjöberg, L.E. 1989. The secular change of gravity and the geoid in Fennoscandia. In: S. Gregersen, and P.W. Basham, (eds), *Earthquakes at North-Atlantic passive margins: neotectonics and postglacial rebound,* pp. 125–39, Kluwer Academic Publishers.

Slunga, R.S. 1991. The Baltic Shield earthquakes. *Tectonophysics* **189**: 323-331.

Smith, A. G. 1971. Alpine deformation and the oceanic areas of the Tethys, Mediterranean, and Atlantic. *Geol. Soc. Amer. Bull.* **82**: 2039–70.

Smith, D.E., Kolenkiewicz, R., Dunn, P.J., Robbins, J.W., Torrence, M.H., Klosko, S.M., Williamson, R.G., Pavlis, E.C., Douglas, N.B. and Fricke, S.K. 1990. Tectonic motion and deformation from Satellite Laser ranging to LAGEOS. *J. Geophys. Res.* **95**: 22013–41.

Snieder, R. 1988. Large-scale waveform inversions of surface waves for lateral heterogeneity, 2: Application to surface waves in Europe and the Mediterranean. *J. Geophys. Res.* **93**: 12067–80.

Snoke, A. W. and Miller, D. M. 1987. Metamorphic and tectonic history of the northeastern Great Basin. In: W. G. Ernst (ed.), *Metamorphism and crustal evolution of the western United States,* pp. 607–48, Ruby Volume VII, Prentice Hall, Englewood Cliffs, N. J.

Soffel, H.C., Bachtadse, V., Böhm, V. and Franke, W. 1992. Palaeomagnetism of Ordovician rocks from southern Scandinavia, the Bohemian massif and the Massif Central. *Geophys. J. Int.,* in press.

Sokolov, V.A. and Heiskanen, K.J. 1985. Evolution of Precambrian volcanogenic–sedimentary lithogenesis in the south-eastern part of the Baltic Shield. *Bull. Geol. Surv. Finl.* **331**: 91–106.

Solomon, S. C. and Sleep, N. H. 1974. Some simple physical models for absolute plate motions. *J. Geophys. Res.* **79**: 2557–67.

Sommermann, A.E., Meisl, S. and Todt, W. 1990. U/Pb-Alter von Zirkonen aus Metavulkaniten des Südtaunus. *Europ. J. Mineral.* **2**: 244.

Soper, N.J. and Woodcock, N.H. 1990. Silurian collision and sediment dispersal patterns in southern Britain. *Geol. Mag.* **127**: 527–42.

Sørensen, K. 1986. Danish basin subsidence by Triassic rifting on a lithospheric cooling background. *Nature* **319**: 660–3.

Sorgenfrei, T. and Buch, A. 1964. Deep tests in Denmark 1935–1959. *Dan. Geol. Unders.* 3, No.36.

Spakman, W. 1986. The upper mantle structure

in the central European-Mediterranean region. In: R. Freeman, St. Mueller and P. Giese (eds), *Third EGT Workshop: The Central Segment*, pp. 215–21, European Science Foundation, Strasbourg, France.

Spakman, W. 1988. Upper mantle delay time tomography with an application to the collision zone of the Eurasian, African and Arabian plates. *Gelogica Ultraiectina* **53** Rijksuniversiteit te Utrecht, Utrecht, The Netherlands.

Spakman, W. 1990a. The structure of the lithosphere and mantle beneath the Alps as mapped by delay time tomography. In: R. Freeman, P. Giese and St. Mueller (eds), *The European Geotraverse: Integrative Studies*, pp. 213–20, European Science Foundation, Strasbourg, France.

Spakman, W. 1990b. Tomographic images of the upper mantle below central Europe and the Mediterranean. *Terra Nova* **2**: 542–53.

Spakman, W. 1991. Structure of the European–Mediterranean mantle to a depth of 1400 km: Preliminary results from P delay time tomography. Unpublished poster, EUG VI Meeting, Symposium 13: The European Geotraverse Project (EGT), Strasbourg, France.

Stampfli, G.M. and Marthaler, M. 1990. Divergent and convergent margins in the northwestern Alps: confrontation to actualistic models. *Geodinam. Acta* **4**:159–84.

Stangl, R. 1990. Die Struktur der Lithosphäre in Schweden, abgeleitet aus einer gemeinsamen Interpretation der P- und S-Wellen Registrierungen auf dem FENNOLORA-Profil. PhD Thesis, University of Karlsruhe.

Starmer, I.C. 1991. The Proterozoic evolution of the Bamble Sector shear belt, southern Norway: correlations across southern Scandinavia and the Grenville controversy. *Precambrian Res.* **49**: 107–40.

Stäuble, M. and Pfiffner, O. A. 1991. Evaluation of the seismic response of basement thrust and fold geometry in the Central Alps based on 2-D raytracing. *Ann. Tectonicae* **5**: 3–17.

Stein, S., Cloetingh, S., Sleep, N.H. and Wortel, R. 1989. Passive margin earthquakes, stresses and rheology. In: S. Gregersen, and P.W. Basham, (eds), *Earthquakes at North-Atlantic passive margins: neotectonics and postglacial*

rebound, pp. 231–59, Kluwer Academic Publishers.

Stephansson, O., Ljunggren, C. and Jing, L. 1991. Stress measurements and tectonic implications for Fennoscandia. *Tectonophysics* **189**: 317–22.

Stephenson, R.A. and Cloetingh, S. 1991. Some examples and mechanical aspects of continental lithosphere folding. *Tectonophysics* **188**: 27–37.

Stets, J. 1990. Ist die Wittlicher Rotliegend-Senke (Rheinisches Schiefergebirge) ein 'pull-apart' Becken?. *Mainzer Geowiss. Mitt.* **19**: 81–98.

Stille, H. 1929. Der Stammbau der Gebirge und Vorländer, 16th International Geological Congress (1926) Madrid Comptes-Rendues

Stosch, H.G. 1987. Constitution and evolution of subcontinental upper mantle and lower crust in areas of young volcanism: differences and similarities between the Eifel (F.R. Germany) and Tariat Depression (central Mongolia). *Fortschr. Mineral.* **65**: 49–86 .

Stosch, H.G. and Lugmair, G.W 1984. Evolution of the lower continental crust: granulite facies xenoliths from the Eifel, West Germany. *Nature* **311**: 368–70.

Stosch, H.G., Lugmair, G.W. and Seck, H.A. 1986. Geochemistry of granulite-facies lower crustal xenoliths: implications for the geological history of the lower continental crust beneath the Eifel, West Germany. *Geol. Soc. Lond. Spec. Publ.* **24**: 309–17 .

Suhadolc, P., 1990. Fault-plane solutions and seismicity around the EGT Southern Segment. In: R. Freeman and St. Mueller (eds), *Sixth EGT workshop: Data compilations and synoptic interpretation*, pp. 371–82, European Science Foundation, Strasbourg, France.

Suhadolc, P., Panza, G.F. and Mueller, St. 1990. Physical properties of the lithosphere-asthenosphere system in Europe. *Tectonophysics* **176**: 123–35.

Suppe, J. 1985. *Principles of Structural Geology*, Prentice-Hall, New York.

Tapponnier, P. 1977. Evolution tectonique du systeme alpin en Mediterranee: poinconnement et ecrasement rigide-plastique. *Bull. Soc. Géol. France* **19**: 437–60.

Thybo, H. 1990. A seismic velocity model along the EGT profile from the North German Basin into the Baltic Shield. In: R. Freeman, P. Giese, and St. Mueller (eds), *The European Geotraverse: Integrative Studies*, pp. 99–108, European Science Foundation, Strasbourg, France.

Thybo, H., Kiorboe, L.L., Moeller, C., Schönharting, G. and Berthelsen, A. 1990. Geophysical and tectonic modelling of EUGENO-S profiles. In: R. Freeman and St. Mueller (eds), *Sixth EGT workshop: Data compilations and synoptic interpretation*, pp. 93–104, European Science Foundation, Strasbourg, France.

Thybo, H, and Schönharting, G. 1991. Geophysical evidence for Early Permian igneous activity in a transtensional environment, Denmark. *Tectonophysics* **189**: 193–208.

Tollmann, A. 1982. Großräumiger variszischer Deckenbau im Moldanubikum und neue Gedanken zum Variszikum Europas. *Geotekton. Forsch.* **64**: 1–91.

Torske, T. 1977. The South Norway Precambrian region – a Proterozoic Cordillean-type orogenic segment. *Nor. Geol. Tids.* **57**: 97–120.

Torsvik, T.H. and Trench, A. 1991. The Ordovician history of the Iapetus Ocean in Britain: new palaeomagnetic constraints. *J. Geol. Soc. Lond.* **148**: 423–5.

Torsvik, T.H., Smethurst, M.A., Briden, J.C. and Sturt, B.A. 1990a. A review of Palaeozoic palaeomagnetic data from Europe and their palaeogeographic implications. In: W.S. McKerrow and C.R. Scotese (eds), Palaeozoic Palaeogeography and Biogeography. *Geol. Soc. Lond. Mem.* **12**: 25–41.

Torsvik, T.H., Olsesen, O., Ryan, P.D. and Trench, A., 1990b. On the palaeogeography of Baltica during the Palaeozoic: new palaeomagnetic data from the Scandinavian Caledonides. *Geophys. J. Int.* **103**: 261–79.

Torsvik, T.H., Ryan, P.D., Trench, A. and Harper, D.A.T. 1991. Cambrian-Ordovician paleogeography of Baltica. *Geology* **19**: 7–17.

Torsvik, T.H., Smethurst, A., Van der Voo, P., Trench, A., Abrahamsen, N. and Halvorsen, E. 1992. Baltica–a synopsis of palaeomagnetic data and their palaeo-tectonic implications: a rewiev of Palaeozoic palaeomagnetic data from Europe and their palaeogeographic implications. *J. geol. Soc. Lond.* **149**: in press.

Trappe, H. 1989. Deep seismic profiling in the North German Basin. *First Break* **7** (5): 173–84.

Trench, A. and Torsvik, T.H. 1991. A revised plaeozoic apparent polar wandering path for southern Britain (eastern Avalonia). *Geophys. J. Int.* **104**: 227–33.

Trommsdorff, V. and Nievergelt, P. 1983. The Bregaglia (Bergell) Iorio intrusive and its field relations. *Mem. Soc. Geol. It.* **26**: 55–68.

Trommsdorff, V., Dietrich, V. Flisch, M., Stille, P. and Ulmer, P. 1990. Mid-Cretaceous, primitive alkaline magmatism in the Northern Calcareous Alps: significance for Austroalpine geodynamics. *Geol. Rundschau* **79**: 85–97.

Trümpy, R. 1990. *Geology of Switzerland, Part A: an outline of the geology of Switzerland*, Schweiz. Geol. Komm. (ed.), Wepf and Co. Publishers, Basel, New York.

Tullies, J. and Yund, R.A 1980. Hydrolytic weakening of experimentally deformed Westerly granite and Hale albite rock. *J. Struct. Geol.* **2**: 439–51.

Turcotte, D. L., 1983. Driving Mechanisms of Mountain Building. In: K.J. Hsu (ed.), *Mountain Building Processes*, pp. 142–6, Academic Press.

Vaasjoki, M. and Sakko, M. 1988. The evolution of the Raahe–Ladoga zone in Finland: isotopic constraints. *Bull. Geol. Surv. Finl.* **343**: 7–32.

Valasek, P. A. 1992. The tectonic structure of the Swiss Alpine crust interpreted from a 2D network of deep crustal seismic profiles and an evaluation of 3D effects. PhD Thesis, Swiss Fed. Inst. Technology Zürich.

Valasek, P., Mueller, St., Frei, W. and Holliger, K. 1991. Results of NFP 20 seismic reflection profiling along the Alpine section of the European Geotraverse (EGT). *Geophys. J. Int.* **105**: 85–102.

Van der Beek, P.A. and Cloetingh, S. 1992. Lithospheric flexure and the tectonic evolution of the Betic Cordilleras, SE Spain. *Tectonophysics*, in press.

Vandenberg, J. and Zijderfeld, H. 1982. Paleomagnetism in the Mediterranean area. In: H. Berckhemer and K. Hsu (eds), Alpine–Mediterranean Geodynamics. *Amer Geophys. U. Geodynam. Ser.* **7**: 83–112.

Vejbœk, O.V. 1985. Seismic stratigraphy and tectonics of sedimentary basins around Bornholm, Southern Baltic. *Dan. Geol. Unders.* Ser. 1, **8**: 1–28.

Vejbœk, O.V. 1989. Effects af asthenospheric heat flow in basin modelling exemplified with the Danish Basin. *Earth Planet. Sci. Lett.* **95**: 97–114.

Vialon, P., Rochette, P. and Menard, G. 1989. Indentation and rotation in the western Alpine arc. In: M. P. Coward, D. Dietrich and R. G. Park (eds), Alpine Tectonics. *Geol. Soc. Lond. Spec. Pub.* **45**: 329–38.

Vivallo, W. and Claesson, L. Å. 1987. Intra-arc rifting and massive sulphide mineralization in an early Proterozoic volcanic arc. In: T.C. Pharaoh, R.D. Beckinsale and D. Rickard (eds), Geochemistry and Mineralization of Proterozoic Volcanic Suites. *Geol. Soc. Lond. Spec. Publ.* **33**: 69–80.

Volbers, R., Jödicke, H. and Untiedt, J. 1990. Magnetotelluric study of the earth's crust along the deep seismic reflection profile DEKORP 2-N. *Geol. Rundschau* **79**: 581–601.

Volker, F. and Altherr, R. 1987. Lower Carboniferous calc-alkaline volcanics in the northern Vosges: evidence for a constructive continental margin. *Terra cognita* **7**: 174–5.

Voll, G. 1976. Recrystallization of quartz, biotite and feldspars from Erstfeld to the Levantina nappe, Swiss Alps, and its geological implications. *Schweiz. Miner. Petrogr. Mitt.* **56**: 641–7.

Voll, G. 1983. Crustal xenoliths and their evidence for crustal structure underneath the Eifel volcanic district. In: K. Fuchs *et al.* (eds), *Plateau uplift, the Rhenish Massif–a case history*, pp. 336–42, Springer, Berlin, Heidelberg, New York.

Von Hoegen, J., Kramm, U. and Walter, R. 1990. The Brabant Massif as part of Armorica/Gondwana: U–Pb isotopic evidence from detrital zircons. *Tectonophysics* **185**: 37–50.

Voshage, H., Hofmann, A.W., Mazzuchelli, M., Rivalenti, G., Sinigoi, S., Raczek, I. and Demarchi, G. 1990. Isotopic evidence from the Ivrea Zone for a hybrid lower crust formed by magmatic underplating. *Nature* **347**: 731–6.

Wahlgren, C.H. and Stephens, M.B. 1990. Post-Sveco-Fennian plastic and brittle–plastic shear zones in westernmost Bergslagen, Sweden. *Geol. Fören. Stockh. Förh.* **12**: 204–5.

Ward, P. 1987. Early Proterozoic deposition and deformation at the Karelian craton margin in southeastern Finland. *Precambrian Res.* **35**: 71–93.

Ward, P. 1988. Early Proterozoic Kalevian lithofacies and their interpretation in the Hammaslahti–Rääkkylä area, eastern Finland. *Geol. Surv. Finl. Sp. Pap.* **5**: 29–48.

Warner, M. R. 1986. Deep seismic reflection profiling the continental crust at sea. In: M. Barazangi and L. Brown (eds), Reflection seismology: a global perspective. *Amer. Geophys. U. Geodynam. Ser.* **13**: 281–6.

Warrik, R.A. and Oerlemans, H., 1990. Sea level rise. In: J.T. Houghton, G.J. Jenkins and J.J. Ephraums (eds), *Climate change: the IPCC scientific assessment,* pp. 257–81, Cambridge University Press.

Weber, K. 1978. Das Bewegungsbild im Rhenohercynikum–Abbild einer varistischen Subfluenz. *Z. Deutsch. Geol. Ges.* **129**: 249–81.

Weber, K. 1981. The structural development of the Rheinische Schiefergebirge. *Geol. Mijnbouw* **60**: 149–59.

Wedepohl, K.H. 1987. Kontinentaler Intraplatten-Vulkanismus am Beispiel der tertiären Basalte der Hessischen Senke. *Fortschr. Miner.* **65**: 19–47.

Weissert, H.J. and Bernoulli, D. 1985. A transform margin in the Mesozoic Tethys: evidence from the Swiss Alps. *Geol. Rundschau* **74**: 665–79.

Welin, E. 1987. The depositional evolution of the Sveco-Fennian supracrustal sequence in Finland and Sweden. *Precambrian Res.* **35**: 95–113.

Welin, E. and Stålhös, G. 1986. Maximum age of the synmetamorphic Svecokarelian fold phases in south central Sweden. *Geol. Fören. Stockh. Förh.* **108**: 3.

Wenzel, F., Brun, J.-P. and the ECORS-DEKORP Working Group 1991. A deep reflection seismic line across the Northern Rhine Graben. *Earth Planet. Sci. Lett.* **104**: 140–50.

Werner, D. 1985. A two-dimensional geodynamic model for the southern segment of the EGT. In: D.A. Galson and St. Mueller (eds), Second EGT Workshop: the Southern Segment, pp. 65–9, European Science Foundation, Strasbourg, France.

Werner, D. and Gudmundsson, H. 1992. *Four geodynamic profiles crossing the Alps,* in prep.

Werner, D. and Kissling, E. 1985. Gravity anomalies and dynamics of the Swiss Alps. *Tectonophysics* 117: 97–108.

Werner, D. and Kissling, E. 1988. A geodynamic model for the lithosphere of the Swiss Alps. *Phys. Earth Planet. Int.* 51: 153–4.

Wernicke, B. P. 1985. Uniform-sense normal simple shear of the continental lithosphere. *Canad. J. Earth Sci.* 22: 108–25.

Wessel, P. and Husebye, E.S.1987. The Oslo Graben gravity high and taphrogenesis. *Tectonophysics* 142: 15–26.

Wever, Th., Meissner, R., Sadowiak, P. and the DEKORP Group 1990. Deep reflection seismic data along the central part of the European Geotraverse in Germany: a review. *Tectonophysics* 176: 87–101.

White, S. 1975. Tectonic deformation and recrystallization of oligoclase. *Contr. Mineral. Petrol.* 50: 287–304

Wildi, W. 1985. Heavy mineral distribution and dispersal pattern in Penninic and Ligurian flysch basins (Alps, Northern Apennines). *Gior. Geol., Bologna, Ser. 3a* 47: 77–99.

Wilson, J. T. 1965. A new class of faults and their bearing on continental drift. *Nature* 207: 343–7.

Wilson, M. and Downes, H. 1991. Tertiary–Quaternary extension-related alkaline magmatism in Western and Central Europe. *J. Petrol.* 32: 811–49.

Wilson, M.R., Hamilton, P.J., Fallick, A.E., Aftalion, M. and Michard, A. 1985. Granites and early Proterozoic crustal evolution in Sweden: evidence from Sm–Nd, U–Pb and O isotope systematics. *Earth Planet. Sci. Lett.* 72: 376–88.

Wilson, M.R., Sehlstedt, S., Claesson, L.-Å., Smellie, J.A.T., Aftalion, M., Hamilton, P.J. and Fallick, A.E. 1987. Jörn: an early Proterozoic intrusive complex in a volcanic-arc environment, North Sweden. *Precambrian Res.* 36: 201–25.

Wonik, T. and Hahn, A. 1989. Karte der Magnetfeldanomalien 000F, Bundesrepublik Deutschland, Luxemburg, Schweiz und Österreich (westlicher Teil) 1:1000000. *Geol. Jahr.* 43: 1–21.

Woollard, G.P. 1975. Regional changes in gravity and their relation to crustal parameters. Bureau Gravimetrique International. *Bull. d'Inform.* 36: 106–110.

Worner G., Schmincke, H.U. and Schreyer, W. 1982. Crustal xenoliths from the Quaternary Wehr volcano (East Eifel). *N. Jb. Mineral. Abh.* 144: 29–55.

Yanovskaya, T. B., Panza, G. F., Ditmar, P. D., Suhadolc, P. and Mueller, St. 1990. Structural heterogeneity and anisotropy based on 2-D phase velocity patterns of Rayleigh waves in Western Europe. *Rend. Fis. Accad. Nazionale Lincei Ser. IX* 1: 127–35.

Ye, S. 1992. Crustal structure beneath the central Swiss Alps derived from seismic refraction data, PhD Thesis, Swiss Fed. Inst. Technology, Zürich.

Zeis, S., Gajewski, D. and Prodehl, C. 1990. Crustal structure of southern Germany from seismic refraction data. *Tectonophysics* 176: 59–86.

Ziegler, P.A. 1982. *Geological Atlas of Western and Central Europe*, Shell Internationale Petroleum Maatschappij, Elsevier, Amsterdam.

Ziegler, P.A. 1987a. Compressional intra-plate deformations in the Alpine foreland – an introduction. *Tectonophysics* 137: 1–5.

Ziegler, P.A. 1987b. Late Cretaceous and Cenozoic intra-plate compressional deformation in the Alpine foreland – a geodynamic model. *Tectonophysics* 137: 303–28.

Ziegler, P.A. (ed.) 1987c. Compressional intra-plate deformations in the Alpine foreland. *Tectonophysics* 137: 1–420.

Ziegler, P.A. 1988. Evolution of the Arctic–North Atlantic and the Western Tethys. *Amer. Assoc. Petrol. Geol. Mem.* 43: 1–198.

Ziegler, P.A. 1990. *Geological Atlas of Western and Central Europe-2nd ed.*, Shell Internationale Petroleum Maatschappij B. V., Geol. Soc. Lond., Elsevier, Amsterdam.

Zoback, M.L. and World Stress Map Team 1989. Global pattern of tectonic stress. *Nature* 341: 291–8.

Zoetemeijer, R., Desegaux, P., Cloetingh, S., Roure, F. and Moreti, I. 1990. Lithospheric dynamics and tectonic–stratigraphic evolution of the Ebro Basin. *J. Geophys. Res.* 95: 2701–11.

Zuber, J.A. and Öhlander, B. 1990. Geophysical and geochemical evidence of Proterozoic collision in the western marginal zone of the Baltic Shield. *Geol. Rundschau* 79: 1–11.

Zulauf, G. 1990a. Zur spät- bis post-Variszischen

Krustenentwicklung in der nördlichen Oberpfalz. *KTB-Rep.* **91-1**: 41–62.

Zulauf, G. 1990b. Spät- bis post-Varisvariszische Deformationen und Spannungsfelder in der nördlichen Oberpfalz (Bayern) unter besonderer Berücksichtigung der KTB-Vorbohrung. *Frankfurt. Geowiss. Arb. Ser. A* **8**: 1–285.

INDEX

Aar massif 53, 189
Aar-Gotthard area *87*, 89
accretion, of island arcs 143
accretionary belt, Caledonian 162, 164
accretionary wedges 11, 23, 26, 141, 143, 152, 184
   Neogene 197
   Penninic 199
   polyphase 192–3
   Saharan 201
active margins 22, 178, 179
Adriatic crust 53, 55, 56, 189
Adriatic margins 124, 183–4, 187
Adriatic microplate *31*, 53, 56, 192, 193, 198,
   205, 211, 226
   anticlockwise rotation of 28, 30, 184, 189, 205
   collision
      with Corsica–Sardinia block 225–6
      with Variscan Europe 30
   fragmented 195
   imbrication of upper crustal flakes 186–7
   overriding lower European crust 189
   a rotating indentor 186
Adriatic Sea 124
Adriatic–African plate 182
Adula block 182
Adula nappe 186, 187, 189
African craton 200
African plate, movement relative to Europe 189
African–Eurasian plate interaction 124
Aiguilles Rouges massif 189, 190
Airy-type compensation 132
Alboran extension site 211
Alboran Sea 31, *210–11*
Algeria 200, 201, 202
Alnö volcanic province 93
Alpine arc, seismicity in 118
Alpine belt 30–1, 83, *87*
Alpine foreland 173
Alpine nappes 28, 89, 188
Alpine orogeny 173–6, 180–90
   Cretaceous convergence 184–6
   foreland volcanic phase 174–5
   Mesozoic rifting phase 181–4
   pre-Triassic evolution 180–1
   Tertiary collision 173, 186–90
Alpine south front 190
Alpine–Mediterranean area 4
   heat flow 75–6, 136

Moho temperature 75
   seismicity in 118–19
   thermal data 73
Alpine–Penninic basins 214
Alpine–Zagros chain 32
Alps 30, 47, 51–5, 68, 78, 136, 209
   compressional regime 190
   crustal core 196
   crustal thickening beneath 102
   geodynamics of 219–24
   lithospheric cross section 107–8
   seismic structure *52*
   thrust fronts of 204
   uplift in 131–2
   wedge structure in deep crust of 30
   see also Central Alps; Southern Alps; Western
      Alps
amphibolites 99–100, 140, 142, 143, 145, 149
Anatolian fault system 31–2
angular unconformities 180, 181, 182, 186
anorthosite 143, 148
Apennine deformation front 193
Apennine foredeep 193
Apennine orogen 192–3
Apennine-Betic chain 32
Apennines 30, 124, 136, 190, 192–3, 204, 211
   see also northern Apennines
Appalachian–Variscan system, bivergent 214
Appalachians 27, 181
applied stress 82
Aquitaine basin 68
arc terranes 11, 20, 23
   see also terranes
Archaean terranes 17, 18, 39, 139, 140–3, 145,
   152, 229
Archaean–Proterozoic palaeoboundary 146
arcs 24, 229
   see also island arcs; magmatic arcs
Ardennes 23
Ardennes–Brabant Massif, detrital zircons 150–1
Argentura–Mercantour massifs 119
'Armorica', rifted from Gondwana 26
Armorican massif 26, 165–6
asthenosphere 1, 32, 103, 203, 230
   heat advection from 170
   lateral flow in 227
   low-viscosity 129, 131
   mixing by convection 2

thin, Blatic Shield 226, 227
  upwelling of 223
asthenosphere push 208–9
Atkanric Ocean 28, 29, 181, 183
Atlantic Ocean, see North Atlantic
Atlas Mountains 30
Atlasian foreland 199
Atlasian system, Algeria 210–11
Atlasian terrane 60
Austroalpine nappes 184, 186
Avalonia 14, 21, 162, 165. 167
  docking of 24, 164
Avalonia–Baltica Suture 21
Avalonia–Laurussia suture 21, 27, 28
azimuthal anisotropy 68
Azores triple junction 204

BABEL Line A 155, 160, 162
BABEL reflection profiles 102, 104, 147
BABEL Working Group 9, 18, 104
  marine seismic reflection survey 42, 43–4. 47
backarc environment 147
backarc basins 11, 20, 145, 205
  expanding 209–11
backarc spreading 24, 32, 166, 193, 202, 203, 204,
  210–11
backfolds, Penninic nappes 189
backthrusting 179, 191
Balearic basin 136
Balearic extension site 211
Balearic Islands 31
Baltic Sea 41, 43–4, 47
Baltic Shield 4, 11, 20, 64, 73, 102, 132, 227
  Archaean nucleus 3
  crust
    structure in northern Shield 37–9
    thick 61, 103
    and upper mantle structure 61, 63
  electric conductivity 79, 80
  faulting 120–1
  geoid anomalies 78–9
  gravity studies 77
  lithosphere
    cross section 103–6
    mechanically strong 86
    thermal lithosphere 75
    thick 65
  Moho temperature 75
  palaeomagnetic data 152
  Precambrian drift history 12–13, 13
  seismicity 114–17
    crustal 88, 114, 116
    distribution of with depth 88, 115–16
    intraplate 85, 86, 87, 88, 89
  stability of 228–30
  strength profile 83, 84, 85
  stress in 121–2, 122
  tectonic evolution 139–50
Baltic Shield/European thrust boundary 106
'Baltic' trilobite fauna 22
Baltic–Bothnian megashear zone 41

Baltica 19–20, 27, 158–9, 165
  anticlockwise rotation of 13, 14, 19, 20
  collision with Laurentia 20, 153
banded iron formations 20, 141, 146
Barents Sea 39, 88
basalt 142
  calcalkaline 134
  intra-plate 178
  MORB-type 177, 178
  ocean floor 99
  tholeiitic 143
base metal mineralisation 147
basement 162–3, 185
  pre-Alpine, Alpine outcrops 180, 181
  pre-Mesozoic 176–80
  Variscan 166, 167, 173, 179
  Proterozoic 177
  Precambrian 20, 44, 152, 153, 155, 158, 160
  Archaean 37
basement high (ridges) 59, 184
basement massifs, Cadomian 21
basement nappes 58, 191
basement uplift 189
basin closure 165
basin inversion 160–2, 173
basins
  Late Cretaceous–Early Tertiary 160
  Triassic 173
  Variscan 172
  Cambro–Ordovician 165, 166, 168
  see also basin types
batholiths 23, 148, 149, 150
Bavaria 150, 179
Belmorian terrane 141–2, 145
Belmorides 145
Benioff zones 196, 207, 210–11, 212, 212
Bergslagen 147
Bernhard nappe 185
Betic Codillera 31
BIRPS (UK) 33
Biscay–Asturia loop 214
Black Forest 26, 132, 179
black shales, conductive 163
Blekinge, S Sweden 149
Bohemian loop 214
Bohemian massif 26, 68, 165–6, 169
Bohemian terrane 179
Børglum fault 158
Bornholm island 45, 149, 160
Bothnia, Gulf of 43–4, 104, 115
Bothnian zone 116
Bouguer gravity anomalies 43, 71, 130, 132, 163,
  193–5
  positive 153, 161, 172
  see also gravity anomalies
Brabant massif 22, 23, 178
Bregaglia intrusion 187
Brenner line 189
Briançonnais swell 182, 184
Briançonnais zone 118
brittle faulting 112

brittle fracture 81
brittle regime 82–3
brittle strength 81–2
brittle-plastic transition 112, 114, 119
brittle/ductile transition 175

Cadomian orogeny 19, 178
Calabrian arc 61, 135, *212*, 212
Caledonian front 21, 44, 45, 153
    defined *154, 157*, 162
Caledonian orogen 41, 139, 162–4
Caledonian orogeny 41, 153
Camargue graben 224
Campidano graben 58, 136
Cape Corse 58
Carpathian north front 204
Carpathians 30
Central Alps 51, 220
    zircons, Lepontine area 151
central European rift system 49
Central Massif *see* Massif Central
central and southern Europe, tectonic evolution
    of 150–1
Central–Viking Graben rift system 23
channel flow 129, 131, 227
Chiasso formation 188
clastic wedges 29
collision 212
    Tertiary 186–90
collision tectonics
    Early Proterozoic 143–6, 229
    Variscan belt 101
    see also orogenic cycles, Mediterranean
collision zones 11–12, 43, 229
    beneath Gulf of Bothnia *42*, 43, 228
collisional belts 149
Cologne basin 117, 122
compression 120, 186, 191, 220
compressional belts 202
Conrad discontinuity 49, 158
conserving mass, principle of 217
Constance, Lake 47, 73, 176
continent–continent collision 217
continental break-up 11
continental crust 61, 103, 182
continental evolution, and plate tectonics 1
continental lithosphere 78, *82*, 83, 217
    strength profiles 83–4, *85, 86, 87*
convergence
    Alpine, plate paths 203–4
    continental 203–4
    Cretaceous 184–6
    crustal 136
    oblique 186
    see also Europe–Africa convergence; plate
        convergence
convergence and extension, local vectors of 204
Corsica 57, 58, 62, 73, 103, 133, 193, 194
    anticlockwise rotation 224, 225
Corsica–Sardinia block 31, 108, 224–5
Corsica–Sardinia channel 225–6

corundum 99
creep 83, 218, 219
creep constants *84*
creep strength 84
critical taper concept 206, 207
CROP (Italy) 33
crust 35–60, 78, 170, 224
    Archaean evolution 12
    Cadomian 26
    European 196
        evolution of 17–32
    Moho as base of 61
    Rheno–Hercynian 169
    Saxo–Thuringian 169
    Sveco–Fennian 148
    three-layered 43, 45, 154–5
    two-layered 19, 44, 45, 154–5
    see also lower crust; lowermost crust; middle
        crust; upper crust
crust-mantle boundary 37, 39, 61, 170
crust-mantle transition 39, 97, 161
crustal attenuation 183
crustal densities 78
crustal extension 23, 26, 158, 166, 197, 229, 230
crustal flexure 30
crustal movements, recent 124–32
crustal reflectors 156
    bivergent 43
crustal roots 27, 61, 201, 207, 217, 218, 219, 230
crustal shortening 27, 31, 230
crustal slab 193
crustal strength, minima 84
crustal structures 76
    Precambrian, contrasted 154–6
crustal thickening 20, 148, 217, 220, 230
    Bothnian collision zone 229
    orogenic 101
    Tunisia *60*, 200
crustal thickness
    and geoid lows 128–9
    sharp changes in 43
crustal thinning 158, 166
    beneath Norwegian–Danish basin 159–60
    beneath Sardinia Channel 200
    of Precambrian basement 153
crustal xenoliths 93, 163
crystalline massifs, aseismic 118
crystalline plasticity 112
Cu–Mo–Au deposits 148
Cu–Pb–Zn deposits 147
Cu–Zn–Pb–Ag–Au mineralisation 148
Cu-sulphide deposits 146
cumulates 100

Dal Group 150
Dalsland thrust 150
Danish basin 44, 106
Danish foreland 153
Danish–Scanlan area 27
Dauphinois–Helvetic shelf 184
    see also Helvetic shelf

decay of convergence 212
décollement, basal 145
décollement zone, mid-crustal 178
decompression melting 148
decoupling 132
    intracrustal 89
    in the lower crust 183–4
    of upper crust, Alps 189
deep flow 129, 131
deep seismic reflection profiles 33, *34*
deformation 83, 88, 89, 150, 170
    Carboniferous 179
    intraplate 118
    polyphase 180
    Sveco–Norwegian 149
    syn- and post-collisional 148
    Variscan 177
dehydration melting 148
DEKORP programmes 8, 33, 49, 51
DEKORP-1 reflection profiles 171, 178
DEKORP-2N 106, 163, *178*
DEKORP-2 102
delamination 101
    of Alpine lower crust 222
    of lithospheric mantle 170
    of thinned European crust 186–7
density contrasts 217
depocentres, Permian 172
detachments 119, 190, 202, *210–11*
    base of Aar massif 196
    compressional 201
    deep 193
    intracrustal 206
    mid-crustal 175
diapirs 201
    granite 147
    mantle 209
    salt 159, 172
differential strength 81
Dinaric orogen, compression in 186
Dinarides *210–11*
Dinarides-Hellenides 196
diorite 149
dip slip 208
discontinuities 102
dislocation climb 83
dislocation glide 83
Dolerite Group 18
Dolomites, thrusting in 186
Dora Maira massif 118, 185
Dorn creep 83
downwarp, double 196
downwelling, Po basin 132
ductile flow 83
dyke swarms 18, 27, 153, 158
dykes
    basic 145, 148, 149
    lamprophyre 148
    Permo-Carboniferous 161

Earth Science Study Centres 5–6, 98

earthquakes 118
    Baltic Shield 114–17
    continental, depth of 84–5
    distribution in Europe *112–13*, 114
    intraplate 111–12
    lower crustal 119
    Southern Alps 191
East Avalonia *see* Avalonia
east European Platform 19–20, 61
East Silesian massif 163
eclogite 26, 105, 106, 169, 179, 230
    in lower crust 105
ECORS (France) 33, 178
effective elastic thickness (EET) 80, 88, 89–90,
    132
Eger graben 224
EGT workshops 5, 8
EGT-S86 33, *34*, 102
EGT-South 33, *34*, 51–60, 102
Eifel region *86*, 88, 133
Eifel volcanic field 94, 224
    crustal structure *96*, 97, 100
    regional heat flow 95
    xenoliths 95, 99–100, 164
elastic properties, of deep crustal rocks 93
elastic response 80
Elbe line 106, 163, 173
Elbe massif 162
electrical conductivity 79–80, 105
    anomaly, S of Elbe line 163
    high 107, 178, 179
electromagnetic studies 79
Embrunais folds 118
English–North German–Polish Caledonides
    21–3, 162
eo-Alpine chain 28–9
eo-Alpine orogeny 184–6
erosion 72
escape 11
    lateral 189, 203
    strike-slip 20, 26
escaping material 209
Eu-anomalies 98, 99, 100
EUGEMI 33, *34*, 47–51, 55, 102, 106
EUGENO-S 33, *34*, 44–7, 102, 106, 150, 153,
    155, 158, 160–1, 162
Eurasian plate 11
Europe 61
    crustal evolution 17–32
    observed stress distribution 222
    terranes and crustal domains 15–15, *16*
    Palaeozoic development 165
    Variscan, plate tectonic affinities *165*
    Caledonian 13, *14*
    Precambrian, tectonic evolution 139–52
    Proterozoic 17–18, 19
    Archaean 17
Europe, central 73
    alkaline volcanism 132–3
    Phanerozoic structures and events
        164–80

seismicity in 117–18
  Variscan 4, 79, 83, *86*, 88, *90*
Europe, central and southern 150–1
Europe, northern, seismicity of *115*
Europe–Africa convergence 17, 30, 58–9, 118,
  184, 186, 202
  controls asthenosphere supply 203
  Neogene, amount of 197–8
Europe–Africa plate boundary 199, 202
European Geotraverse (EGT) 3–9
  central section, xenolith evidence 100–2
  seismic exploration of upper mantle 60–9
  seismic structure
    central segment 47–51
    northern segment 37–47
    southern segment 51–60
European plate, subduction of 55
European Science Foundation (ESF) 3
European Science Research councils (ESRC)
  working group 3, 4, 5–6
evaporites 159, 172
extension 189, 202, 217, 220, 226
  Aegean area 120
  Apennines 120
  behind orogens 203
  by rising viscous pillows 208
  Devonian to Early Carboniferous 169
  hanging wall 208
  high heat flow 136
  lithospheric 207
  polyphase 193
  synorogenic 207
  and unroofing of Tauern thermal dome 189
  west Mediterranean 209, 211
  see also crustal extension
extension–compression model *225*, 226
extensional basins 28,32, 219
extensional collapse 20

Faroe–Rockall rift 214
fault splays 27, 157–9
fault zones 147, 172, 184
faulting/faults 150
  aseismic slip 116
  listric 106, 158, 178
  low-angle 189
  normal 122, 158, 183
  postglacial 120–1
  reverse *174*, 189
  strike-slip 26, 27, 32, 47, 102
  synsedimentary 183
  transcurrent 170, 182, 189
  transform 183, 214
felsic rocks 95, 97
FENNOLORA 18, 33, *34*, 39–44, 61, 102, 104–5,
  105, 147, 149
Fennosarmatia *see* Baltica
Fennoscandia 128
Fennoscandian uplift 125, *127*, 127–31, 226–7
ferropicrite 144
finite element analysis 217

Fjerritslev fault 117, 158, 159
flake structure 164
flexural modelling, of Alpine region 132
flexural rigidity 227
flower structures 157, 189
fluids, in the crust 176
fluids and fractures, and conductivity 79
flysch 26, 169, 177, 178–9
flysch basins 15, 17, 184, 188, 189
flysch wedge 192–3
focal depth distribution 118, *119*
focal mechanisms, northern and western Alps
  123–4
fold belts, Alpine 11
fold-thrust belts 199, 203, *210–11*, 211, 212
  and backarc extension 204
  Maghrebide 202
  Numidian-Atlas 200–1
  physics of 206
  south-Alpine 191
  see also thrust belts
folds
  box and open 201
  overturned 160
  reclined, large-scale 141
  recumbent 143
fore-arc basins 11
foredeep, Po basin 188
foreland, common, Alps and Apennines 191
foreland basement flexure 207
foreland basins 26, 30, 90, 187–8
  Silurian 21, 27
foreland flexure, Molasse basin 89
fractionation 100
fracturing 82, 182–3
Free Air gravity anomalies 128, 129, 131
Fynen 153

gabbro 148, 149
Genova, Gulf of 224
geodynamic modelling 215–19
geodynamic processes 228–30
geoid anomalies 78–9, 125–6
Germany, eastern 39
Gibraltar, Straits of 31
Giudicarie transpressive belt 186
glaciation 176
  and deglaciation 127
Global Positioning Systems (GPS) 124
Glückstadt graben 158
gneiss 98, 143, 148
  alumino-silicate 141
  calc-alkaline 145
  leuco-granitic 96
  mylonitic 143
  tonalitic 98, 140
Gondwana 19, 26, 151, 165
Gondwana–Laurasia (Laurussia) separation 28,
  181
Gonfolite Lombarda group 188
Gosau basins 186

Gothian domains 18
Gothian orogen 41, 139, 149, 154, 229
Gothian terrane boundary 153
Gotthard massif 189
graben areas, high heat flow 136
graben structures 19, 23, 58, 116, 132, 136, 150,
    160, 173, 224
    Early Tertiary 173
    see also named grabens and graben systems
Grand Paradiso block 185
granite 41, 193
    alkali 143
    crustal melt 145, 148
    diapiric 147
    post-tectonic 145, 149, 150
    rapakivi 18, 148, 149, 152
    S-type 26
granite-gneiss terranes 12, 17
granitoids
    I-type 148, 179
    Late-Variscan 180, 181
    Moldanubian 95, 98
granodiorite 140
granulite 95, 97, 98-9, 100
Granulite Belt of Finland 145
granulite-eclogite conversion 222-3
gravity 76-9, 129
gravity anomalies 105, 117, 132, 217, 219, 220
    see also Bouguer gravity anomalies
gravity modelling 76, 77
greenstone belts 141, 142-3, 145
greenstone terranes 12, 17
Grenville Front, extension of 150
Grenville orogeny 18
Grenville thrust belt 18
Grimmen Achse (Axis) 160

Haparanda monzonites 145
Harz Mountains 173, 177, 178
Harzburger gabbro 163
harzburgite 93
heat flow 18, 170
    positive anomalies 136
Hegau volcanics 93, 95
Heldburg Gangschar 93
    crustal structure 96, 97
    regional heat flow 95
    xenolith suite 94, 95
Hellenic arc 61, 63
Hellenic trench 204
Hellenide/Tauride south front 204
Hellenides 210-11
Helvetic Dauphinois domain 182
Helvetic nappes 53
Helvetic overthrust 55
Helvetic Shelf 30
    see also Dauphinois-Helvetic Shelf
Helvetic zone 51, 52-3, 187
Hesse Basin 172, 173
Hessian Depression 49, 224
    see also North Hessian Depression

Hetta granodiorites 145
High Atlas 30, 211
hinge advance 209, 212
hinge retreat 191, 203, 211, 212
    extensional 209
Holy Cross Mountains, Poland 23, 160, 161
Horn graben 158, 162
hydrocarbon traps 172

Iapetus Ocean, closure of 20
Iberia 26, 29, 30, 31, 203, 204
ILIHA experiment 34, 35, 69
imbrication 102-3, 107, 146, 191, 192, 195, 200,
    201, 206
Inari terrane 37, 39, 77-8, 103, 141, 143, 144
indentation 26, 203, 205, 212
    and escape 208-9
Insubric line 51, 53, 55, 188, 189, 191, 196,
    209
intrusions 144, 147, 172, 180, 187
    beneath Silkeborg gravity high 158
    layered 158
inversion 27, 30, 153
inversion structures 47
island arc collision 106, 230
island arc systems 100
island arcs 11, 20, 99, 143, 144, 145, 146, 147
isostatic compensation 125, 129
isostatic equilibriuum, deviation from 131, 132
isostatic uplift 207
Ivrea basement block 183
Ivrea body 132, 193
Ivrea zone 170, 195

Jatulian Group 146
Jatulian shelf 152
Jeffreys-Bullen model 61
Joint Programme 4, 6, 7
Jormua ophiolite 146, 147
Jothnian basin 116
Jura Mountain 87, 88, 89, 119
Jutland 153, 159, 160

Kalix volcanic province 93
Karasjok greenstone belt 145
Karelia 17, 37, 142
Karelian terrane 142-3, 145-6
Kattegat 35, 36, 156, 158, 161
Kattegat platform 44
Kazakhstan plate 20
Keiv Group 143
Keiv-Porosozero suture 143
Kittilä greenstone belt 142
Knipovich Ridge 115
Kola Peninsula 17
Kola suture zone 143
    see also Polmak-Pasvik-Pechenga belt
Kola-Karelian orogen 17-18, 139, 140-6, 152,
    229
Kongsbergian orogen 149
Kraichgau basin 172

La Galite island 200
Lapland 115
Lapland Granulite belt 37, 39, 77–8, 103, 104, 145
Lapland–Kola orogen 104
Lapponian Supergroup 142
Laurentia 14, 165
Laurentia-Baltic suture 21, 22
Laurussia 20, 165, 167
Lausitz 173, 224
lava plateaux 142
Leine graben 132
lenses
    mafic–ultramafic 148
    tectonic, anorthosite 143
leucogranite, crustal-melt 149
Lewisian Complex 139
Ligurian basin 136, 226
Ligurian coastal zones 194
Ligurian rift 190, 224
Ligurian Sea 52, 56, 73, 103, 108, 136, 193, 194
    rifting in 57, 58
    seismic activity 118, 119
    subsidence 220
Ligurian–Alboran Sea closure 31
Liguride units 193
linear magnetic anomaly, negative 143
lithosphere 1
    depth-dependent rheology of 81–4
    hetereogeneities, dynamic role of 217
    modelling of 81–2
    physical properties 71–80
        electric conductivity 79–80
        gravity 76–9
        thermal structure 72–6
    radial convergence of 67, 223
    seismic 2
    stacked slabs of 195–7
    stratification, rheological 194–5
    strength of 83
    structure
        differences in 65, 67
        layered 61, 63, 229
        mechanical 80–91
        thermal 72–6
    subcrustal, strong 84
    subducted 63, 134
    thermal structure along EGT 73–6
    thickening of 217, 230
    thin 62, 136
    three-layered 208
    two-layered response to stress, Baltic Shield 88
    see also continental lithosphere; lower
        lithosphere; oceanic lithosphere
lithosphere–asthenosphere boundary 75
lithospheric collision 114
lithospheric cooling 160
lithospheric cross section, integrated 102–9
lithospheric roots 218, 229, 230
    Alps 65. 103, 119, 136, 219–20
        asymmetrical 61–2, 107–8

beneath Po basin 67, 220, 222, 223
    displaced 218
    downbuckling over 219
lithospheric shortening 118
lithospheric structures, Caledonian 164
lithospheric tectonics, and the orogenic cycle 211–12
lithospheric thinning 223
    and stretching 181
    and young volcanism 224
lithostatic pressure 82–3
Lizard–Giessen–Harz nappes 24
loading and unloading, glacial 125, 127, 227
Loissin well 163
Loke shear 22–3, 163, 164
Lombardian basin 183
Lombardian Flysch basin 186
Lopian 142
lower crust 39, 49, 148, 164, 170
    anomalous structure 160, 161
    European, subduction of 53
    layered 41, 170–1
    Lüneburg massif 163
    Rheno–Hercynian belt 178
    Scandinavia 149
lower lithosphere 198
    compositional banding 65
    and isostasy 128
    layered structure 64
    S-wave velocity 65
    thickening of 217, 220
Lower Rhine graben 122, 132, 173
    crustal model for 175–6
Lower Saxony basin 173
lowermost crust 43, 47
Lugano line 183
Luleå–Kuopiu suture zone 146–7
Lüneburg massif 21, 24, 162, 163

Maghreb 30
Maghrebide fold-thrust belt 202
magmatic arcs
    Andean-type 24, 143, 144
    calc-alkaline 11, 145
    Sveco-Fennian orogen 147, 152
magmatic rocks, pre-Devonian 26
magmatism 99, 133, 136
    arc 11, 178, 191
    late- and post-collisional 170
    post-orogenic 230
    rapakivi granite 148, 149, 152
    Tertiary 175
    Mid-Cretaceous 186
    Permo-Carboniferous 27, 158
magnetic anomalies 71, 105, 159
magneto-telluric (M-T) measurements 71, 153
Mandel–Ustaoset fault 150
mantle 227
    see also upper mantle
mantle flow patterns 202

mantle material, as a Newtonian viscous fluid
   220
mantle xenoliths 93
Marsili basin 133–4
Martegnas mélange 183
mass balance arguments 217–20, 222–3
Massif Central 26, 93, 150, 224
Massif Central ocean 166
mechanically strong crust (MSC) 81, 86, 87, 88,
   90, *91*
mechanically strong lithosphere (MSL) 2, 81, 84,
   88, 90, *91*
mechanically weak lithosphere 89–90
Mediterranean 17, 32, 73, 108–9
   orogenic cycles 211–13
   orogenic loops 30–1, 32, 213–14
   tectonics, recent 203–14
Mediterranean basin 31
mélange zone 147
melts 11, 12
Meseta loop 214
Messina, straits of 30
metamorphic core complexes 207–8, 219
metamorphic haloes 207
metamorphism 170, 180
   amphibolite facies 99, 141, 148, 150, 208
   eclogite facies 20
   and extension 207–8
   greenschist grade 178
   high-pressure 26, 185
   low-pressure 26
   mid-Cretaceous 184–5
   pre-Devonian 26
   Saxo-Thuringian belt 179
microplate movements 124
microplates 28, 169, 217
mid-Atlantic ridge 115
mid-German Crystalline high 24, 26, 49, 178–9
middle crust 49, 56
   Aar massif 53
   Baltic Shield 38–9, 87–8
   Ligurian Sea 58
middle Rhine graben 173
Midlands massif 21
migmatites 140, 141, 143
Milan fold belt 189
Mohn's Ridge 115
Moho 37, 49, 56, 102, 104, 158, 160–1, 191, 208
   asymmetric 219
   Atlasian foreland 199
   beneath the Alps 53–5, *54*, 107, 222
   beneath central Finland 148
   beneath European Variscan crust 27
   beneath Norwegian–Danish basin 160
   changes in velocity contrast at 41, 43
   mapping of 60–1, *62*
   and P-wave velocity 60
   reflection at 47
   rise and fall of 44–5
   shallow 136
   structure below European unit 196

temperature at 75
updoming of 38–9
   see also seismic Moho
Moho offsets 55, 104, *210–11*
Moho overlap 190, 197
Moho stacking 196, 197, 199
Molasse basin 30, 47, 49, 51, 52, *87*, 88–9, 107,
   119, 136
molasse basins 17, 26
Moldanubian terrane 230
Moldanubian zone 93, 166, 169, 177, 179–80
   see also Hegau; Urach
Monferrato complex 197
Mont-Blanc massif 189, 190
Moravo–Silesian unit 179–80
Morcles nappe 189
mountain belts, geodynamic modelling of 217–19
Mte Rosa massif 185
Münchberg Klippe 26
Murmansk terrane 140–1, 143
mylonites 143, 145, 191

nappes 24, 30, 53, 146, 150, 179, 182, 185, 187,
   189
   advancing 184
   Alpine 89, 188
   Austroalpine 184, 186
   basement 58, 191
   Caledonian 39
   exotic 26
   gravitational 20
   Penninic 186, 188, 189
NARS portable seismic stations 35, 63, 65, 69
Navier–Stokes equation 217, 220
Neiden granitic pluton 141
Neseuretus province trilobites 166
NFP-20 (Switzerland) 8, 33, 51, 52, 102, 107
Ni–Cu deposits 148
Norbotten arc 147
Nördlinger Ries crater 176
normal incidence seismic reflection techniques 33,
   35, 43, 49
Norrland, Sweden 105
North Africa 204
north Alpine crustal front 199
North America–Greenland continent 18
North Atlantic, opening of 173
North Atlantic Ridge 222
North Cape, Norway 39
North German Basin 44, 45, 47, 73, 153, 157,
   172, 173
north German foreland 177–8
north German line 172
North German–Polish Caledonides 15–16, 162,
   165
   collapse of 24, 164
North Hessian Depression (NHD) *86*, 88, 93,
   106, 223
   crustal structure 96–7
   regional heat flow 95
   xenoliths 95, 98–9, 100, 163

North Sea 30, 114, 115, 173
North Sea coast, subsidence of 176
North-Helvetic Flysch 188
northern Alpine foreland 117, 220
northern Apennines 55–6, 57, 58, 107, 118, *210–11*, 220
northern Carpathians *210–11*
Northern Phyllite zone 24
Norwegian Caledonides 115, 229, 230
Norwegian Sea 114, 115
Norwegian–Danish basin 30, 116–17, 153, 159–60, 162
Novate intrusion 187
Numidian series 201
Numidian terrane 201

obduction 20, 193
ocean spreading 224
ocean-bottom seismometers 58, 59
oceanic basins
  local, deep 202
  small, opening of 31
  unrecognised 198
oceanic crust 100, 164, 190, 193
  Devonian 169
  formation of 136
  Mediterranean basins 61
  subducted 24, 169, 187
  two-layer 224
oceanic lithosphere 23, *82, 83*
  Ionian, subducted 134–5
  juvenile 229
oceanic regions, structure of 1
oceanization 56–7
oceans, Devonian 25–6
Odenwald 49, 178
Old Red Continent 20, 24, 156, 167, 169, 177
olistostromes 24
olivine, creep in 83
Onega, Lake 17
ophiolite 31, 146, 148, 183, 192
Oran Meseta 200, 201, 202
Orijärvi island arc 147
orogenic activity, post-collisional 26
orogenic belts *210–11*, 211–12
orogenic cycles, Mediterranean 211–13
orogenic float 23, 164
orogenic loops 30–1, 32, 213–14
orogenic plateaux 207
orogenic regimes, differing 190–3
orogens, collapse of 15–17, 219
orthogneiss 141, 143, 145
Oslo graben 116, 150
Oslo region 115
Oslo rift 153
Oslo-Skagerrak graben system 158
Outokumpu nappe 146
overthrusting 55, 101

P-wave velocities 97, 107, 162
  eclogites/peridotites 101

Sardinia-Sahara 199, *200*
  xenoliths 92–3, 95
P-waves 33, 41, 52, 60–3
palaeo-poles, Baltic Shield 12–13
palaeogeographic misfits 170
palaeomagnetism 12–15
Pan-African orogeny 19
Pangaea 24, 26–7, 28
Pannonian basin 30, 68, 207, 209, *210–11*
Pantellaria rift 136
paragneiss 143, 145, 149, 150
Paris basin 68
partial melting 98, 99, 148
passive margins 22, 178, 182–3, 201
Pävie fault 120–1
Pechenga Series, thrust bound slices of 143–4
Pelagian Sea 60
pelites 143, 144
peneplane, post-Sveco-Norwegian 19
Penninic collision suture 191
Penninic front 55
Penninic nappes 186, 188, 189
Penninic zone 53
Penninic-Austroalpine nappes 30
perched basins 192
peridotite, upper mantle 105
Permian basins 230
phyllite 143
phyllite zone 178
Piedmont zone 118
Piemont ocean 28, 182, *183*
  opening of 183
pillow lavas, andesite 144
plastic flow 112
plastic shearing 112
plate collision 118
plate convergence 17, 20, 21, 169, 222
  see also convergence
plate margins, convergent 134
plate movements, driving force for 216–17
plate tectonic processes 104, *112–13*, 114
  recent, Mediterranean 202–14
plate tectonics 1, 11–12
plate units, Alps–Apennines 195–7
  Alpine unit 196–7
  European unit 196
  Northern Apennines unit 196–7
  Po Plain unit 196
  sub-Monferrato unit 197
plateau collapse 207, 213
plutons 141, 145, 148, 149, 179
Po basin 52, 55–6, 78, 188, 189, 191–2, 220
Po Plain 51, 55, 75, 118, 125, 131
Pogallo fault 183
POLAR profile 33, *34*, 37–47, 61, 77, 79, 102, 103, *104*
Polish Trough, inversion of 160
Polmak–Pasvik–Pechenga (PPP) belt 104
  see also Kola suture zone
Pomeranian-Kujawic Wall 160
pop-up structures 189

pore fluid pressure 83
pore pressure 83
postglacial rebound 120, 226–7
power law creep 83
pre-collisional shortening 170
Pritzwalk massif 162
Protogine zone 116, 150
Provençal basin 136, 193, 226
Provençal Sea 194
pull-apart basins 28, 157, 182
pull-apart structures 158, 172
Pyrenean orogeny 30
Pyrenees 68, 118

Raahe–Ladoga fault zone 147
radioactive decay 2, 73
rapakivi granites 148, 149, 152
rapakivi massifs 18
ray tracing 37, 43, 200
Rayleigh waves 33, 35, 64–5
reactivation, in Tornquist fan 157, 158
red beds 18, 181
reflection, sub-Moho 161
restites 99, 101–2
restorations 197, *198*, 199
Rhenish massif *50*, 51, 106, 117, 122, 132, 167,
    169, 170, 177, 178
Rheno-Hercynian terrane 230
Rheno-Hercynian units 26
Rheno-Hercynian zone 24, 94, 106, 163,
    176–8
  base of crust 100
  Ordovician sequences 167
  origin as rift basin 169, 177
  see also Eifel; North Hessian Depression
rheology
  depth-dependent 81–4
  and intraplate seismicity 84–91
Rhine graben system 122, 132
  seismicity in 117–18
  see also lower Rhine graben; middle Rhine
    graben; upper Rhine graben
Rhön mountains 49, 224
Rhone–Rhine–Eiger rift system 30
ridge push 120, *122*, 202, 203, 207, 219, 222
Rif, Morocco 30
rift basin 90 169, 176–7, 178–9
rift zones, thinned crust 193
rifting 19
  active 56–7
  Atlantic 29
  differential, incipient 224
  oceanic, propagation of 226
  Red Sea 32
  Tyrrhenian Sea 226
  Oligocene–Miocene 224
  Tertiary 175
  Jurassic 28
  Mesozoic 27, 159–60, 181–4
  Devonian 167–9, 178
  Silurian 169

Ordovician–Silurian 25–6
Cambrian–Ordovician 166
rifts 148
  aborted 158
  failed 196
  intra-arc 147
rigidity 2, 227
Ringkøbing-Fyn high 44, 45, 106, 153, 158, 173
Rømø fracture zone 158
Rönne graben, inversion of 160
rotation, en-bloc 172
Rotliegend troughs 214
Russo-Baltic Platform *see* Baltica

S-wave velocities 65
S-waves 33, 41, 63–4, 65
Saar/Nahe basin 171
sag basins 162
Sahara Atlas 201, 211
Sahara platform 60, 199
Saharan glaciation 19, 25, 26, 166
salt, Messinian 32
salt pillows 159
Saorge-Taggia line 119
Sardinia 57, 58, 62. 73. 79, 93, 103, 133, 194, 200
  anticlockwise rotation 224, 225
Sardinia Channel 56–9, 73, 103, 108, 136, 200,
    202
Sardinia rift 224
Satellite Laser Ranging (SLR) 124
Savo schist belt 147
Saxo-Thuringian basin 179
Saxo-Thuringian terrane 24–5, 230
Saxo-Thuringian zone 25–6, 93, 166, 177,
    178–9
Saxony 179
Scandian orogeny 20
Scandinavian Caledonides 20, 114
Scania 27, 160
Scania volcanic province 93
Schams nappes 182
schist 148
Scientific Coordinating Committee (SCC) of
    ESRC 4, 5–6
Scottish–Norwegian Caledonide orogeny 165
sea floor, new 209, 211
seafloor spreading 11, 25, 208
Sealand 153, 159
sedimentary basins 19, 20, 44, 45
  Early Permian *171*
  negative geoid anomalies 125
sedimentation, Norwegian–Danish basin 159
sediments 19
  Phanerozoic 153
  post-Palaeozoic 60
Seiland volcanic province 93
seismic Moho 97, 100–1
seismic refraction surveys 35, 51
seismic tomography 34, 61, 202
seismic wave paths 33–5, 37
seismic waves 2

seismicity 111–19
  intra-crustal, Kattegat 162
  intraplate, and rheology 84–91
seismogenic layer 112, 114, 119
serpentinite 146, 183
Sesia-Lanzo block 185
shear waves 37
shear zones 104, *112*, 116, 150, 153, 161, 175–6
shears 196
Silesia 26
Silkeborg gravity high 153, 158
sillimanite 96
Silmano nappe 186
Simplon line 189
Sirkka Thrust 145
Skagerrak graben 156
Skagerrak–Oslo rift system 27, 153
Skellefte field, collision zone in 43
Skellefte island arc 146, 147
slab pull 202, 203
small plate-motions, intra-Mediterranean 197
Solway line 21
Sorgenfrei Tornquist zone 44, 106, 115, 117, 153,
    158, 160–1
  development of 27–8, 47
Sörvaranger terrane 141, 143
south German basin 173
south-Alpine backthrust (thrustbelt) 187, 196
Southern Alps 56, 75, 191
space geodetic techniques 124–5
Spessart Mountains 178, 179
spinel peridotite 97
spinel-lherzolite 93
spreading 124, 183
spreading sites 203, 212
Steinheim crater 176
STREAMERS 9
strength envelopes *82*, 83, *84*
strength profiles 83–4, *85, 86, 87*
stress 29–30, 90, 124
  compressional 122
  in continental lithosphere 111–12
  deviatoric 222
      compressional 218
      horizontal 217, 218, 219
    horizontal 120, *221*
  compressional 172, 173, 219, 220
  extensional 220
  internal 217
  relaxation of 83
  state of 120–4
  tectonic, Baltic Shield 87–8
  tensional 29, 217–18
  vertical 120
stress patterns *225, 226*
  regional 120
stress release 120
stress systems 225, 230
stretching and subsidence, post-Variscan 181
strike-slip 209
    Avalonia-Laurussia suture 21

dextral 173
  sinistral 122, 181, 189
  sinisatral systems 28
Strona-Ceneri basement block 183
stylolites 173
subcrustal reflectors 56
subducted slabs 56, 134, 187, 190, 199, 212
  from the African plate 61, 63, 109, 199, 202
  and hinge retreat 203
  of Piemont oceanic crust 185
subduction 11, 104, 114, 180, 211, 229
  A-type 26
  Alpine 196
  Apennine 196
  asymmetric model of 198
  B-type 26, 195
  with backarc spreading 203
  Calabro-Panormide front 211
  crustal 222
  Dinaride 196
  during Tertiary collision 187
  of jagged crustal edge 208–9
  at Kola suture zone 143–5
  of lithosphere 119, 133, 134
  oblique, of Tornquist ocean 164
  of oceanic layering 229
  of Saxo-Thuringian basin 179
  Silurian/Early Cambrian 169–70
  Sveco-Fennian 147
subduction hinge retreat *see* hinge retreat
subduction zones 24, 28, 43, 68, 143, 147, 184–5,
    196, 202, 206
subsidence 182, 183
  Ligurian Sea 220
  northern Apennines 220
  Norwegian–Danish basin 160
  Po basin 188, 220
  Po Plain 31, 125
  stepwise 182
Sumi-Sariola Group 145
Suretta nappe 186, 189
surface heat flow 71, 72, 73, 223
surface heat flow density 72–3
surface waves 34–5, 64–9
suspect terranes 15
sutures 11, 143, 150
Sveco-Fennian orogen 18, 43, 139, 146–9, 152,
    229
Sveco-Fennian province 104, 105, 154
Sveco-Norwegian orogen 18, 19, 139–40,
    149–50, 154, 229
Sveco-Norwegian orogeny 18–19
Swabian Jura 117, 122
Sweden 39, 121, *122*
  earthquakes 115, *116*

Tambo nappe 186, 187
Tampere island arc 147
Tanaelv belt 39, 103–4, 145
Tauern window 189
Taurides *210–11*

tectonic activity, Permo-Carboniferous 27
tectonic collage 230
tectonic processes, driving forces 215–27
tectonic units complex 30–1
tectono-metamorphic activity, Devonian to
    Early Carboniferous 169–70
Teisseyre Tornquist zone 28, 41, 45, *46*, 47, 68,
    157, 160
Telemark supracrustals 150
telescoping, of terranes 11, 18, 23
Tell Mountains 30, 118
Tellian-Numidian terrane 60
Tellian-Numidian thrust zone 59–60
Tello-Rifian system *210–11*
Tellorifian fold-thrust belt 211
temperature field, EGT, modelling of 74–5
temperature gradients, near Earth's surface 72–6
terranes 60, 200, 205, 230
    Archaean 17, 18, 39, 139, 140–3, 145, 152, 229
    island arc 20, 141
    tracing origins of 13, *14,* 14–15
    see also arc terranes; named terranes
Tethyan basin 182
Tethyan-Alpine collision suture 196
Tethys 28, 181
Teutoburger Wald 173
thermal activity, Tertiary to Recent 176
thermal assimilation 198
thermal conductivity 72–4
thermal equilibrium 136
thermal instabilities, mantle 215–17
thermal softening, of root terranes 207
thermal updoming 180–1
thrust belts 162
    architecture 206
    bivergent *187*
    Early Proterozoic 39
    Kola Peninsula 17, 18
    north German Caledonides, composition of
        162–3
    piggyback 212, *213*
    see also fold-thrust belts
thrust sheets, stacked 145, 192–3, 197, 201
thrust slices, Luleå–Kuopiu suture zone 146
thrust systems, shallow, internal *210–11*
thrust wedge 145
thrust zones 59–60, 143, 150
thrusting
    Eastern Alps 185–6
    in-sequence 186–7
    South Alpine 191
thrusts 104
    listric 189
    post-collisional 145
    ramped 195, 206
Thuringia 179
Thuringian Forest 178
tillites 26
Toce-Lepontine area, updoming of 188–9
Tornquist fan 157–9
Tornquist Sea 19, 23, 164, 165, 169

Tornquist zone 27, 30, 106, 120, 153–62, 172
    Early Permian basins 172
    see also Sorgenfrei
        Tornquist zone; Teisseyre
        Tornquist zone
Trans-European fault 21, 27, 28, 44, 65, 106, 153,
    157
Trans-Scandinavian Batholith 149, 150
transform zone, dextral 181
transient heat flow 135–7
transient thermal phenomena 76, 108
transitional crust 58, 59
    two-layered 57
transpression 122, 161, 179, 181, 182
    dextral 56, 179, 179–80, 184, 188
transtension 28, 30, 172, 226
triple junctions 21, *22,* 158, 204
tuffs 142
Tunisia 59–60, 73, 152, 200–1
    crust and lithosphere 108–9
    seismic activity 118
Tunisian Atlas 201
turbidites 24, 141, 143, 146
Tuscan region 134
Tuscanides 192
Tuscany geothermal area 136
Tyrrhenian basin 31, 136
Tyrrhenian Sea 119, 193, 225–6
    volcanism in 133, *134,* 134

U-Pb zircon dating 148, 150, 151
Ukranian massif 20
ultracataclasites 150
Umbrian series 192
unconformities, sub-Permian 160
underplating 170, 185, 189, 229, 230
underthrusting 169, 230
uniform elastic plate concept 80
uplift 72, 136, 208, 217, 219, 230
    Alps 125, 220
    Fennoscandia 125, *127,* 127–31, 226–7
    Late Tertiary 20
    northern Apennines 220
    and subsidence, linked 222
    vertical 188–9
upper crust 45
    Aar massif 53
    Baltic Shield 37, 87–8, 115
    elastic 217
    interleaved high- and low-velocity layers 41,
        106
    layering of 58
    Sveco-Fennian orogen 146
upper lithosphere, rheologically zoned 81
upper mantle 199
    asthenospheric 223
    exploration of 33–5
    seismic exploration along EGT 60–9
    structure beneath Baltic Shield 105–6
    upper Rhine graben 117, 122–3, 132–3, 136,
        173, 175

upwelling, active 132
Urach *86*, 88
 xenoliths 107
Urach field 73
Urach volcanics 93, 95
Urach/Hegau 101
 crustal structure *96*, 97–8
 xenolith suite 94, 95
 see also Hegau
Urach/Hegau province, regional heat flow 95
Uralides 20, 27
Urals 20, 181

Vainospää granite 145
Valais trough 182
Valencia, Gulf of 224
Variscan belt 101
Variscan
 central Europe 4
 crustal seismicity 88, *90*
 electrical conductivity 79
 strength profile 83, *86*
Variscan domain 15, 16–17
Variscan evolution, pre-collisional 24–6
Variscan fold belt, degradation of 27
Variscan orogeny 26–7, 95
Variscan region 102, 106–7
Variscan roots, disappearing 230
Variscan tectonic front 180
Vaskojoki anorthosite 145
Vättern graben system 19
Vavilov basin 133, 134
velocity-depth distribution,
 lithosphere/asthenosphere 37, 39–44, 47
Very Long Baseline Interferometry (VLBI) 124
Viking graben 116
Vinding fracture zone 158
viscoelastic layer 217
viscoelastic properties, of Earth's surface 72
viscoelastic theory 131
viscosity
 upper and lower mantle 131
 variations of with depth 220
viscous pillows, rising 208, 209
viscous response 80

Vogelsberg 224
volcanic arcs 24
volcanic provinces, northern EGT 93
volcaniclastics, Permo-Carboniferous 180, *181*
volcanism
 alkaline 224
 bimodal 149, 170
 calc-alkaline 178
 intra-plate 172–4
 island arc 133–4
 ocean island basalt-type 133
 Permian 170, 172
 recent 132–5, 136
 Sardinia 58
 young, and upwelling of asthenospheric mantle 224
volcanoes, active 202
Voronech uplift 20
Vosges 26, 132, 179

Wales 26
waveform inversion 35
 techniques 65, 67–8
websterite 97, 99
west Mediterranean terranes 205
Western Alps 118
Westerwald 224
White Sea 145
wide-angle seismic reflection techniques 33, 35
World Stress Map 120
wrenching 11, 28, 157
 dextral 172, 173

xenolith investigation, conceptional limits of 100
xenoliths 72, 106–7, 164, 174
 amphilobite-facies 99–100
 evidence for lithospheric composition 91–8
 mafic 99
 petrology of 95
 populations along the EGT 93–4

Zechstein 47, 159
Zechstein Sea 172
zircons 26, 141, 146, 150–1

Printed in the United States
By Bookmasters